Charles Seale-Hayne Library
University of Plymouth
(01752) 588 588

LibraryandITenquiries@plymouth.ac.uk

A CRITICAL REVIEW OF VAN

Earthquake Prediction From Seismic Electrical Signals

A CRITICAL REVIEW OF VAN

Earthquake Prediction From Seismic Electrical Signals

Editor

Sir James Lighthill

University College London

World Scientific
Singapore • New Jersey • London • Hong Kong

Published by

World Scientific Publishing Co Pte Ltd

P O Box 128, Farrer Road, Singapore 912805

USA office: Suite 1B, 1060 Main Street, River Edge, NJ 07661

UK office: 57 Shelton Street, Covent Garden, London WC2H 9HE

A CRITICAL REVIEW OF VAN
Earthquake Prediction from Seismic Electrical Signals

ISBN 981-02-2542-3

Printed in Singapore.

PREFACE

JAMES LIGHTHILL

*University College London, Department of Mathematics, Gower St.,
London WC1*

This book summarises the proceedings of the meeting A CRITICAL REVIEW OF
VAN, held jointly by the International Council of Scientific Unions (ICSU) and the
Royal Society at the Society's London premises on 11-12 May 1995. It was attended
by ICSU's Executive Director, Julia Marton-Lefèvre, by the Royal Society
Treasurer, John Horlock, and by 38 other participants from 9 different countries.

I presided over this meeting in my capacity as Chairman of ICSU's Special
Committee for IDNDR (the UN-designated International Decade for Natural Disaster
Reduction, 1990-99). I began by thanking the Royal Society for its greatly
appreciated offer to be a "neutral venue"— a highly distinguished venue —for a small
but select scientific meeting (with attendance by invitation only) devoted entirely to
just one single subject: A CRITICAL REVIEW OF VAN.

The VAN approach to earthquake prediction was named from the initial letters
of the surnames of its distinguished originators (Professor P. Varotsos, K.
Alexopoulos and K. Nomicos) all of whom had accepted invitations to the meeting
—although in the end ill health sadly prevented Professor Alexopoulos from
attending. The meeting would concentrate, in an appropriately critical spirit, on
asking 3 questions:

(a) What is VAN?

(b) Do claims for its value rest on scientifically sound bases?

(c) Do there exist for VAN scientifically credible mechanisms?

This book faithfully reflects a wide diversity of views on those issues that were
represented at the meeting.

The International Council of Scientific Unions, besides comprising different large
Unions like the International Union of Theoretical and Applied Mechanics (IUTAM)
which I represent, includes also many inter-Union committees. For IDNDR (the
1990's) ICSU set up the Special Committee in that field which I chair, and which has
already fostered major practical programmes on atmospheric and oceanic disasters

such as hurricanes and typhoons, on river flooding, on droughts and associated famine disasters, and on great volcanic eruptions. All of these are fields where big proven successes in disaster reduction came from increasing the reliability and timeliness of forecasts.

Yet this doesn't at all imply that the same may be true for seismology! Hitherto, in fact, our work in that field was mainly concentrated on the Global Seismic Hazard Assessment Program (GSHAP), aimed at targeting high-priority areas for earthquake-resistant construction.

But the recent stimulus provided when the International Union of Geological Sciences (IUGS) nominated to the Special Committee a new representative, Seiya Uyeda, led to our deciding to try to evaluate from an objective scientific point of view the VAN approach (from seismic electrical signals) to earthquake prediction. It was seen as essential to attempt such a critical review in collaboration with another ICSU body: the Sub-Commission on Earthquake Prediction – chaired by Professor Max Wyss – of the International Association for Seismology and Physics of the Earth's Interior (IASPEI); itself lying within that International Union of Geodesy and Geophysics (IUGG) whose representative on the Special Committee, Academician V.I. Keilis-Borok, would join with me in offering concluding reviews of the meeting.

So the first and second sessions, including a description of VAN and some arguments for its effectiveness, together with a critical review of mechanisms which might underlie observations made in the course of VAN experiments, would be opened by Professor Uyeda and continued by the originators of the VAN method and some of their close collaborators; while the third and fourth sessions would be devoted to counterarguments against VAN, with Professor Wyss opening the fourth session. Moreover the opener of the third session would be Professor Robert Geller, currently engaged in editing for Geophysical Research Letters (GRL) a major volume of correspondence about VAN.

Apparently the present timeliness of an objective review of this subject is re-emphasized by the concurrent appearance of that written debate – to which a wide variety of earthquake-prediction specialists have been contributing – alongside the present proceedings of the ICSU/RS meeting, attended by many such specialists together with several distinguished geophysicists who had agreed to contribute from an initially neutral standpoint to A CRITICAL REVIEW OF VAN. Discussions in our meeting's fifth and sixth sessions benefited very particularly from penetrating questions by these essentially neutral geophysicists and from their contributions to narrowing divergences of view in this field.

I offer my warm thanks to all participants for giving this review meeting such a satisfactorily objective and friendly character. And, at the conclusion of this Preface, I emphasize too that my own commitment from the outset has been, not only to chair the meeting, but also to edit its proceedings, from just such a neutral and objective standpoint.

CONTENTS

Part IV. Arguments in Favour of the VAN Approach

Part V. Some Related Experimental Programmes

Part VI. Reactions to the Review Meeting

Part I
What is VAN?

INTRODUCTION TO THE VAN METHOD OF EARTHQUAKE PREDICTION

S. UYEDA

Earthquake Prediction Research Center, Tokai University, Shimizu 424, Japan and Geodynamics Research Institute / Department of Geology and Geophysics, Texas A&M University, College Station, Texas 77843, USA

Earthquake prediction has been in practice in Greece for more than a decade by the so called VAN method. The method is based on detection of characteristic changes in the geoelectric potential, the so called Seismic Electric Signals (SES) that appear prior to earthquakes. SES are distinguished from noise through a set of criteria based on simple physical principles. It has been found that SES are observed only at particular locations (sensitive sites) and that a sensitive site is selectively sensitive to SES from particular seismic source area(s), making epicentral prediction possible to within about 100km. Magnitude of an impending earthquake is predicted, to within 0.7 units, from its relationship, also empirically discovered, with epicentral distance and intensity of SES.

The main published counter-arguments against the effectiveness of VAN are these: 1) SES are noise, 2) the success and alarm rates are not as high as the VAN group claims, 3) such a success rate can be achieved by chance and 4) The VAN method lacks physical mechanism. Some facts and thoughts on these objections are presented.

1   Introduction

In the 1970's, optimism prevailed in geoscientific community that successful earthquake prediction was in sight. The gap strategy, Haisheng prediction, and the dilatancy model, among others, contributed to that optimism [1, 2]. However, subsequent events yielded only discouraging results, so that the present consensus appears to be that earthquake prediction, in particular short-term prediction, is still beyond the foreseeable future.   Some theoretical seismologists even believe that earthquake prediction would be impossible in principle because of intrinsic chaotic nature of fracture process [3, 4]. However, there is an outstanding exception to the general pessimism, which is the success of the VAN method in Greece. The VAN method is unique in that it has been successfully predicting earthquakes for more than a decade [5-8]. Naturally, the VAN method has been highly controversial.   Some of the criticism and confusion, however, seems to be rooted on misunderstanding. It is the intent of this paper to provide a general introduction to the VAN method and to try to help removing misunderstanding.

Thanks to seismology, we now know that earthquakes are caused by sudden fault displacements, which in turn are due to stress accumulated at plate boundaries and in plate interiors by plate motions. However, their prediction is still very difficult. For earthquake prediction, one has to specify three elements, i.e., when, where and how large the impending earthquake will be, with precision useful for human society. With regard to "when", it is customary to classify prediction into three categories, namely the long-term (more than tens of years), intermediate-term (one to tens of years) and short-term (less than one year) prediction.

Former two classes of predictions, mainly based on the past history of fault movement and seismicity, can be useful for long-term planning for mitigation of earthquake hazard. On the other hand, short-term prediction is based on precursory phenomena immediately before earthquakes. A wide variety of phenomena, including seismological, geodetic, geochemical, hydrological, electro-magnetic or even meteorological and biological phenomena, has been postulated to be potential precursors[1, 9]. Except for a few cases like Haisheng prediction, however, there have practically been no success but frequent failures [10, 11]. Although some success may be hoped in the future, owing to the development of new technologies and data handling techniques, such as GPS, satellite interferometry, water table monitoring, so far, at best, most of the reported precursory phenomena have been noticed only after earthquakes. They are called post-predictions.

The VAN method, named after the initials of Professors P. Varotsos, K. Alexopoulos and K. Nomicos, stands out as a notable exception in that it has been actually making short-term predictions before, not after, earthquakes. Soon after the disastrous earthquake in Athens area in 1981, the VAN group started monitoring geoelectric potential changes, because solid state physicists Varotsos and Alexopoulos anticipated theoretically that some electric current would be generated in the earthquake source region just prior to earthquake. Nomicos developed the necessary data acquisition system.

The VAN group now claims that earthquakes in Greece with Ms(ATH) (magnitude announced by the Seismological Institute of the National Observatory of Athens, SI-NOA) greater than 5 can be predicted within the error of 100 km in epicentral location and 0.7 unit in magnitude. The time of earthquake occurrence is claimed to be from several hours to 11 days after detecting the signals, but it can be several weeks for repeated and prolonged signals (SES activity and GVEF, see below).

2 Outline of the VAN Method

2.1 What they measure

In the VAN method, one monitors continuously the geoelectric potential changes and their EW and NS gradients. At each station, several short dipoles with different lengths (50-200 m) in both EW and NS directions and a few long dipoles (2 - 20 km) in appropriate directions are installed (See Figs. 2a and 2b of Ref. [12]). Such a multiple dipole system is necessary for noise rejection as will be explained later. Digitally recorded data on geoelectric potential changes are transmitted in real time to the central station in Athens through public telephone lines to be stored digitally and displayed on analog pen recorders.

2.2 Seismic Electric Signals (SES)

The VAN scientists claim that some of the observed geoelectric potential gradient changes with amplitude of the order of 1mV/100m are indeed precursors to earthquakes and call them Seismic Electric Signals (SES). There are, according to the VAN group[7], three types of precursory signals: namely single SES, Electrical or SES Activity and Gradual Variation of Electric Field (GVEF).

Single SES is a signal isolated in time, having a duration ranging from half a minute to a several hours, and precedes a single earthquake, whereas in SES activity a number of SES appear within a short time, such as in a few hours or a day, and are followed by more than one seismic events. Example shown in Fig. 1 is a typical SES activity. SES activity can have varied wave forms, some recent examples of which are shown in Figs. 5a, 5b, 6a, and 6b of Ref. [12].

GVEF, having a duration of many hours to days, is reported to appear weeks before a large (Ms(ATH) $\geq$ 5.5) earthquake[7]. GVEF has amplitude an order of magnitude larger than usual SES which appear during the ending phase of GVEF as shown in the conceptual view held by the VAN-group (Fig. 2): electrical signals of different frequencies are emitted at different stages during the earthquake preparation

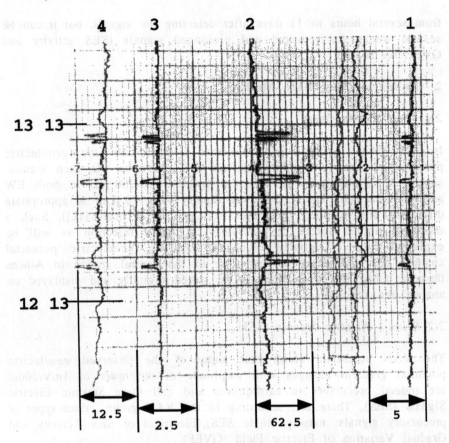

Fig. 1. An example of SES activity recorded at Ioannina (IOA) with 4 SES of the same polarity within 50 minutes on August 31, 1988 [13]. (The apparent reversed polarity of the channel 2 with regard to other channels is due simply to connections intentionally reversed for easier SES recognition.) Two lines without SES activity are from Nauplio-station. Based on this SES activity data, the VAN-group issued a prediction telegram on September 1, 1988, stating "SIGNIFICANT ELECTRICAL ACTIVITY WAS RECORDED AT IOA-STATION ON AUGUST 31 1988   EPICENTER AT N.W. 300 OR W 240 WITH MAGNITUDES 5.3   AND 5.8". Together with two other telegrams based on later series of SES, this telegram was followed by the Killini-Vartholomio events [13].

Numbers such as 12.5 show scales for SES in mV.
Channel 1: EW dipole, L=47.5 m
Channel 2: NNE dipole, L=2.5 km
Channel 3: NS dipole, L=48 m
Channel 4: EW dipole, L=181 m

period and the higher frequency signals generally come closer to the fracture. As the VAN group admits[7], however, GVEF has been observed only rarely, so that its physical nature, such as its duration, lead time, intensity and selectivity (see section 2.4) is largely still unknown in the present author's view. Similar long variations were observed earlier by Sobolev in Kamchatka[15] and more recently by Nagao and others in Japan[16].

Mention must be made here that recently there have been a number of reported occurrences of electromagnetic precursors from various parts of the world including Japan, Russia and USA, covering a wide range of frequency from DC-ULF, ELF, VLF, LF to MHz[17-19]. Time sequence deduced from these various observations seems to agree qualitatively, but not necessarily quantitatively, with the idea represented in Fig. 2. Comprehensive understanding of these observations in terms of physics will be an important research target in the years to come and a path to achieve more efficient earthquake predictions as will be advocated in the next section.

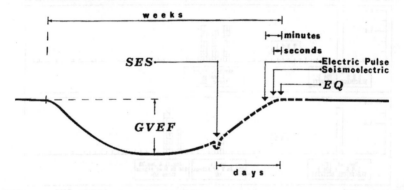

Fig. 2. Conceptual, not to scale, diagram showing the possible time sequence of the three types of electrical precursors (GVEF, SES and electric pulse) based on observations in Greece [7].

2.3 The lead time

The time lag between signals and earthquakes is different for different types of signals, apparently causing some confusion among critics. It is between several hours and two weeks (11 days according to VAN) for single SES. For SES activity, the sequence of events seems more complicated as seen in the time charts such as Fig. 3 (reproduced from Fig. 22 of Ref. [7], which is often quoted in the recent VAN

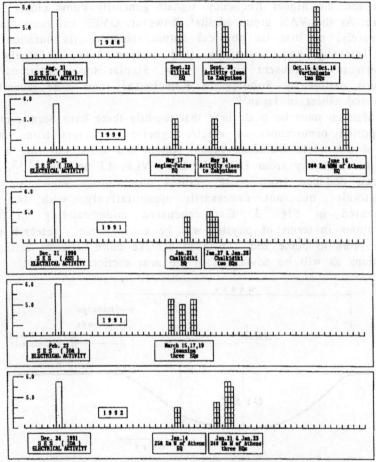

Fig. 3.   Time sequences of events associated with five significant SES activities recorded during the period 1988 - 1992 [7].   Similar time sequences have been observed many more times in subsequent period as seen in Figs. 7a and 7b of Ref. [12].

predictions). Although not shown in the figure, small magnitude earthquakes start within about two weeks. Then, the first strong earthquake comes at about three to four weeks. Moreover, in several recent cases, the largest event took place at about six to seven weeks after the initiation of SES activity. In most such cases, but not always, there were additional SES activities during the course of events.

It seems that due to this intriguing course of events, VAN's recent prediction often states "the seismic activity may follow the time charts of Fig. 28 of Varotsos and Lazaridou, 1991[13] or Fig. 22 of Varotsos et al, 1993a[7]". It may be noted also that in all cases of SES activity, no strong event has been observed during the first three weeks and hence the time window for large events is restricted to almost one month starting at the end of the three weeks. It seems unwise to the present author to ask for a rigid rule of game regarding the lead time of prediction at this stage. The VAN method is not a game with fixed rules but a developing science to which new discoveries are constantly added.

If we establish in future how much shorter lead times are associated with signals with specific higher frequencies, VAN measurement combined with higher frequency measurements would greatly help improving the time resolution of predictions.

2.4 The Selectivity (the way to predict epicenter)

On top of the establishment of noise rejection criteria to be mentioned later, the VAN research has made two major discoveries that have made the present art of prediction possible. One is the so called selectivity, which consists of two important contents. The first is that there are only selected sites which are sensitive to SES (sensitive sites). Sensitive sites had to be found through painstaking testing at many sites by the method described in Ref.[7]. This fact seems to give a partial reason why previous efforts to catch precursory electric signals have failed. Lately it has become gradually clear that sensitive sites are of heterogenous geologic structure, and close to faults, each possibly providing heterogeneous electrical resistivity structures, which make the search for sensitive sites more efficient.

The second, which is even more striking, is the fact that a sensitive site is sensitive only to SES from some specific focal area or areas that are not always close to them. A map identifying those focal area(s), SES from which are sensed by a site, is called the selectivity map of the site (Fig. 4). Conversely, once the selectivity map has been established, it provides the means to predict the epicenter of the impending earthquake. It is to be noted that in VAN's prediction telegram, epicenter of the impending earthquake is given in terms of distance in km and direction reckoned from Athens, an example of which is given in the caption of Fig. 1. The present author share the opinion of some critics that this may not be the best way for the purpose and welcomes the presentation in terms of longitude and latitude adopted in Table 1 of Ref. [12].

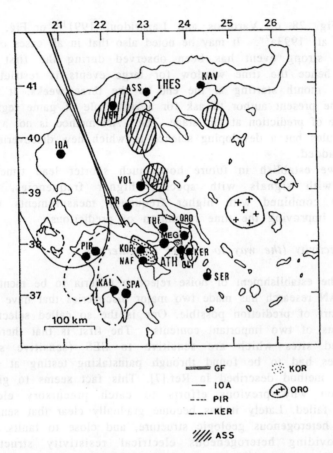

Fig. 4. Distribution of VAN stations and "selectivity map" of several stations as of 1989 [20]. Since the end of 1989, only four stations, i. e., IOA, ASS, KER and PIR, have been in operation due to lack of funds. "Selectivity map" now is more refined than this figure by new data. Double line GF indicates the major fault separating Greece into outer accretionary zone and inner zone. It also seems to form a major barrier for transmission of SES.

Selectivity maps are always incomplete and are constantly improved with accumulation of new data. When they are poorly established or the sensitive focal area is large or there are more than one sensitive focal areas, epicentral estimate is difficult. Double predictions are often issued under such circumstances. But, it has often been found that the vector direction (the polarity and ratio of NS and EW components) of the changing electric field varies from one source area to another, providing additional means for epicentral prediction. When the selectivity maps of different sites overlap, naturally SES are observed simultaneously at these sites and help narrowing down the epicentral area more convincingly.

This selectivity rule, which is highly reproducible, had to be discovered purely empirically. In physical sense, the sensitive sites and corresponding focal area(s) have to be somehow selectively electrically connected, which is highly unlikely unless the uppermost part of the earth is extremely heterogeneous in electrical resistivity, but if it is so, it would not be categorically impossible[7,21]. It would even be consistent with the view that the amount of electric current density at the source would have to be unrealistically huge for SES to travel hundreds of km in a homogeneous medium. Simple calculation indicates that 10^{-9} A/cm^2, which is compatible with laboratory data[19,22], would suffice if the observation site and focal area are connected by a narrow conductive channel[23,7].

2.5 The VAN-relation (the way to predict the magnitude)

The other important discovery of the VAN group is the relationship among the focal distance, r, magnitude, M, and the observed intensity of SES, $\Delta V/L$, where ΔV is amplitude of SES and L the dipole length. It would be natural that there is a relation among r, M and the strength of the signal, and the formula empirically postulated by VAN [5] is:

$$\log(\Delta V/L \times r) = aM + b \qquad (1)$$

where a is a constant 0.34-0.37 and b is a site-dependent constant. Once the epicentral area is estimated from the selectivity rule, both $\Delta V/L$ and r are known, so that the expected value of M can be derived from the above formula.

From the onset of the project in early 1980's, the VAN group has consistently calibrated their system with the Ms(ATH) scale as it used to be the only magnitude immediately available to them for their day to day practice. Ms(ATH), because of the subscript s, is sometimes mistaken as a surface wave magnitude, but it is simply defined as Ms(ATH)= ML + 0.5, where ML is the local magnitude. Although the use of Ms(ATH) by VAN apparently has caused some confusion in the seismological community[24], the problem should be non-essential as long as one convertible scale has been consistently used. Nevertheless, it is a good move on the VAN side to have started to use also the more common magnitude scales such as those from USGS and ISC, that are just as readily available as Ms(ATH) nowadays.

Equation (1) says that for the same earthquake, SES amplitude decays as 1/r, which provided another way to locate the epicenter as the intersecting point of Apollonian circles when SES is detected at more than two stations[6]. After the end of 1989, however, such a practice has virtually been stopped because of only four working stations.

The 1/r decay of SES seems to give some clue to the physical mechanism of SES. For a homogeneous isotropic medium, this would suggest a long line source, which is unlikely. Again more or less a point source connected with linear conduits seems to be compatible with the observation. Another important message of the above formula is that M is related with the maximum amplitude of SES only and that neither its wave form nor duration is involved.   Physical meaning of it is intriguing to the present author.

2.6 Practice of Prediction and Results

When an SES is observed at some station(s), the epicentral area(s) estimation is first done by the aid of the selectivity rule. Then, the magnitude is assessed using the VAN relation (equation (1)). Time of the impending earthquake depends on the type of signals as explained in 2.3. Only when the assessed magnitude Ms(ATH) is equal or greater than 5, which is likely to have serious consequences in Greece, does the VAN group send a prediction to the Greek government. The VAN group has generally taken the stance that warning to the general public and taking necessary countermeasures belong to the responsibility of the government, not to individual scientists whose public statements would cause unwanted social panic.

At the same time, the VAN group considers it essential that the scientific community is informed that the prediction is made before the earthquakes. For this reason, they send the prediction simultaneously, on request, to some 30 foreign scientific institutions, with the hope that they freely acknowledge the prior receipt of prediction and yet they do not disclose it to the public before the event.

The results of predictions have been reported in a series of publications (i.e. Refs.[5-8, 12-14 and others]). These publications together with catalogues of earthquakes allows one to assess the success rate and alarm rate of the VAN predictions given a set of success/failure criteria. Applying the criteria set forth by the VAN, the success rate and alarm rate (for Ms[ATH] ≥ 5.3 earthquakes) given by Uyeda [20] were both about 60% (for May, 1988 - Oct., 1989) and those given by Hamada[25] were about 50%. (for Jan., 1987 - June, 1989). On the other hand, Drakopoulos and others[26] gave practically zero % values for two selected three month periods in 1986 and 1988. Similarly Wyss and Allmann[27] gave 9% for the success rate. Although there is a certain room for individual judgment in evaluation, such a wide discrepancy is striking and will be discussed in 3.2. Readers are advised also to examine the VAN's replies[12,28] to the latter evaluations.

3 Major Criticisms and Some Related Aspects

Major objections to the VAN method have been voiced as follows: 1) SES are noise, 2) the success and alarms rates are not as high as the VAN group claims, 3) such low rates can be achieved by chance and 4) the VAN method lacks physical mechanism.

3.1 Are SES noise?

Noise rejection is the very first necessary step in any physical measurements. Noise can be generated from two major sources: i.e., those generated in hardware and those received at stations. In the data acquisition, transmission and recording system developed mainly by Nomicos[29], extreme care has been taken to eliminate the possibilities of instrumental noise. Moreover, in the recent years, data are transmitted by two entirely independent means from four stations (IOA, KER, ASS. and PIR), one continuous realtime transmission and the other daily transmission of the same data stored in data loggers at stations. SES received through these independent methods have been proven always identical[30], excluding the possibility of their being noise picked up through transmission.

As to noise received at stations, they have developed a set of rules as follows:
1. Changes with magneto-telluric origin can be removed because they are recorded at all dipoles at all the stations simultaneously.
2. SES must appear simultaneously at all of short and long dipoles at a station or stations concerned.

3. SES must satisfy the $\Delta V/L=$ constant relation for short dipoles in both EW and NS directions (in a homogeneous area).

4. The polarity and amplitude of SES of short and long dipoles must be compatible with the assumption that the source is distant compared with the dipole lengths: projection of SES vector calculated from short dipoles on to long dipoles must have same polarity and comparable amplitude with observed change of the latter.

Rules 2 and 3 are used to reject the effects of electrode instabilities due to such factors as rainfall and noise from nearby (within a few hundred meters) sources, whereas rule 4 eliminates noise from sources at distances comparable to the length of long dipoles. The physical meaning of these rules are straightforward. In actual practice of rules 3 and 4, some allowances for measurement errors and local resistivity heterogeneity has to be given. This is the subjective part of the rules.

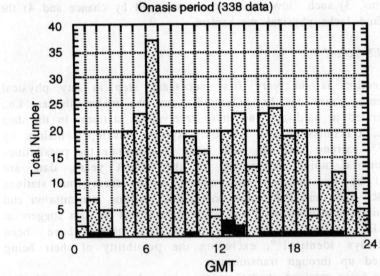

Fig. 5. Frequency histogram of occurrences of 338 anomalous changes (gray) and 8 SES (black) against time of day (GMT), recorded at IOA during July 19 - October 18, 1988 [31]. SES lasting more than one hour are given more than one count.

In order to examine the overall objectivity of these rules, we have made an exhaustive check of the original VAN data by performing the SES selection process independently of the VAN group[31]. We first examined the raw records of Ioannina (IOA) station for a three month period (July 19 - Oct. 18, 1988) for preliminary check and then for a longer period of 28.5 months (April 30 1988 - Sept. 15, 1990).

The data for the three-month preliminary check were those from all the 8 (7 short and 1 long) dipoles working at IOA at that time. The number of all the anomalous changes picked up and tabulated by us was 338, excluding those due to magneto-telluric origin (rule 1). Gray columns in Fig. 5 show the frequency histogram of these changes against time (GMT) of the day. Clearly, the changes are more frequent in the day time, indicating that most of them are noise related to human activity. Then the rules 2, 3, and 4 were applied in this order to these 338 changes. 184 changes were recorded only on the long dipole and 44 only on short dipoles. Among these, 11 were judged as due to rainfall. Thus the rule 2 eliminated 228 changes, leaving 110, of which only 13 passed the rule 3. Finally, 8 changes survived the rule 4.

They are 1 isolated SES on July 27, 4 distinct SES of August 31, 2 continuous multi-peaked SES of Sept. 29 and similar one on Oct. 3, all in 1988. The latter three cases can be viewed as SES activity. Frequency histogram of these 8 changes are indicated by black columns in Fig. 5. 7 out of 8 SES, i. e., on August 31. Sept. 29 and Oct. 3 were found to be exactly the same changes on which basis the VAN group issued the successful predictions for the Killini-Vartholomio earthquakes[13]. One SES on July 27 apparently did not prompt the VAN group to issue a prediction for some reason unknown to us.

We then proceeded to check the longer period data of 28.5 months. The total number of anomalous changes, excluding the magneto-telluric ones, was over 3,000, but the number of changes that survived the rules 2, 3, and 4 was only 38, as shown in Fig. 6(b), where the vertical axis is amplitude. During the study period, the VAN group issued 18 predictions based on SES observed at IOA station (Refs. [8, 13, 32]) as shown in Fig. 6(a), where the vertical axis is predicted Ms(ATH). As they send their predictions only when the predicted Ms(ATH) is equal to or greater than 5, if everything goes well, 18 predictions in Fig. 6(a) should represent a subset of our 38 SES in Fig. 6(b). In fact, 14 of the 18 VAN predictions were found to be based on the same changes that are included in the list of our SES. The 4 cases of disagreement are VAN prediction telegrams 13, 16, 18 and 20. In each case, we noted the changes but discarded them as noise. In retrospect, we could have dealt with them differently if we were more experienced.

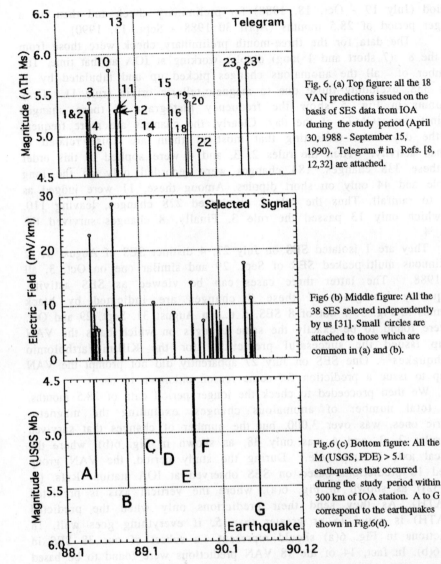

Fig. 6. (a) Top figure: all the 18 VAN predictions issued on the basis of SES data from IOA during the study period (April 30, 1988 - September 15, 1990). Telegram # in Refs. [8, 12,32] are attached.

Fig6 (b) Middle figure: All the 38 SES selected independently by us [31]. Small circles are attached to those which are common in (a) and (b).

Fig.6 (c) Bottom figure: All the M (USGS, PDE) > 5.1 earthquakes that occurred during the study period within 300 km of IOA station. A to G correspond to the earthquakes shown in Fig.6(d).

In Figs. 6(a) and 6(b) , small circles are attached to the SES common to the two sets. It may be noticed that generally higher bars correspond better than shorter ones. Only exception is the case of telegram 13, predicting the largest Ms(ATH) = 6.4 - 6.5 during the study period, which we rejected as noise by the rule 4. Earthquakes

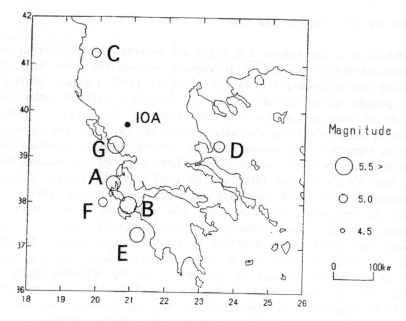

Fig. 6 (d) Epicenters of the earthquakes A - G in Fig. 6(c.).

did follow the prediction, but M(USGS) of the largest event was 4.9. As stated already, our SES on July 27, 1988 has no corresponding VAN prediction. From September, 1989 to June, 1990, we found many possible but weak SES, but the VAN group did not issue any predictions, except for the two strong ones (23 and 23' in Fig. 6(a)). In fact, the seismicity was low during the period as seen in Fig. 6(c), which shows all the M(USGS PDE)>5.1 earthquakes that occurred within 300 km of IOA station (see Fig. 6(d)).

Despite a few cases of disagreement, it appears impressive that the two independent operations led to such a close overall match in picking up 14 common SES out of over 3,000 anomalous changes. This result seems to confirm that the SES identification process of VAN has high degree of objectivity and , at the same time, gives another partial explanation on why attempts to predict earthquakes over 30 years by electric precursors have failed and Kagan and Jackson[33] had to state "prediction based on electric and other precursors is a difficult or maybe an impossible task".

3.2 Are the VAN success and alarm rates low?

As mentioned at the end of 2.6, there are striking discrepancies in the estimates of VAN's success. It is striking because the same data and the same success/failure criteria are given to every assessor, although there always exists certain room for judgment left to individual assessor in interpreting the data and applying the criteria. Criteria are set by the VAN group themselves based on their own performances. It seems therefore only natural that the criteria were so designed that reasonable success rates   would result. Then, what are the reasons why some say VAN's success rates are so low?   In the present author's understanding, the VAN considers that double prediction, 10-20   or so km of epicentral deviation from 100 km limit, a few days of deviation in origin time - namely not strictly box-type criteria - , for instance, are permissible. Negation of such common sense judgments would contribute to give rise to lower success rates.   Among other common confusions is that of the lead time for different types of signals. As explained in 2.3, lead time related to SES activity seems still to be learned from experience, so that the assessors are advised to be careful not to deny the success by simply applying wrong criteria[26,27,34]. However, there seem to be more serious origins of the discord.

One is related to confusion in magnitude scales. As explained in 2.5, the VAN has been using Ms(ATH) scale for their predictions, for the simple reason that it was the only available one in day to day prediction practice until rather recently. Unfortunately, however, Ms,(ATH) sounds like a surface wave magnitude which it is not. Mixed use of this scale with other scales can lead to wrong conclusions (e.g. Ref.[27]).

The other more grave origin of discord is concerned with the difference in the number of the earthquakes that should be predicted. The VAN group states that their targets are Ms(ATH)≥5.0 earthquakes and their permissible error range in magnitude is 0.7. If the predicted Ms(ATH) was 5.0 and the actual Ms(ATH) was 4.3, that prediction is naturally scored as successful.   The alarm rate in this case, however, remains the number of successfully predicted Ms(ATH)≥5.0 earthquakes   / the total number of Ms(ATH)≥5.0 earthquakes.

Some assessors[27,34], however, contend that the VAN in such a case should target all the Ms(ATH)≥4.3 earthquakes and the alarm rate should be the number of successfully predicted Ms(ATH)≥4.3

earthquakes / the total number of Ms(ATH)≥4.3 earthquakes. Because of the Gutenberg-Richter relation, the number of earthquakes increases drastically with lowering magnitude threshold, giving rise to a drastic lowering of the alarm rate with the latter definition. It seems to the present author that the latter definition is unreasonable in the true sense of experimental error and certainly not in accord with VAN's criteria to which each assessor is bound to adhere.

To the present author, however, to know if the VAN predictions meet the VAN's own criteria is a rather trivial exercise and the essential purpose of evaluation of the VAN method should be to find out whether the SES has physically linked with earthquakes and the VAN predictions are meaningful to society and these points should be evaluated by assessor's own criteria.

3.3 By chance?

A very popular argument against the VAN prediction is that their success can be achieved by chance, the so called null hypothesis. Again the authors of the above cited papers are vocal in this argument in the present volume. With the low rates of success, which were incorrectly derived in the present author's view as explained in the preceding section, success by chance may seem plausible. However, their arguments are not without other serious errors also. Leaving detailed discussion on their errors to Refs. [35, 36], only the simple but persuasive comment of Hamada[25,37] that the alarm success rate of the VAN prediction goes up dramatically with higher magnitude threshold is mentioned here. Such is simply impossible if the predictions had been cast randomly into a population that consists of Gutenberg-Richter type distribution. This point will be substantiated further in the following.

Concentrating the discussion to large earthquakes has definite advantages in assessing the possible physical link between SES and earthquakes[12,38]. First one can avoid the non-essential confusion due to various uncertainties inherent to small events, such as the detectability in different magnitude scales and possibilities of chancy success due to too many earthquakes. Moreover, it is the success and alarm rates for large events that really matters in earthquake prediction.

We have examined the problem using the less ambiguous magnitude scale, i.e. MB given in USGS NEIC PDE[39]. During the period January 1,

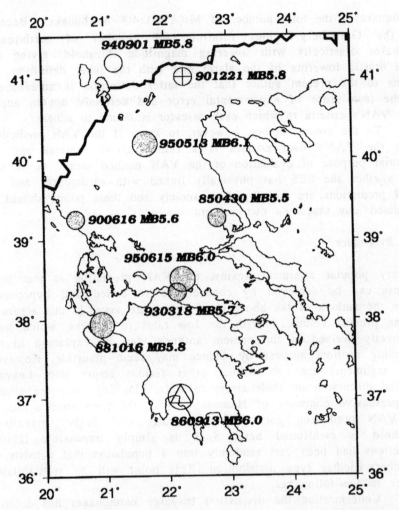

Fig. 7. All the earthquakes with NEIC MB ≥ 5.5 during January 1, 1984 - June 30, 1995. Shaded circle: successfully predicted, circle with triangle: unsuccessfully predicted, circle with cross: missed, open circle: out of Greek territory as indicated by the national border line.

1984 - September 10, 1995, 9 MB ≥ 5.5 earthquakes occurred in the region as shown in Fig. 7. Among these, one event denoted by white circle does not count as it occurred out of the Greek territory. The shaded circles denote the events that were successfully predicted by VAN, whereas the circle with a triangle is unsuccessfully predicted event and the circle with a cross missed event. Except the 1986 Kalamata earthquake (MB=6.0), for which epicenter could not be predicted in advance, and 1990 MB=5.8 earthquake in northern Greece, all other 6 earthquakes were successfully predicted, including the recent two M≥6.0 events [12]. Although the sample numbers are too small for any meaningful statistics, the face value of the alarm rate for MB≥5.5 earthquakes is 75%.

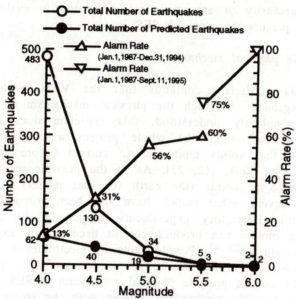

Fig. 8. Diagram showing how the alarm rate increases with the magnitude threshold. Open circle: the total number of earthquakes with MB equal or greater than the threshold values on the abscissa according to the USGS NEIC PDE catalogue for Jan.1, 1984 - Dec. 31, 1994. Black circles: the total number of successfully predicted earthquakes derived from checking VAN predictions against the PDE catalogue. The triangles are their ratio, namely the alarm rate or the rate of success. Inverted triangles show the values when the two M≥ 6.0 events in 1995 are incorporated.

The rise of alarm rate with magnitude can be seen in Fig. 8, in which the total number of earthquakes and total number of successfully predicted earthquakes with magnitude equal or greater than the threshold values on the abscissa are shown. In this case, the period January 1, 1987 - December 31, 1994 was selected because the complete list of predictions are available in Refs. [12, 32, 13, 7, 17], so that anyone can examine our results by checking against the USGS NEIC PDE Catalogue. The triangles are their ratio, namely the   alarm rate or the rate of success. Inverted triangles show the values when the May 4, 1995 (ML:5.5) event and the two MB>=6.0 events (May 15 and June 15, 1995) are incorporated. Since the NEIC catalogue is currently available only to Dec. 31, 1994, two alarm rate curves are not to be linked, but MB>=5.5 portion will not change even after the MB≤5.5 portion has been redrawn with updated data, because there was no large earthquake in the late 1994. These results seem to decisively rule out the necessity of any statistical discussion in evaluating the validity of the precursory nature of SES.

3.4   *Lacking in physical mechanism?*

Another popular criticism contends that the VAN method lacks physical background. Although the physical mechanism of it is far from being completely understood, this criticism does not seem necessarily warranted since the whole project was initiated by a physical model that solids emit electric current before fracture as explained in e. g., Refs. [12, 23]. As to the possible mechanism of current generation in solids (the earth for that matter) during pre-fracture stage, several other models have also been proposed[40-44] and some remarkable laboratory experiments supporting the hypothesis that dislocation motion can produce current necessary to explain SES have been conducted[22, 45]. Possible role of electrokinetic effect has also been proposed for the source of the current[46,47]. Thus there seems to be no lack of possible physical mechanism in SES generation. How are these possible mechanisms related with the tectonic process in the real earth is a first class challenge to geoscientists.

An important problem in the VAN method, other than current generation, is how SES is transmitted from the source to sensitive sites hundreds of km away, namely the physical and geological nature of current conduit(s) and what makes a site sensitive. Recently, active research has been initiated to solve these problems. For example, Fig. 9 shows the NE - SW cross section including the IOA station, indicating the underground apparent resistivity structure obtained by magneto-telluric sounding carried out under a Japan-Greece cooperation program[48]. A structure with extremely low specific resistivity was found under the northern side of Mt. Mitsikeli, that might act as a conduit for SES transmission. As to the SES conduit(s), narrow continuous water channel(s) might also be speculated since a cave with running water exists in the valley west of the VAN station.

Furthermore, magneto-telluric studies at the IOA station[12,48] have revealed highly anisotropic nature of the underground electrical structure, that are responding differently to the effects of ionospheric origin, thunderbolts and SES, providing useful clue for understanding the mechanism of SES transmission and reception (see Figs. 3 and 4 of Ref. [12] for example).

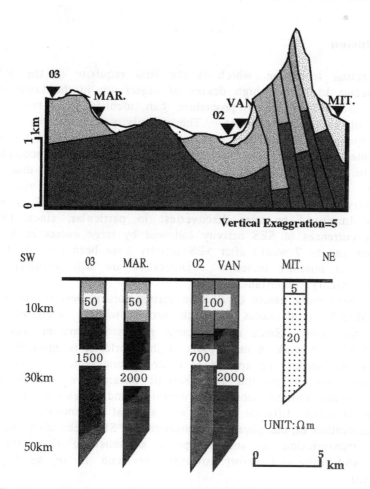

Fig. 9. Apparent resistivity cross section derived from MT observations at five sites, including the IOA station, along the NE - SW line which is perpendicular to the strike of local geologic structure [48]. MAR: Marmara Village, VAN: IOA station, MIT: Mt. Mitsikeli.

There seems to be no particular reason to suspect that the VAN method would not work in some of the world's seismic regions other than Greece and it is urgently needed to apply it elsewhere to save human lives from earthquake disasters. Clarification of problems addressed in this section will be highly valuable if the VAN technique is to be transferred to other regions, where different natural and societal conditions may require substantial improvements of the method.

4 Conclusion

Noise rejection technique, which is the first requisite of the VAN method, seems to have a high degree of objectivity in the sense that anybody who follows their procedure can identify SES from the multitude of anomalous changes. The selectivity rule discovered by the VAN group has a high degree of repeatability so that it can be used for epicenter prediction. The selectivity maps are always incomplete and are to be constantly improved through experiences, so that the precision of epicentral prediction is continuously refined. Predictability of time sequence of seismic events after SES is also constantly improved by new discoveries. In particular, since 1988, repeated occurrences of SES activity followed by large events in 3 - 4 weeks after and 6- 7 weeks after SES activity have been noticed. Their generality and possible tectonic significance are still unknown but potentially highly important.

The precursory nature of SES to earthquakes is best revealed for larger (M≥5.5) earthquakes by high scores without need of any statistical treatment. Since no particular reason is apparent why the method should not work in other parts of the world, it is advisable and urgent that the method be applied to other seismic regions to save as many human lives as possible from seismic hazard, even though such an effort would require substantial devotion and ingenuity of those concerned to adapt different natural and societal circumstances.

Clarification of the physical mechanisms of SES generation and its selective transmission to and reception at sensitive sites are in progress and enhanced effort towards this end is to be highly encouraged.

Acknowledgments

I would like to thank Sir James Lighthill for inviting me to deliver the present introductory paper at the ICSU / Royal Society meeting "Critical Review on VAN" and also for giving me many constructive comments on the manuscript.

Appendix

For the same reasons as ours, Mulargia and Gasperini[49] have looked at the largest earthquakes and claimed that backward associations outnumber forward associations. Since some of the statements in the version available to the present author[49] are literally incomprehensible, e.g., the first paragraph of the section "Scoring VAN prediction: the alarm rate for large earthquakes in 1987-1992", some conjecture was needed to decipher what they mean.

That the aforementioned claim of Ref.[49] is wrong and how such an erroneous claim came about has been thoroughly shown by the VAN [12], so that I will not go in any details here. They also used the magnitude scales given in the NEIC PDE catalogue. While we used strictly MB to avoid any ambiguity, they used what they call Mave. Then, they put the threshold values Mave $\geq$ 4.9 and Mave $\geq$ 5.1, so that the number of earthquakes to be examined became 27 and 11 (these numbers should be 26 and 10 [12]). These procedures appear reasonable. What is not reasonable is that in making associations between predictions and earthquakes, the authors of Ref. [49] disregarded VAN's "basic rules".

First, they score as correct association regardless of DM. This looks generous to VAN, but it is not. Putting threshold of M as above and no threshold to VAN's predictions generally kills about half of forward associations when the predicted magnitude is 5.0, for instance. This way the authors of Ref. [49] are not comparing relevant sets of data to state with. Second, by allowing 22 days uniformly, again looking generous to VAN superficially, they are simply doing some exercise almost irrelevant to VAN problem as elaborated in Ref. [12].

Without going in any detail, therefore, their conclusion seems to be of no value, except showing that almost any conclusion can be derived by playing, violating for that matter, the very basic principles.

Two interpretations suggested by Ref. [49], which the authors admit not entirely satisfactory, merit some comments, however. That VAN's "sensitivity" unconsciously increases after a large earthquakes is simply untrue. This I can testify only from my personal experience in working with them on real time basis. Much worry is always before the events. The other point that the signals may appear more frequently because the stress changes may be more vigorous after the earthquake than before may theoretically be a possibility. In fact, many signals are observed corresponding to large aftershocks. What is significant here, however, seems to me that the authors of Ref. [49] are now taking the view of at least assuming that VAN signals have a physical basis which in the present author's view is a welcome progress on the part of the strong opponents.

References

1. T. Rikitake, *Earthquake Prediction*, Elsevier, Amsterdam, 357pp (1976).
2. Z. Ma et al, *Earthquake Prediction: Nine Major Earthquakes in China (1966-1976)*, Seismological Press Beijing and Springer-Verlag Berlin Heidelberg, 332pp (1989).
3. R. Geller, *Nature*, **352**, 275 (1991).
4. R. Geller, in *Proceedings of Earthquake Prediction Research Symposium*, National Comm. for Seismology and Seismological Society of Japan (1994).
5. P. Varotsos and K. Alexopoulos, *Tectonophysics* , **110**, 73 (1984a).
6. P. Varotsos and K. Alexopoulos, *Tectonophysics*, **110**, 99 (1984b).
7. P. Varotsos et al, *Tectonophysics*, **224**, 1 (1993a).
8. P. Varotsos et al, *Tectonophysics* , **224**, 269 (1993b).
9. T. Rikitake, *Tectonophysics*, **54**, 293 (1979).
10. M. Wyss and M. Baer, *Nature*, **289**, 785 (1981).
11. R. Andrews, *Earthquakes and Volcanoes*, **23**, 4, 170 (1992).
12. P. Varotsos et al, in *Critical Review on VAN* (this volume) (1995).
13. P. Varotsos and M. Lazaridou, *Tectonophysics*, **188**, 321 (1991).
14. P. Varotsos and K. Alexopoulos, *Tectonophysics*,**136**, 335 (1987).
15. G. A. Sobolev, *Pageoph*, **113**, 229 (1975).
16. T. Nagao et al, in *Critical Review on VAN* (this volume), (1995).

17. *Electromagnetic Phenomena related to Earthquake Prediction*, eds. M. Hayakawa and Y. Fujinawa, Terra Scientific Pub. Co., Tokyo, 677 pp (1994).

18. Y. Fujinawa, in *Critical Review on VAN* (this volume) (1995).

19. Y. Enomoto, in *Critical Review on VAN* (this volume) (1995).

20. S. Uyeda, *Zisin*, **44**, 391 (1991).

21. H. Utada, *Tectonophysics*, **224**, 153 (1993).

22. V. Hadjicontis and C. Mavromatou, *Geophys. Res. Lett.*, **21**, 1687 (1994).

23. P. Varotsos and K. Alexopoulos, *Thermodynamics of Point Defects and their Relation with Bulk Properties*, North Holland, Amsterdam, 474pp (1986).

24. M. Wyss, *Geophys. Res. Lett.*, to be published (1995).

25. K. Hamada, *Tectonophysics*, **224**, 203 (1993).

26. J. Drakopoulos et al, *Tectonophysics*, **224**, 223 (1993).

27. M. Wyss and A. Allmann, *Geophys. Res. Lett.*, to be published (1995).

28. P. Varotsos, *Geophys. Res. Lett.*, to be published (1995).

29. K. Nomicos and P. Chatzidiakos, *Tectonophysics*, **224**, 39 (1993).

30. P. Varotsos et al, *Jishin Journal*, **17**, 18 (1994).

31. T. Nagao et al, *Geophys. Res. Lett.* , to be published (1995).

32. E. Dologlou, *Tectonophysics*, **224**, 189 (1993).

33. Y. Kagan and D. Jackson., *Geophys. Res. Lett.*, to be published (1995).

34. F. Mulargia and P. Gasperini, *Geophys. J. Intern.*, **111**, 32 (1992).

35. P. Varotsos et al, *Geophys. Res. Lett.*, to be published (1995).

36. H. Takayama, *Geophys. J. Intern.*, **115**, 1197 (1993).

37. K. Hamada K., in *Critical Review on VAN* (this volume) (1995).

38. H. Kanamori, in *Critical Review on VAN* (this volume) (1995).

39. K. Al-Damegh and S. Uyeda (in preparation) (1995).

40. L. Slifkin, *Tectonophysics*, **224**, 149 (1993)

42. D. Lazarus, *Tectonophysics*, **224**, 265 (1993).

43. L. Slifkin, in *Critical Review on VAN* (this volume) (1995).

44. D. Lazarus, in *Critical Review on VAN* (this volume) (1995).

45. V. Hadjicontis and C. Mavromatou, in *Critical Review on VAN* (this volume) (1995).

46. I. P. Dobrovolsky et al, *Phys. Earth Planet. Inter.*, **57**, 144 (1989).
47. N. Gershenzon and M. Gokhberg, *Tectonophysics*, **224**, 169 (1993).
48. M. Uyeshima et al, (in preparation) (1995).
49. F. Mulargia and P. Gasperini, in *Critical Review on VAN* (this volume) (1995).

SHORT TERM EARTHQUAKE PREDICTION IN GREECE BY SEISMIC ELECTRIC SIGNALS

P. VAROTSOS, M. LAZARIDOU, K. EFTAXIAS, G. ANTONOPOULOS, J. MAKRIS AND J. KOPANAS

Solid State Section, Department of Physics, University of Athens, Greece.

More than fifteen years ago, the study of Thermodynamics of point defects in solids led to the conclusion that transient currents are emitted from a solid containing electric dipoles upon a gradual variation of pressure (stress). This emission occurs when the stress σ reaches a critical value σ_{cr} which does not have to coincide with the fracture stress σ_{fr}. Therefore this (transient) current emission can be considered as a precursor of the fracture of the solid. As similar conditions might occur before an earthquake, field experiments started in 1981 in Greece aimed at detecting such an earthquake precursor.

After preliminary experimentation, a continuous monitoring of the electric field of the earth at various sites in Greece started in 1982 and a telemetric network was completed in 1983. The data are transmitted on a real time basis to a central station located at Athens. Transient variations of the electric field of the earth, i.e. the so called Seismic Electric Signals, SES, have been actually found to precede earthquakes (EQs). Their physical properties were described in a series of papers and can be summarised as follows: The SES detection is possible only at certain points of the earth's surface (sensitive points). Each sensitive point, however, can collect SESs only from a restricted number of seismic areas (selectivity effect). For a given pair "sensitive point - seismic area" the SES amplitude scales with the earthquake magnitude M.

A proper combination of SES physical properties can lead to the estimation of the parameters (time-window, epicentre and magnitude) of the impending EQ(s). As the telemetric network was operating on a real time basis, predictions (addressed to the Greek authorities) could be issued well in advance. During the last eight years these predictions are also sent to a number of foreign Institutes and hence an independent evaluation of the results can be made.

As regards VAN predictions of the strongest EQs, i.e. those announced from the Athens seismological Institute as having magnitudes $M_s(ATH) \geq 5.8$, the following results have been obtained: During the period January 1, 1987 to June 15, 1995, fourteen EQs within the area $N_{36}^{41} E_{19}^{25}$ occurred. By accepting as successful prediction a case when the deviations are $\Delta r \leq 100$ km and $\Delta M \leq 0.7$ in the determination of the epicenter and magnitude respectively, ten EQs (out of 14) have been successfully predicted. For three EQs no predictions were issued while in one case the deviations were $\Delta r > 100$ km and $\Delta M > 0.7$-units.

A number of papers by experts on statistics have shown that our predictions cannot be ascribed to chance. On the other hand Mulargia and Gasperini (1992) claim the opposite. This disagreement initiated a debate (Geophysical Research Letters, to be published) and was also discussed in the Royal Society Meeting. It is shown that the latter calculation is not correct; the fact that the success (and alarm) rate dramatically increases for larger magnitude thresholds provides a clear proof that our predictions cannot be ascribed to chance. Furthermore the recent claims by Mulargia and Gasperini that VAN signals are postseismic effects, are shown to be invalid.

The predictions issued for the last three big EQs in Greece [i.e. $M_s(ATH)$=6.0 on May 4, 1995 close to Thessaloniki; $M_s(ATH)$=6.6 on May 13, 1995 in northern Greece; $Ms(ATH)$=6.1 on June 15, 1995 in central Greece] are described in this paper in separate Appendices.

1 Introduction. Motivation of the field experiments

For reasons of electrical neutrality, when aliovalent impurities are added to an ionic solid, vacancies (and/or interstitials) are formed. As these vacancies have "opposite" effective charge to that of the impurities, they are attracted by the impurities and hence electric dipoles are formed. For example, in the case of pure NaCl when some tens of ppm Ca^{+2} are added, an equal number of cation vacancies (in addition to the intrinsic cation and anion vacancies generated from Thermodynamic reasons) are formed thus leading to the formation of electric dipoles of the form "Ca^{+2} + cation vacancy" and hence to a drastic increase of the static dielectric constant. (These vacancies are usually called "bound vacancies"). The dipoles can change their orientation usually through jumps of the neighbouring cations (Na^{+1}) to the vacancy. The relaxation time τ of this dipole is given by:[1-3]

$$\tau = (\lambda v)^{-1} \exp\left(\frac{g^m}{kT}\right) \qquad (1)$$

where $\lambda =$ the number of jump paths accessible to the jumping species with an attempt frequency v and g^m denotes the Gibbs energy for the corresponding migration (m) process.

Equation (1) shows that the quantity τ decreases exponentially with temperature T. Furthermore it can be shown that the time τ also changes with pressure (for T=constant) mainly because the so called migration volume v^m ($\equiv \partial g^m / \partial P \mid T$) is different than zero. (An additional pressure variation of τ comes from the pre-exponential factor in eq(1), but -as explained by Varotsos and Alexopoulos[1,3]- is very small due to the fact that the frequency v varies only slightly with pressure as it results from the small values of the Gruneisen constant). The value of τ can alternatively increase or decrease with pressure depending on the sign of the volume v^m.

Pressure stimulated currents (PSC) are transient currents that are emitted from a solid containing electric dipoles (of the type described above) upon a gradual variation of pressure. These currents can be emitted either by increasing or by decreasing the pressure depending on whether v^m is negative or positive. From physical point of view, polarisation PSC arise when the pressure reaches a critical value P_{cr} in the neighbourhood of which the relaxation time becomes (under gradual variation of pressure) sufficiently small so that the dipoles re-align from a random orientation into the direction of a continuously acting (external or internal) electric field (under analogous conditions depolarisation PSC can alternatively be observed)[3]. It can be shown that the polarisation PSC maximum occurs when the following thermodynamical condition holds[3]:

$$\frac{bv^m}{kT} = -\frac{1}{\tau(P_{cr})} \tag{2}$$

where $b \equiv dP/dt$ the rate of the pressure variation and $\tau(P_{cr})$= the value of the relaxation time when $P = P_{cr}$. Varotsos et al[4] have shown that the derivation of eq(2) is quite general and holds irrespective of whether the rate b is constant or not.

The following point should be clarified: whenever v^m is negative the (polarisation) PSC is released upon increasing pressure [i.e. $v^m<0$ and $b>0$ in eq(2)]. Thus transient signal is emitted when the pressure P approaches (from smaller values) P_{cr} and diminishes when P becomes a little bit larger than P_{cr}. When the pressure still continues to increase it reaches later the fracture stress P_{fr}. Therefore the PSC emission can be considered as a precursor of the fracture of the solid; in case that b=const the lead time is given, of course, by $\Delta t = (P_{fr} - P_{cr}) / b$.

In addition to point defects line defects are present in the solids. We emphasize that the interpretation of slip can only be achieved through the consideration of the latter (i.e. the dislocations) which usually move through the migration of point defects. Ionic electrical charge resides on dislocations in non-metallic crystals as a consequence of the establishment of equilibrium between the dislocation jogs and the point defects in the bulk of crystals.[5] (We clarify that only the jogs of edge dislocations can be ionically charged). An abrupt increase or decrease in stress results in bowing out of dislocations segments between pinning points and hence effective "electric dipoles" are formed as the Debye-Huckel clouds (consisting of vacancies with opposite charge than the dislocation core) do not instantaneously follow the core.

By summarising we can state that solid state physics considerations indicate that, upon a gradual stress variation of a solid, transient electric signal should be emitted before the rupture. As we thought that similar conditions might occur during the preparation process of an earthquake, field experiments started in Greece in 1981 towards detecting precursor electric signals and continued up to date.[6-17]

2 Configuration of the network

Various experimental details were published in two recent reviews, i.e. by Varotsos and Lazaridou[11] and by Varotsos, Alexopoulos and Lazaridou.[4] Here only a brief description will be given.

Since 1982, continuous measurements of the electric field of the earth are carried out in Greece. Transient variations of the electric field of the earth (usually called Seismic Electric Signals, SES) have been found to precede earthquakes (EQs). A telemetric network of eighteen stations (via leased telephone lines) was completed in the beginning of 1983. Data are transmitted to a central station located at an Athens suburb, labelled GLY (Fig. 1) and hence the SES can be recognised on a real time basis. Technical details on this transmission can be found in an accompanied paper by Nomicos, Makris and Kefalas.

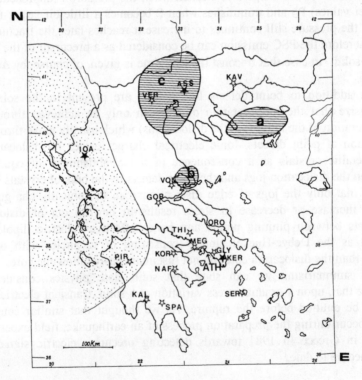

Fig. 1. Map showing the sites of the 18 VAN stations in Greece. The solid stars show the stations that operate up to date while the open stars correspond to the stations which stopped in the end of 1989. The station VOL was reactivated (but at a new site and hence is not yet calibrated) very recently, i.e. September 1994.

The sites of these stations are shown in Fig. 1. The open stars depict those stations the telemetric connection (and the operation) of which stopped (due to the lack of funds) in the end of 1989, while the solid stars indicate the four stations that are still in operation. (Therefore the predictions issued during the last five years -except that discussed in Appendix III- are based on the data collected exclusively from these 4 stations).

Non polarized electrodes Pb/PbCl$_2$ at a depth of 2m are used at all stations. A minimum of eight dipoles is installed at each station; some of these dipoles have lengths (L) between 50m and 400m and are called "short dipoles", while the others have appreciably larger lengths (L≈ usually between 2 and 20 km) and are called "long dipoles". A minimum of four short dipoles is usually installed in perpendicular directions (usually in EW and NS), e.g. two parallel short dipoles with unequal lengths in the EW direction and two others in the NS direction. We emphasize that no common electrode should exist among the dipoles. As for the long dipoles (which again do not have any common electrode) their sites should be carefully selected (as explained by Varotsos and Lazaridou[11] or in the Appendix II of Varotsos, Alexopoulos and Lazaridou[4]) so as to allow the distinction between the true precursory signals and the noise emitted from sources lying several kilometers away from the station.

Varotsos et al[4] explained in detail why a combination of short and long dipoles is absolutely necessary for the recognition of true precursors in the sense of their discrimination from noise. An electric disturbance is classified as a SES after it has met *all* four criteria mentioned in Appendix 2 of Varotsos and Lazaridou.[11]

In the following we shall briefly describe how we discriminate the SES from noise of various sources:

Possibility of picking up noise in the telemetering process. The reliability of the "real-time" telemetric system has been repeatedly checked by comparing the data collected in situ and the data recorded at the central station. An independent check came from the following experiment: within the frame of the EPOC-0045 project of the European Communities dataloggers (model 21X of Campell Scientific Inc) have been installed at the four remote stations labelled with solid stars in Fig. 1 (i.e. IOA, ASS, KER and PIR) and the data collected in situ are transmitted to Athens once or twice per day via switched telephone lines. Thus data from each of these stations were collected in two quite independent ways: (i) through the "real time" telemetric system and (ii) with the dataloggers. By comparing systematically the SESs collected in both ways no difference was noticed.

Discrimination of SES from noise of electrochemical origin. This type of noise, which is usually ascribed to a change in the contact potential between the electrodes and the ground, e.g. due to rain, can easily be recognised when parallel dipoles for each measuring direction are installed.[6-8,11] This noise is usually not recorded simultaneously at the parallel dipoles as they have independent electrodes; in the rare case that it is simultaneously recorded, the variations ΔV of the potential difference do not scale with the length of the (short) dipoles in a given direction, i.e. $\Delta V/L \neq$ constant. (We recall that one of the four criteria for the SES recognition is the requirement that $\Delta V/L \approx$ constant, in a given direction, when heterogeneities do not exist). Note also that the long dipoles are less affected by the electrochemical disturbances when compared to the short dipoles.[3.11]

Discrimination of SES from artificial (man-made) noise. We distinguish two cases:

In the case of a signal emitted from a nearby artificial source (i.e. that lying at a distance comparable with the length of the short dipoles) the criterion $\Delta V/L \approx$ constant, for a measuring direction, is violated.

In the case of a signal from a distant noise source (i.e. that lying at a distance of several kilometers from the sites at which the short dipoles are installed) the distinction between noise and SES cannot be made on the basis of the aforementioned criterion because may be $\Delta V/L$ is constant for the short dipoles. In this case a comparison is necessary between the recordings of the short dipoles and the long dipoles in a way precisely explained by Varotsos and Lazaridou[11] or in the Appendix 2 of Varotsos et al.[4]

The following point should be added: when a noise source operates continuously it becomes necessary to eliminate it from the records. A technique towards this direction has been developed[11] provided that this noise source remains permanently installed at the same site.

Discrimination of SES from magnetotelluric changes. The subtraction of magnetotelluric disturbances (MT) from the records can be easily done when measuring at a given site both the electric field E and the magnetic field H (the latter is measured at IOA with coil 3-component magnetometers). By following the usual procedure, as explained by Hadjioannou et al,[14] the residual E-ZH (where Z is the impedance tensor) is estimated and hence the SES is better defined.

The following remark should be made: when a network of 18 electrotelluric stations is operating is very difficult for the magnetotelluric changes to be misinterpreted as SES. This is so because MT appear simultaneously at all the stations in contrast to the SESs which are recorded only at a restricted number of them[7] (e.g. the maximum number of stations at which an SES was simultaneously recorded was 4 while MT were recorded at *all* the 18 stations). Note that during the period 1989-1994, when only four stations were in operation, the SESs were recorded *only* at one station.

As it will be explained below, an SES is better defined at the direction in which the MT noise becomes less. Figure 2a depicts the sites of the electrodes of the three long dipoles at IOA. (Note there are two long dipoles with independent electrodes that connect Perama-village with the site of the station, i.e. the site at which the short dipoles have been installed; an additional long dipole connects the station to Ioannina city). The sites a, b, c of the short dipole arrays at IOA are depicted in Fig. 2b. (In addition to these short dipoles, 10 other short dipoles are currently operating there). Simultaneous measurements of the magnetic and electric field (Figs 3a and 3b respectively) at site b indicate that there is a strong polarisation of the electric field along the direction $N_{22}W$. This effect reflects the following: let us compare the electric field variations measured along the directions NS, EW at site b (Fig 4A) and those resulted after rotating the axes NS and EW at site b by 22 degrees anticlockwise $\{E(\partial)=R(\partial)\ E(t)\}$. It is obvious that the MT-noise on NS(-22) at b (Fig. 4B) is very small (and hence the SES marked is clearly defined). Detailed experimentation has also led to the determination of the site "c" in which the direction NS was found to be the direction of local channelling (and hence the direction at which MT noise is less). It is evident that the SES can be easily recognised either at NS at site "c" (Fig 4A) or at NS(-22) at site "b" (Fig 4B).

3 Summary of SES physical properties

Lead time. The various types of electrical precursors detected in Greece are summarised in Fig. 19 of Varotsos, Alexopoulos and Lazaridou.[4] Three of them, which are used for actual prediction purposes, can be summarised as follows:

(a) Gradual variation of the electric field of the earth (GVEF) which is a long duration anomaly and has a lead-time Δt of the order of 1 month; it has an amplitude one order of magnitude larger than that of the SES[3,7].

(b) Single SES with Δt around 11 days[15] (or smaller[6]).

(c) SES electrical activities (i.e. many SESs within a relatively small time, e.g a recent example is given in Figs 5,6). This case seems to follow the time evolution depicted in Fig 7. This evolution can be summarised as follows: a time period of at least 3 weeks is elapsed between the initiation of the SES activity and the occurrence of strong EQ (small shocks may start earlier). The strongest EQ usually occurs during the fourth week; otherwise smaller EQ(s) with magnitude around 5-units appear during this week and the strongest EQ occurs after an additional period of 2-3 weeks. The latter behaviour, observed for the first time in 1988 (first time-chart in Fig. 7a)[11,16], has the advantage that the epicentre of the strongest EQ becomes known with good accuracy as it

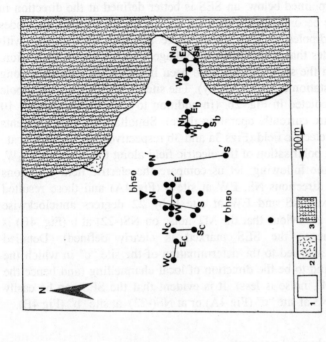

2a. Long dipoles; 1: quaternary sedimentary deposits; 2: alpine formation of the Ionian Unit (mainly limestones).

2b. Short dipoles in the region of Fig. 2a shown by a square; 1: alluvial deposits; 2: flysch of the Ionian Unit; 3: limestones of the Ionian Unit. The triangles stand for the two boreholes (bh).

Fig. 2. Configuration of electrodes at IOA station. The geological map is made after a recent reconnaissance by D. Papanikolaou and E. Logos (1993) (unpublished data) (it is slightly different than that published by Varotsos, Alexopoulos, Lazaridou and Nagao (1993) which was based on the official IGME geological map):

will lie in the vicinity of the preceding smaller EQ(s) which occurred a few weeks earlier.

SES amplitude. For SESs registered at a given station and originating from a given seismic area, the SES amplitude expressed as $\Delta V/L$ (for a dipole with given orientation) scales with the magnitude according to the relation

$$\log(\Delta V/L) = (0.34\text{-}0.37)\, M + b$$

The plots for dipoles of the two orientations have the same slope but different intercepts.

SES polarity and the ratio of two SES components. For a dipole of given orientation (e.g. EW or NS) a SES arriving from a given seismic area always has the same algebraic sign.

For each pair "seismic area-station" the ratio of the amplitudes in the EW and NS- directions, i.e. $(\Delta V/L)_{EW} / (\Delta V/L)_{NS}$, remains the same for various EQs from the same seismic area.

Selection of sites appropriate for SES collection. Experiments show that SES cannot be observed at all points of the earth's surface but only at certain points called *"sensitive points"*. (We have found other sites, called *"insensitive"*, operating for a long time, e.g. more than 5 years, in which no SES was recorded). Therefore the installation of a (permanent) station appropriate for SES collection (which is called "sensitive") should necessarily be preceded by a tedious experimentation described by Varotsos, Alexopoulos and Lazaridou:[4] In short, a number (e.g. 10) of temporary low noise stations is installed and only after a long period, i.e. after the occurrence of several significant EQs from a given seismic area, can we select the most appropriate (if any) of these temporary stations that happen to collect SES. Although general rules cannot be drawn, it seems that the areas which have the most potential to be sensitive are: (1) the vicinity of a major fault, (2) regions of crystalline rocks close to large heterogeneities, such as geological contacts with significantly different conductivities, (3) area with strong local inhomogeneities.

We empasize that the stations depicted in Fig. 1 have been selected to be SES sensitive after a long experimentation.

Selectivity effect. *Selectivity* refers to the experimental fact that a (sensitive) station is capable to collect SES only from a restricted number of seismic areas. The following may happen: a given station S_A situated at A may be able to record SES originating at seismic area B but not at area C, even if $r_{AC} < r_{AB}$.

A map showing the seismic areas that emit SES detectable (for EQs above a magnitude threshold of course) at a given station is called *"selectivity map of this station"*. In Fig. 1 the western large dotted area corresponds to the selectivity map of IOA. (There is a suspicion that this selectivity map extends more to the south, not including KAL area, and hence IOA can collect SES

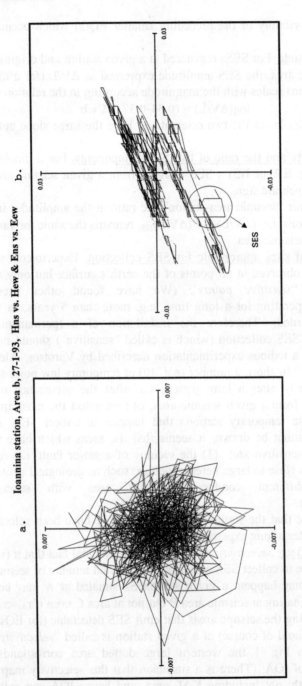

Fig. 3.   Variations of the magnetic field (Fig. 3a) and the electric field (Fig. 3b) at the site b of IOA station. The mark SES corresponds to the SES depicted in Fig. 4 and indicates that its polarisation is different from that of the MT. The electric field (see the text).

a.   Ioannina station, Area b, 27-1-93,   Hns vs. Hew & Ens vs. Eew   b.

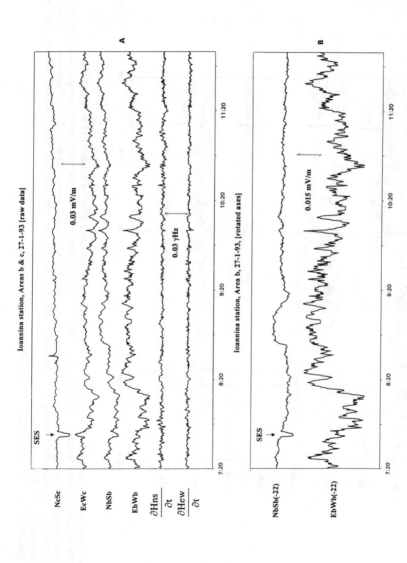

Fig. 4. SES collected at IOA station on January 27, 1993. A: the two channels at the bottom refer to the two horizontal magnetometers and the other channels to the electric measurements in EW and NS directions at the areas b and c shown in Fig 2b B: Calculated values of the electric field at site b after rotating the axes EW and NS by 22 degrees anticlock-wise.

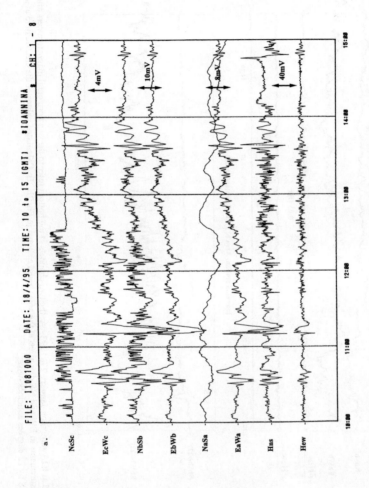

Fig 5. SES (electrical) activity recorded at IOA on April 18, 1995 which led to the prediction of the 6.6 EQ described in Appendix II.

a. The two channels at the bottom correspond to recordings of the time derivative of the magnetic field measured along the directions EW and NS which indicate that no significant change is associated with the SES activity. The other six channels correspond to the electrical variations recorded at sites a, b and c shown in Fig. 2b. The scale "8mV" corresponds to EaWa and NaSa (the latter is out of order), the scale "10mV" to EbWb while the scale "4mV" to the last three channels NbSb, EcWc and NcSc.

b. Recordings of the three long dipoles the sites of which are shown in Fig. 2a.

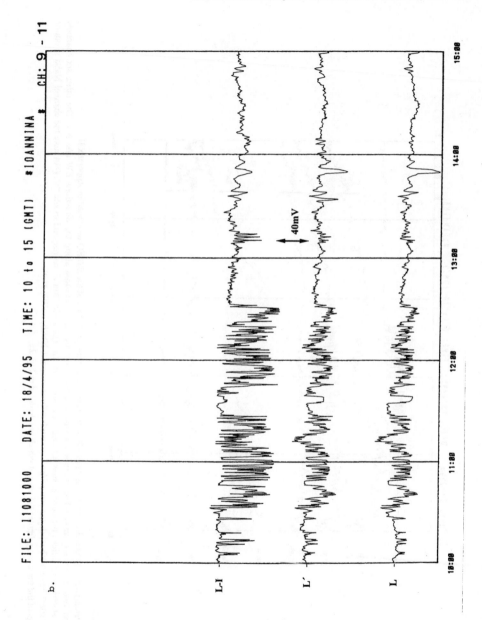

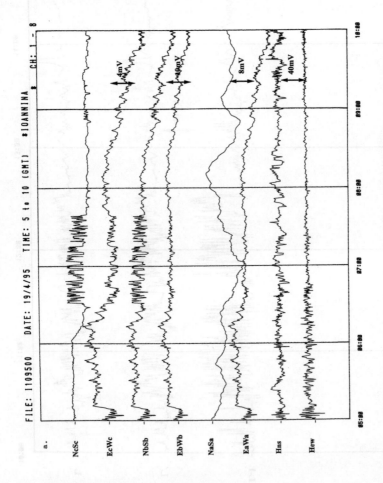

Fig. 6. SES (electrical) activity recorded at IOA on April 19, 1995 which led to the prediction of the 6.6 EQ described in Appendix II.
a. The lower two channels correspond to the time-derivative of the magnetic field measured along the directions EW and NS. The other six channels correspond to the electrical variations recorded at sites a,b and c shown in Fig 2b. The scale "8mV" correspond to site "a" (but note the channel NaSa is out of order), the scale "10mV" to EbWb while the scale "4mV" to the last three channels NbSb, EcWc, NcSc.
b. Recordings of the three long dipoles the sites of which are depicted in Fig. 2a.

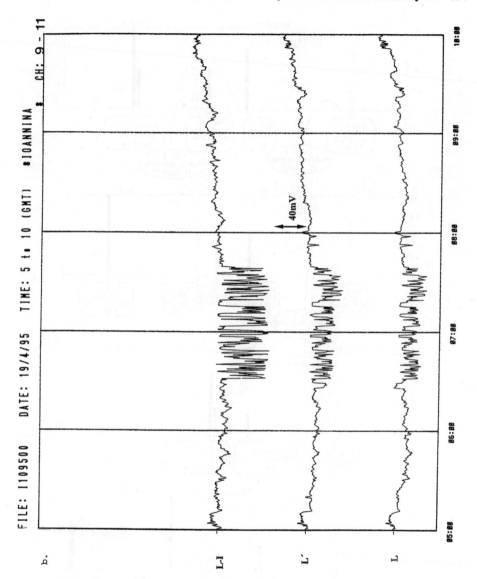

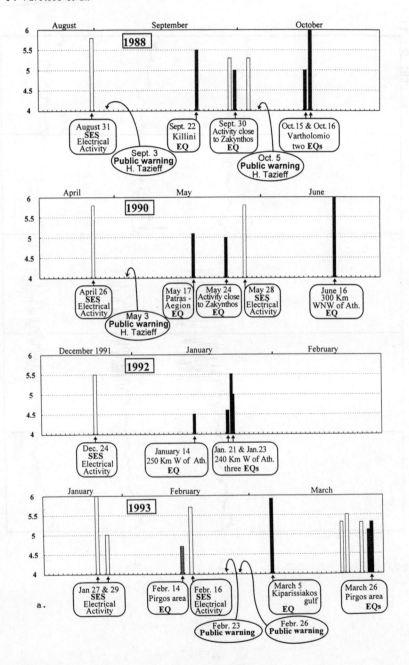

Fig. 7a,b. Time-evolution of a number of SES (electrical) activities observed since 1988. Open bars and full bars correspond to SES activities and EQs respectively.

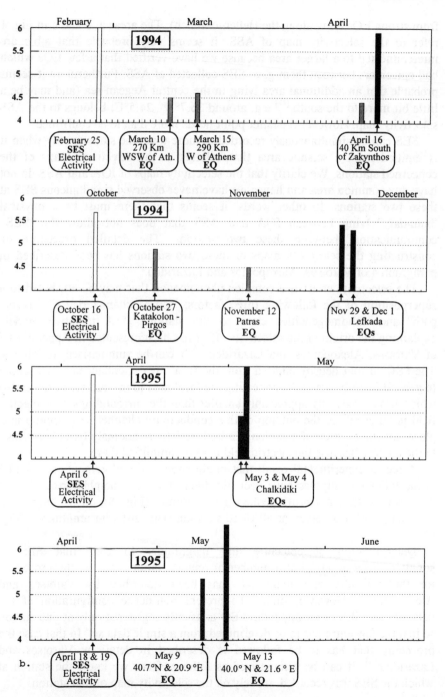

b.

from strong EQs lying along the Hellenic trench). The areas a, b and c in Fig. 1 refer to the selectivity map of ASS. It seems very probable that a,b,c are interconnected to a larger area because we have verified that a few EQs which had epicenters between them gave SES collected at ASS. Furthermore it seems probable that an additional area lying in the central Aegean sea (and may be a little bit more to the south,[17] e.g. around 36.7°N , 24.5°E) belongs to the ASS selectivity map. However the latter point needs more data to be confirmed.

SES can be simultaneously recorded at more than one stations only when it is emitted from a seismic area that belongs to the selectivity maps of the concerned stations. We clarify that the selectivity maps of IOA and ASS do not have any common area and hence we have never observed simultaneous SES at these two stations. In other words, it seems that there must be a physical "boundary" lying between IOA and ASS that does not allow the "SES communication" between these two stations. The detailed procedure of constructing the selectivity maps of these two stations has been described in paragraph 4 of Varotsos, Alexopoulos and Lazaridou.[4]

The selectivity effect is a complex phenomenon that may be attributed to a superposition of the following three factors: "source characteristics", "travel path" and "inhomogeneities close to the station".[11] A tentative model explaining the origin of the selectivity effect has been presented in Appendix IV of Varotsos, Alexopoulos and Lazaridou.[4] It can be summarised as follows: The PSC current density emitted from the focal area is enhanced at a channel (e.g. a sufficiently wet path connecting the focal area to the surface of the earth) which has a resistivity appreciably smaller than the surroundings. The electric field is larger along the outcrop of this conductivity channel. An "appropriate place" for the location of a station is a high resistivity anomaly lying very close to the outcrop because the electric field is again amplified there.

A recent experimental result that might support the aforementioned model is the following: dipoles have been installed at various neighbouring sites at ASS with different (local) geological conditions (Fig. 8); SESs have been (simultaneously) observed at all these sites but with different amplitudes (Fig. 9).

Determination of epicentre and magnitude. In case that SES is simultaneously observed at a number of remote stations (e.g. 4) the procedure for the epicentre and magnitude has been described by Varotsos and Alexopoulos.[7] However, under the present non-dense configuration of the network, the SES of an impending EQ is usually registered only at one station so that the epicentre has to be determined from a single data set. In that case the procedure that has to be followed is described in detail by Varotsos and Lazaridou[11]. It can be summarized as follows: Once we know the station at which the SES was recorded, we rely on the selectivity map of that station.

Then by considering the SES polarity and the exact value of the ratio $(\Delta V/L)_{EW}$ / $(\Delta V/L)_{NS}$ we select as a candidate epicentre the region (s) of the selectivity map which, from earlier experience, is consistent with those SES data. Then by relying on the plot "log $\Delta V/L$ vs M" that corresponds to this concrete pair: "station-candidate seismic area" we estimate the magnitude.

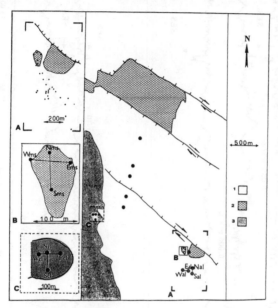

Fig. 8. Configuration of short dipoles at ASS station. The geological map comes from a recent reconnaissance by D. Papanikolaou and D.G. Syskakis (1993, unpublished data). 1: terrestrial deposits; 2: mica gneiss-schists; 3: crystalline limestone.

4 Predictions issued by the VAN telemetric network

The real time collection allows the recognition of the electrical precursor(s) and the analysis well before the EQ occurrence and hence predictions are issued well in advance. Table 1 summarises all the predictions issued during the last 8 years, i.e.from January 1, 1987 until June 15, 1995. Note that since May 15, 1988 all predictions were sent not only to the Greek authorities but also to scientific institutes in other countries (USA, Japan, France, Sweden, etc.).

Three examples of the recent predictions can be found in the Appendices. Each prediction usually contains the following information:

(1) the date and the station(s) at which the SES were recorded (sometimes copies from the original recordings are also attached). It is always clarified whether the prediction is based on single SES or a SES electrical activity (and

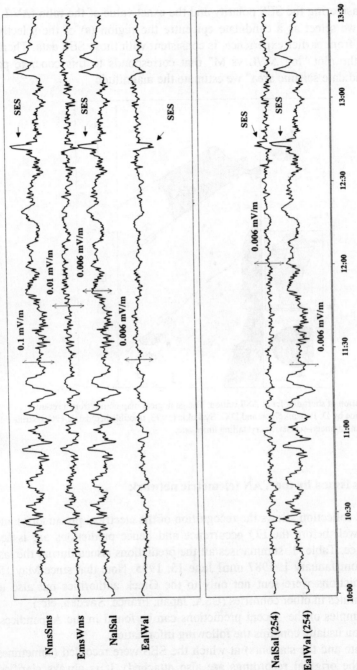

Fig. 9. An SES at ASS station simultaneously recorded at two neighbouring sites: alluvial deposits (al), and mica-schist (ms). Note that this SES was also recorded at the long dipole EmsWcl (L=1.7 Km) and at another long dipole EmsWcl (L=11.4 Km, which has one electrode at Lagada and the other close to the short dipoles of the station. The lower two channels correspond to the calculated values of the electric field at site "al" after rotating the axis EW and NS by 254 degrees (clockwise).

rarely to GVEF). The latter information indicates the time-window of the expected earthquake(s).

(2) estimation of the expected epicentral area.

(3) estimation of the magnitude of the impending earthquake. We emphasize the following fact: since our first publication we clarified that our stations have been calibrated in respect to the magnitude officially announced by the Seismological Instistute of the National Observatory of Athens (SI-NOA). This magnitude labelled Ms in the Preliminary Seismological Bulletin (PSB) of SI-NOA is derived from the approximate formula $Ms=M_L+0.5$ where M_L the local magnitude measured by SI-NOA. As this magnitude does not coincide with the surface wave magnitude Ms reported by USGS and in order to avoid confusion, in the following it will be labelled Ms(ATH).

In accord with an official decision in 1985, predictions were issued *only* when the expected magnitude Ms(ATH) is larger than (or equal to) 5.0-units. The predictions issued during the 3 year period 1987-1989 were published by Varotsos and Lazaridou[11] while those of the subsequent period Febr. 6, 1990 - May 31, 1992 can be found in the paper by Varotsos, Alexopoulos, Lazaridou and Nagao.[17] The texts of the predictions from July 1, 1992 to September 5, 1993 are reported in Table 2 by Varotsos, Uyeda, Alexopoulos, Nagao and Lazaridou[13] while those of the period Sept. 5, 1993 to June 15, 1995 are given in Table 2 of this paper. We recall that the summary of the totality of the predictions issued during the 8-year period 1987-1995 was already given in Table 1.

In Fig. 10 we give time-charts of the totality of the predictions issued during the period 1987-1994 along with the totality of the EQs within the area $N_{36}^{41}E_{19}^{25}$. For the help of the reader we present, for each 2-year period, two time charts: one which includes the totality of predictions along with the totality of EQs with Ms(ATH)$\geq$5.3 and another one depicting the totality of predictions along with the totality of EQs with Ms$\geq$4.3. The scope of these time-charts is to show the following: when we consider a low magnitude threshold for the EQs (i.e. Ms(ATH)$\geq$4.3), there are many events which might arouse arguments concerning "chancy correlations" between some predictions and EQs; on the other hand when increasing the magnitude threshold of the EQs, the number of the EQs drastically decreases and hence the reader can better visualise their correlation with the predictions. In the latter time-charts (i.e. those with Ms(ATH)$\geq$5.3) there are long time periods during which no predictions was issued and no significant EQ occurred. This is a convincing evidence concerning the existence of causality between SES and EQs. A detailed statistical treatment of this problem has been published by Hamada.[18]

Predictions issued for the most significant earthquakes: In Table 3 we give **all** the EQs that occurred within $N_{36}^{41} E_{19}^{25}$ (see Fig. 11) during the eight year

Table 1. Summary of all Predictions issued from January 1st, 1987 to June 15,1995

No.	Date	Lat., N	Lon., E	M	Remarks
a	26/2/1987	37.94	20.32	6.5	
b	27/4/1987	37.7	21.5	5.5	This is a Gradual Variation of the Electric Field of the earth *(GVEF)* which is observed before big events *(M≥5.5)* almost one month before the EQ. (see p. 340 of Ref 11)
c	13/6/1987	38.0	21.5	5.2	
d	1/2/1988	39.3	25.4	5.0	
e	10/3/1988	40.2	20.8	5.0 i	
		or 38.8	21.0	5.0	
f	2/4/1988	38.0	20.9	5.0	
		or 36.0	21.4	5.5	
g	3/4/1988	38.9	23.8	5.0	
h	7/4/1988	38.8	21.1	5.0	
		or 40.2	20.8	5.0	
j	28/4/1988	37.9	20.3	5.0 i	
		or 39.0	20.5	5.0	
1	15/5/1988	37.9	20.3	5.3	
2	21/5/1988	37.9	20.3	5.3 i	
		or 40.2	20.8	5.0	
3	30/5/1988	37.9	20.3	5.4 i	
		or 40.2	20.8	5.0	
4	4/6/1988	37.9	20.3	5.0	
5	10/6/1988	36.7	22.2	5.1	
6	21/6/1988	37.9	20.3	5.0 i	
		or 40.2	20.8	4.8	
7	10/7/1988	37.1	21.2	5.2	
8	13/7/1988	38.0	23.0	5.0	
9	18/7/1988				Uncertain, NNW 80 or SW 100.
* 10	1/9/1988	38.0	21.0	5.8	
		or 39.9	21.3	5.3	
* 11	30/9/1988	38.0	21.0	5.3	
		or 40.0	21.0	5.0	
* 12	3/10/1988	38.0	21.1	5.3	
		or 38.0	23.0	5.0	
13	21/10/1988	38.0	21.0	6.4 i	Several tenths of Km away from this epicenter.
		or 40.5	20.4	5.5	
14	2/3/1989	40.1	21.0	5.0	
		or 37.9	20.3	5.4	
15	3/6/1989	37.9	20.3	5.5	
		or 40.2	20.8	5.0	
16	13/6/1989	38.0	21.5	5.2 i	
		or 40.2	20.8	4.8	
17	23/7/1989	38.2	24.1	5.0	
18	16/8/1989	38.7	21.6	5.0	
19	24/8/1989	38.6	21.7	5.2	
		or 37.1	21.2	5.8	
20	11/9/1989	38.6	21.7	5.2	
		or 37.1	21.2	5.8	
21	15/9/1989				Uncertain, SES from *ASS (THES)*.
* 22	18/10/1989	39.9	21.3	5.3	
		or 38.0	21.0	5.5	
* 23A	26/4/1990	38.0	21.0	5.8	"...a displacement of the epicentral zone of Kilini-Vartholomio (i.e. 38.0° W 21.0°E) by a few tens of Kilometers is expected"
* 23A'	28/5/1990	38.0	21.0	5.8	"

Continuation of Table 1

No.	Date	Lat., N	Lon., E	M	Remarks
24	20/10/1990	38.0	20.8	5.6	
★ 25	2/1/1991	40.7	23.1	5.5	
★ 26	23/2/1991	40.0	21.3	6.0	
★ 27	27/12/1991	38.0	21.0	5.7 i	
		or 40.5	20.0	5.0	
27a	27/1/1992	38.0	21.0	5.2	This is the **largest** expected magnitude.
★ 28A	22/2/1992	38.0	20.5	5.0	"
★ 28B	22/2/1992	38.6	21.7	5.0	"
★ 28C	22/2/1992	37.2	21.4	5.4	"
		or 36.5	21.9	6.0	"
★ 29	13/3/1992	38.0	25.2	5.4	
		or 40.0	26.5	5.4	
30	1/4/1992	38.0	21.0	5.3	
★ 31	22/4/1992	38.2	21.4	5.0	
		or 37.4	21.3	5.5	
32	11&13/11/1992	38.4	22.0	5.0	Issued by PAT.
★ 33	30/11/1992	39.7	20.9	5.5	
★ 34	30/1/1993	38.0	20.9	6.0	
		or 39.7	20.9	5.0	
34b	26/2/1993	37.6	21.0	6.0	Public announcement by P. Varotsos which further clarified (based on additional SES at *IOA* on February 16, 1993) the epicentral area expected by the previous prediction.
★ 35	25/3/1993	38.0	20.9	6.0	Continuation of No. 34 & 34b.
		or 39.7	20.9	5.5	
★ 36A	5/4/1993	37.5	21.3	5.3	
★ 36B	5/4/1993	40.2	25.0	5.5	
		or 40.8	23.1	5.0	
★ 36A'	12/4/1993	39.1	20.4	6.0 ii	
★ 37	14/6/1993	38.4	22.2	5.0	
		or 39.1	20.4	6.0	
38	23/7/1993	37.5	21.3	5.3	
39	2/8/1993	38.0	23.0	5.0	
40	23/8/1993	37.5	21.3	5.3	
41	2/11/1993	38.4	21.5	6.0	
★ 42	28/2/1994	37.4	21.3	6.0 ii	
★ 43A	14/5/1994	37.3	21.1	6.2	
		or 38.5	21.2	5.7	
★ 43B	14/5/1994	38.5	24.3	5.0	
★ 44	16/10/1994	38.9	20.9	5.0 ii	
		or 38.0	21.0	5.7	
★ 45	11/12/1994	40.8	23.1	5.0	
★ 46	15/1/1995	40.8	23.1	5.0	
★ 47	23/2/1995	38.0	23.7	4.5	Quite uncertain as it does not correspond to SES; although *Ms<5.0* it was issued because the epicenter could lie close to Athens.
★ 48	7/4/1995	40.8	23.1	5.2-5.8 iii	
★ 49	30/4/1995	39.8	20.8	5.5-6.0 iii	
★ 50	19/51995				see Appendix III for details

★   Signifies SES electrical activities.

i. This is the most probable solution (between the two alternatives) mentioned in the prediction text.

ii. Case with $\Delta t \cong 7 \pm 1$ weeks (for the last and strongest EQ of the expected seismic sequence) as in Fig. 28B of Varotsos and Lazaridou (1991).

iii. See the detailed text of the prediction (Appendices I and II)

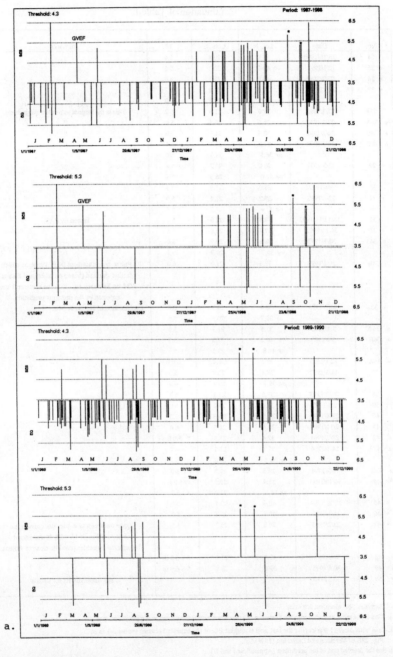

Fig. 10. Time chart showing the totality of predictions (upper) along with the earthquakes (down) during Ms(ATH)≥4.3 or

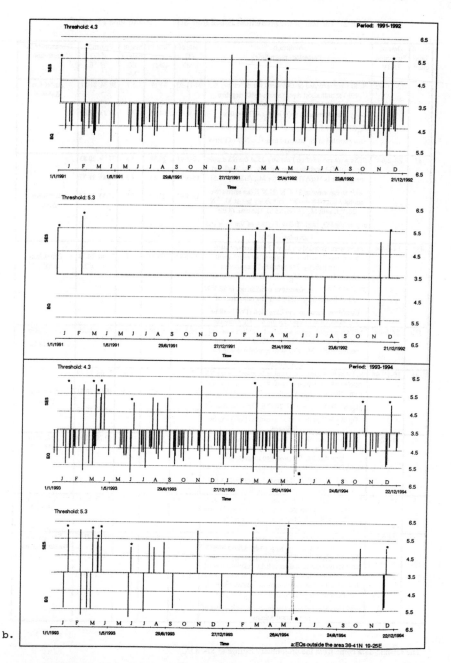

the period 1987-1994. For each two year period we plot two cases: the totality of EQs either with
with Ms(ATH)≥5.3.

TABLE 2: The text of the predictions issued from Sept. 5, 1993 to June 15, 1995[a]

No of Tele-gram	Date of Prediction DMY	Prediction (Epicentre - magnitude)	Date of EQ DMY	Time of EQ GMT	Magni-tude of EQ Ms	Epicentre of EQ °N °E	Deviation of the prediction
41	02-11-93	SESs were recorded on October 23 and October 31, 1993 at both short dipole arrays (with negative polarity) and at the long dipoles of IOA. The predicted magnitude(s) is M_S(ATH)≈6.0 or M_B(USGS)=5.5. The epicenter(s) might be within the area depicted in the attached map (i.e. NW of Patras city).	04-11-93	05:19	5.4	38.34 21.91	Δr≤40Km ΔM=0.6
			22-11-93	14:13	4.8	38.23 22.89	
42	28-2-94	SESs were recorded on February 25 at the NS dipole array of IOA with negative polarity. The predicted magnitude is Ms(ATH) ≅6.0 (or M_B (USGS) =5.5) with an epicenter at 37.4° N 21.3° E (an alternative solution but with smaller probability lies at 39.5° N 20.3° E with M_S (ATH) ≅5.5). The time of the impending EQ_S might follow Fig28A of Varotsos and Lazaridou (1991).	10-3-94	20:39	4.3	37.09 20.83	
			16-3-94	07:15	4.3	36.88 21.77	
			16-4-94	23:09	5.8	37.33 20.63	Δr <80Km Δî=0.2
43A	14-5-94	SESs were recorded on May 12 (the activity started since May 3) at the NS dipole array of IOA with negative polarity. The predicted magnitude is M_S(ATH)=6.2 (or M_B=5.7) with an epicenter at 37.3° N 21.1° E (an alternative solution lies at 38.5° N 21.2° E with M_S(ATH)=5.7).The time-chart of Fig28A by Varotsos and Lazaridou (1991) will be followed, i.e.3-4 weeks for the first big event etc.	23-5-94	06:46	6.0	36.48 24.70 (?)	Δr≅300km Δ M≅0.2
43B	14-5-94	SESs were recorded at KER (not simultaneously with the aforementioned SESs at IOA). They might correspond to M_S ≅5.0 at an epicenter 80Km NE from Athens.	24-5-94	02:05	6.1 or 5.4	38.71 26.29	Δr≅170Km 0.4≲ΔM≲1.1
44	16-10-94	SESs were recorded on October 16 mainly at the NS dipole array with positive polarity. The predicted EQ parameters are M_S(ATH)=5.7 at 38.0° N 21.0° E or M_S(ATH)=5.0 at 38.9° N 20.9° E. The time-chart of Fig28A by Varotsos and Lazaridou (1991) will be followed.	27-10-94	07:02	4.8	37.68 20.99	
			29-11-94	14:30	5.4	38.68 20.51	Δr≅50Km ΔM≅0.4
			01-12-94	07:17	5.3	38.66 20.48	
45	11-12-94	SESs were recorded on December 3 at ASS at both short dipole arrays (with negative polarity) and at the long dipoles. The predicted magnitude is M_S(ATH)=5.0 (or M_B (USGS)=4.5) with an epicenter at 40.8° N 23.1° E. The time-chart of Fig28A by Varotsos and Lazaridou (1991) will be followed.	5-1-95	04:48	4.2 (THES 4.4)	40.70 22.96	Δr<20Km ΔM≅0.7
46	15-1-95	SESs were recorded on January 15 at ASS similar to those reported in the previous prediction (and hence the predicted EQ parameters are the same as previously).	13-2-95	13:16	4.8	40.73 22.72	Δr<50Km ΔM=0.2
47	23-2-95	Unusual electrical variations were recorded on February 20 and 21 at KER (only at three long dipoles) which do not have SES properties, i.e. they do not correspond to electric field variations related with a remote sourse. In case that the SES calibration holds for these unusual electrical variations as well, an EQ with M_B(USGS) ≅4.0 (or M_S(ATH) ≅4.5) might be expected at an epicenter lying a few tens of Km around Athens.	3-3-95	00:36	3.4	38.51 23.74	Exceptional (uncertain) case when Mpred <5.0
			29-3-95	07:17	3.0 (3.5)	38.26 23.67	
			18-4-95	15:32	3.6 (3.9)	38.42 23.61	

The seismic data come from the preliminary listings of SI-NOA.

a. The full text of the additional predictions issued on April 7 (n[0] 48), April 30 (n[0] 49) and on May 19, 1995 (n[0] 50) are given in the Appendices I, II and III respectively.

TABLE 3. All earthquakes with Ms (ATH) ≥ 5.8 within N^{41}_{36} E^{25}_{19}
and the deviations of their predictions. Period: Jan. 1, 1987-June 15, 1995

EARTHQUAKE					PREDICTION	
		EPICENTER				
No.	DD-MM-YY	°N	°E	Ms(ATH)	Δr(Km)	ΔM
1	27/02/87	38.37	20.42	5.9	<50	-0.6
2	18/05/88	38.39	20.44	5.8	<50	+0.5
3	16/10/88	37.90	20.96	6.0	<30	+0.7
4	19/03/89	39.29	23.57	5.8	missed [a]	
5	20/08/89	37.24	21.12	5.9	unsuccessful [b]	
6	16/06/90	39.13	20.38	6.0	~100	+0.2
7	21/12/90	40.95	22.43	5.8	missed [c]	
8	05/03/93	37.07	21.46	5.8	~80	0.0
9	13/06/93	39.26	20.58	5.9	<40	+0.1
10	25/02/94	38.73	20.58	5.8	missed	
11	16/04/94	37.43	20.58	5.8	<30	0.0
12	04/05/95	40.5	23.5	6.0	~40	+0.2 [d]
13	13/05/95	40.0	21.6	6.6	~70-80	+0.6 [e]
14	15/06/95	38.5	22.2	6.1	~40-50	-0.5 [f]

a. Absence in a Conference, see Ref(11)
b. Prediction No. 18 in Table 1.
c. See Ref(17)
d, e, f. Their predictions are discussed in the Appendices I, II, and III respectively
The seismic data comes from SI-NOA monthly catalog except of the epicentres of the last three EQs which
are taken from USGS.

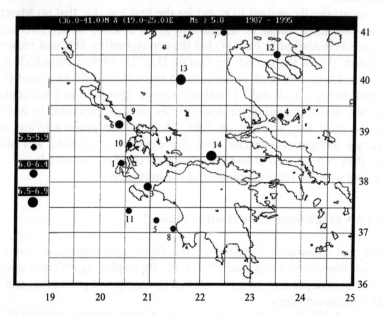

Fig. 11. Map showing the epicenters of the totality of EQs with Ms(ATH)≥5.8 within N^{41}_{36} E^{25}_{19} ,
during the period Jan. 1, 1987 to June 15, 1995. Ten out of these 14 EQs were successfully
predicted (see Table 3).

period Jan 1, 1987 to June 15, 1995 along with their corresponding predictions. An inspection of this Table indicates that for ten out of these 14 EQs successful predictions ($\Delta r \leq 100$ km, $\Delta M \leq 0.7$) have been issued, for one EQ the prediction had larger deviations (and hence it was characterised as unsuccessful) while three EQs were missed (i.e. no prediction was issued). Among these ten successes, for seven of them the accuracy of the prediction of the epicentre was better than 50 km. The following remarks could be added:

(1) the prediction of the 5.8 EQ on May 18, 1988 was confirmed during an International Conference held at Sweden,[10,11]

(2) the two predictions of the EQs on Oct. 16, 1988 and March 5, 1993 have been publicly announced as described in Refs(11) and (12).,

(3) the prediction of the EQ of Jun. 16, 1990 -in view of the impending big magnitude- was also addressed to the Prime-Minister,[17]

(4) the first prediction the correctness of which was officially confirmed by the Government was that of the EQ on April 16, 1994,

(5) for the EQs on April 16, 1994 and on May 4, 1995 the Government not only confirmed their successful predictions but also clarified that countermeasures were taken on the level of the public authorities in the expected areas,

(6) concerning the missed EQs: (a) for the first one, i.e. that on March 19, 1989, no prediction could be issued due to an absence in an International Conference as described in Ref.(11) but the corresponding SES was recognized in retrospect, (b) for the second, i.e. that on Dec. 21, 1990, although clear SES activity was recognized well in advance no prediction could be issued [it was the period during which only 4 stations were in operation and we could not achieve any determination of the epicentre for reasons explained in Ref.(17)], (c) for the third one, i.e. that on Febr. 25, 1994, no prediction was issued because we thought, incorrectly, that the expected magnitude was slightly smaller than 5.0 (due to a bad estimation of the SES amplitude as it was overlapping with the some magnetotelluric disturbances).

5 Can VAN predictions be ascribed to chance?

Varotsos and Lazaridou[11] published the three year continuous sample of predictions that covers the period 1987-1989. By employing different statistical methods, five independent groups evaluated this sample of predictions and drew the following conclusions:

(i) Hamada[18]: *"... with a confidence level of 99.8% it is rejected that this success rate can be explained by a random model of earthquake occurrence taking into account a regional factor which includes high seismicity in the predictions area...".*

(ii) Shnirman, Shreider and Dmitrieva[19]: *"... the earthquakes and the VAN prediction telegrams are in obvious correlation if we select both for strongest magnitudes ..."*

(iii) Nishizawa, Lei and Nagao[20]: *"... results show that the model assuming seismic electric signals as precursors of earthquakes gives the best fit to the data... The present results show the causal relationship between SES and earthquakes without any a priori models associated with the lead time... The results show that SES are not post-seismic events".*

(iv) Uyeda[21]: *"... the actual success rate and alarm rate ... are both estimated to be 60%...".*

(v) Mulargia and Gasperini (MG)[22]: *"... the claimed success can be very confidently ascribed to chance; ... VAN predictions show a better association with the events which occurred before them...".*

It is obvious that MG-claims disagree with the conclusions of the other four groups. MG calculated the significance level (s.l.) -which indicates the risk that one takes when the predictions are regarded as "non-chancy"- and found that for various magnitude thresholds it is appreciably larger than 0.05. An independent investigator, i.e. Takayama,[23] immediately replied that MG-calculation is wrong for the following reason:

"This method (cf. the MG-method) is valid when two event series are geographically fixed. However VAN method predicts an epicentre whose position changes each time. Therefore, the prediction in space domain should also be taken into consideration. ... If seismic activity is geographically homogeneous in the examined domain, the probability that an earthquake occurs in the predicted domain by chance decreases by a factor of $\Delta s/S$ where S is the area of the entire domain and Δs is $\pi(\Delta r)^2$... We concluded that in cases marked by *, the statistical hypothesis that VAN earthquake precursors were observed by chance is rejected with the significance level of 0.05".

The above factor has a drastic influence on the results. For example, for the case $M_{pred} \geq 5.8$, $M_{EQs} \geq 5.1$, $\Delta t \leq 22$ days, $\Delta r \leq 30$ km (Table 1 of MG), the initial MG value was s.l.=0.859 thus ascribing these predictions to chance. On the other hand, when including the above factor, Takayama found for this case s.l.=0.020 to 0.021 instead of s.l.=0.859.

In addition to the above serious deficiency of MG calculation, there are others explained in a paper by Varotsos, Eftaxias, Vallianatos and Lazaridou[24] submitted to Geophysical Research Letters in November in 1993. This paper suggested five Principles which have to be followed in an evaluation of an earthquake prediction method but were violated by Mulargia and Gasperini; it initiated a debate in which a significant number of researchers participated. As the debate will be shortly published in a separate issue we do not refer to its contents but we restrict ourselves here (see Appendix IV) only to Mulargia and Gasperini who participated and expressed their opinion in this Meeting.

6 Conclusions

The SES discrimination from various noise sources is based on a strict application of physical laws and the prediction of the epicentre and magnitude of the impending EQ(s) can be achieved when the SES physical properties are properly combined.

As regards VAN predictions of the strongest earthquakes [e.g. $M_s(ATH) \geq 5.8$] -which can confidently not be ascribed to chance- the results of the last eight years show that the alarm rate is around 70% while the success rate is even larger.

As for the three recent EQs, with $M_s(ATH) \geq 6.0$, which occurred at three different sites of continental Greece during May-June 1995, successful predictions (described in the Appendices I to III) were issued well in advance.

APPENDIX I

Prediction of the Ms(ATH)=6.0 earthquake on May 4, 1995 in Chalkidiki

In Fig. 12 we give a photocopy of the prediction issued on April 7, 1995. It was based on an SES activity collected on April 6, 1995 at ASS station which lies very close to Thessaloniki. The three parameters can be summarised as follows:

Concerning the time-evolution of the activity it was estimated that Fig. 28A of Varotsos and Lazaridou (i.e. the case of Killini-Vartholomio EQs in 1988, see also the first time chart of Fig. 7a of the present paper) or Fig. 22 of Varotsos et al (1993), might be followed. The predicted magnitude was $M_s(ATH)=5.2$ to 5.8 depending on the epicentral distance of the expected EQ from ASS, i.e. 5.2 when the epicentre will lie close to ASS or up to 5.8 when it is close to the southern coast of Chalkidiki peninsula. (We clarify that the prediction text allowed the epicentre to be close to Chalkidiki peninsula,

THE UNIVERSITY OF ATHENS

DEPARTMENT OF PHYSICS
SOLID STATE PHYSICS SECTION
Knossou st. 36, Ano Glyfada
Athens 165 61, GREECE

P. Varotsos
Professor of Physics
Chairman of the Department
of Physics

Athens, April 7, 1995

Dear Prof.

We have recorded a _**SES activity**_ with the following characteristics:*

Date : The activity started on April 6, 1995

Station : ASS (i.e. close to *Thessaloniki*).

Components : Both short dipole arrays (and on the long dipoles).

Polarity : Positive

**Impending earthquake** :

Magnitude : **Ms(ATH)** ≈5.2, if the epicenter lies a few tens of km around ASS (but see below).

MB(USGS) ≈4.7   "   "   "   "   "   "

Epicenter : For the selectivity map of ASS, see Fig. 1 of Varotsos et al, Tectonophysics 224, 269-288 (1993). If the epicenter lies in area "c" (i.e. around *Thessaloniki*) then Ms(ATH)≈5.2. For larger epicentral distances, then M will be larger [e.g. south of, but very close to, Chalkidiki peninsula Ms(ATH)≈5.8]. A more precise determination of epicenter is not possible, because the other stations around Thessaloniki are still out of operation.

Time : According to Fig. 28A of Varotsos and Lazaridou, Tectonophysics 188, 321-347, 1991 or Fig. 22 of Varotsos et al, 1993.

Yours sincerely

Prof. P. Varotsos

* P.S. Another SES activity was recorded at IOA the same day, i.e. on April 6, 1995, which is not simultaneous with the aforementioned activity at ASS. It might correspond to Ms(ATH)≈5.0 at 200 km west of Athens.

Fig. 12. Photocopy of the prediction sheet issued on April 7, 1995 (see Appendix I)

because, as mentioned, the areas labelled a, b and c in the selectivity map of ASS -see Fig 1- might be interconnected).

Actually four weeks later, i.e. at 00:34 GMT on May 4, a Ms(ATH)=6.0 EQ actually occurred with an epicentre at 40.5°N, 23.5°E (USGS), i.e. almost 40 km ESE of ASS. It was preceded by a number of foreshocks on May 3. By comparing to the predicted parameters one verifies a very good agreement.

APPENDIX II

Prediction of the Ms(ATH)=6.6 earthquake on May 13, 1995 in northern Greece

On April 30, 1995 we issued a prediction which consisted of three pages. A photocopy of the first page is given in Fig. 13a and is a short paper under the title *"Recent Seismic Electric Signal activities in Greece"*, the abstract of which stated: *"Three SES activities were recently recorded at IOA station. They might indicate that a pronounced series of EQs will occur in Greece with $M_s(ATH) \approx 6.0$ units."* The two strongest SES electrical activities were recorded on April 18 and April 19. A photocopy of the second page of the prediction is given in Fig. 13b and contains the probable time-chart that will be followed as well as a map indicating the two candidate epicentral areas. We draw the attention to the fact that the prediction text depicted in Fig. 13a clarified that the *epicentral area located close to IOA was indicated as the most probable.* Finally the third page of the prediction was just a photocopy of the SES activity recorded at IOA on April 19, 1995. (i.e. that shown in Fig. 6a)

At 08:47 GMT on May 13, 1995 (i.e. during the fourth week after the strongest SES activities, see the last case in Fig. 7) a $M_s(ATH)=6.6$ occurred. The epicentre announced by USGS was at 40.0°N, 21.6°E, i.e. the EQ occurred at a region which according to the Greek seismologists was previously considered to be aseismic; for more than 1000 years no such EQ had occurred there. Note that the distance between the predicted epicentre (i.e. that close to IOA) and the actual one is around 70-80 km while $\Delta M \approx 0.6$.

APPENDIX III

Prediction of the $M_s(ATH)=6.1$ on June 15, 1995

On May 19, 1995 a prediction was issued which consisted of two pages: a photocopy of the first page is given in Fig. 14a while the second page was just a copy of a map showing the sites of the VAN stations as well as the selectivity maps of the stations IOA and ASS. We clarify that a third page was also sent on May 20, 1995 to the Greek authorities which was a covering letter (in Greek) stating the following (verbatim):

"Secret

3 pages

    To the Excellency

a) Minister of Public Works Mr. K. Laliotis

b) Vice-Minister of Public Works Mr. K. Geitonas
Copy:
c) Sub-Minister Mr. Tsaklidis
d) Earthquake Planning Organisation (OASP) Mr. D. Papanikolaou
>   *Dear Mr Ministers*
>   *I attach a scientific information which probably indicates an impending **new** strong earthquake at a **new (another)** area. This information is not equally reliable (with that which was sent to you for the previous earthquake) because it comes from an experimental station in the area of Volos while all the other stations around this area still remain out of operation.*
>   *As **the situation is critical** and in view of the fact you are not now at your offices the present information is also sent to Mr. Tsaklidis and Mr. Papanikolaou.*
>   *Yours sincerely*
>   *P. Varotsos"*

We turn to Fig. 14a (which is basically a copy of a short paper): The abstract clarified that a new strong EQ might hit Greece at a different epicentral area but with comparable magnitude as compared to the EQ on May 13, 1995.

The strong SES activity was recorded on April 30 at VOL which was reactivated on September 1994 at a new site and hence it was not yet calibrated; however the SES amplitude (10 mV/km) allowed an estimation that, as mentioned in the abstract, the expected magnitude could be around 6.6. Since the selectivity map of VOL was still unknown the epicentre was estimated as follows: in addition to the short dipole arrays the SESs were recorded at two long dipoles (having almost the same direction and lengths $L_1=4$ km and $L_2=22$ km) with the same $\Delta V/L$ value; the latter fact indicated that the impending focal area should lie at a distance r appreciably larger than the dipole lengths, i.e. $r \gg L_1, L_2$ and hence $r/L_2 \gg 1$. As the ratio r/L_2 had to be, at least, around 4-5 the epicenter should lie at a distance more than ~100 km from VOL. Furthermore as the SESs were not recorded at the other four stations, i.e. IOA, ASS, KER and PIR, we excluded as candidates epicentres those seismic areas belonging to their selectivity maps. (Of course we safely excluded the area around 40.0°N, 21.6°E, i.e. the epicentre of the May 13 EQ because it was preceded by SES activities collected at IOA).

Thus the prediction of the epicentre was summarised as follows: *"The new EQ might occur in the remaining part... of **continental** Greece"*. More presisely the following areas were excluded from continental Greece: central western Greece, Chalkidiki area (including Thessaloniki), the aforementioned area within a radius of at least ~100 km around VOL, Peloponese, the neighbouring area around Attica (i.e. Athens) and, of course, the area of northern Greece

April 30, 1995

RECENT SEISMIC ELECTRIC SIGNAL ACTIVITIES IN GREECE

by
P. Varotsos and M. Lazaridou

Dept. of Physics, University of Athens, Greece

Abstract: *Three seismic electric signal activities were recently recorded at IOA station. They might indicate that a pronounced series of earthquakes will occur in Greece with $M_s \approx 6.0$ units.*

The most catastrophic earthquake (EQ) in Greece during the last eight years seriously damaged the village of Vartholomio on October 16, 1988. The epicenter was at 37.90°N, 20.96°E with a surface wave magnitude (according to the National Observatory of Athens) $M_s(ATH)=6.0$. It was preceded by a $M_s(ATH)=5.5$ EQ on Sept. 22, 1988 at 37.99°N, 21.11°E, which destroyed Killini harbour. The prediction of these EQs was publicly announced by Haroun Tazieff (through AFP and Antenne 2 of the French television) after the receipt of related telegrams sent by the first author. Three seismic electrical signals (SES) activities were recorded on Aug. 31, Sept. 29 and Oct. 3, 1988 (at IOA station, located in northwestern Greece) which were extensively discussed by Varotsos and Lazaridou (1991) and by Varotsos et al (1993).

On April 6, 1995, a SES activity was recorded at IOA station with characteristics (polarity, form, etc.) almost similar to those of the aforementioned activities on Sept. 29 and Oct. 3, 1988, but with a smaller amplitude. By following the procedure described by Varotsos and Lazaridou (1991) we concluded that a seismic activity is expected with $M_s(ATH) \approx 5.0$ units at an epicenter 200 km west of Athens. This information was sent on April 7, 1995 to the Greek authorities and to 29 Institutes abroad. In the meantime, however, two new SES electrical activities were recorded (on 18 and 19 April, 1995) at IOA station (see Fig. 1) that have comparable amplitudes and almost the same characteristics with those of Sept. 29 and Oct. 3, 1988. Therefore a $M_s(ATH) \approx 6.0$ EQ might occur close to the aforementioned Vartholomio area. An alternative solution is $M_s(ATH) \approx 5.5$ - 6.0 with an epicenter a few tens of km NW from IOA. Note that the latter solution seems to be more compatible with the experimental fact that $\Delta V/L$ differs between the two long dipoles at IOA having unequal lengths (case of nearby source). The two alternative candidates are shown in Fig. 2. The time evolution of the current activity might follow the same time chart as in Fig. 22 of Varotsos et al (1993).

References

Varotsos P. And Lazaridou M. 1991. "Latest aspects of earthquake prediction in Greece based on seismic electric signals I". Tectonophysics 188, 321-347.

Varotsos P., Alexopoulos K. And Lazaridou M. 1993. "Latest aspects of earthquake prediction in Greece based on seismic electric signals II". Tectonophysics 224, 1-37.

a.

Fig. 13 a,b. Photocopy of the prediction issued on April 30, 1995 (see Appendix II). A third page was also attached that was a copy of the recordings of the SES (electrical) activity recorted at IOA (i.e Fig. 6a)

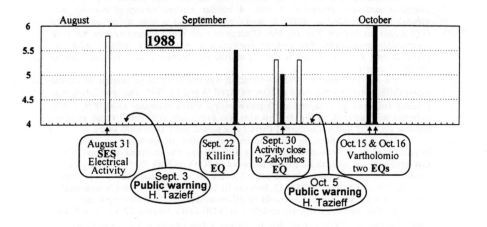

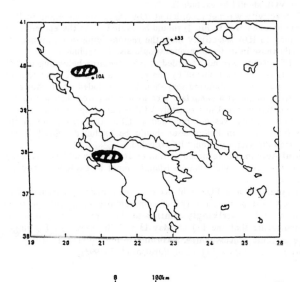

b. Fig. 2. : The two alternative epicenters.

Athens, May 19, 1995
CONTINUATION OF THE SEISMIC ELECTRIC SIGNAL ACTIVITIES IN GREECE

by P. Varotsos and M. Lazaridou

Abstract: *The 6.6 earthquake of May 13, 1995 was preceded by seismic electrical activities recorded at IOA. A similar seismic electrical activity was subsequently recorded at* **VOL.** *It seems probable that a new strong earthquake (EQ) might hit Greece. This EQ should occur at a different epicentral area but with comparable magnitude. The present prediction is not equally reliable with the previous one, as VOL station is not yet calibrated (because it is operating during the last 6 months).*

On April 27 and 30, 1995 a prediction was issued based on seismic electrical signal (SES) activities recorded at IOA on April 18 and 19, 1995. This prediction was actually followed by a M_s=6.6 earthquake with a USGS epicenter at 40.0°N, 21.6°E (labelled with asterisk in Fig. 1). This area was previously considered to be *aseismic* because such EQ had not occurred there for a period more than 1000 years.

Since last September an experimental station is operating at VOL. The inspection of its records indicated that an SES activity was recorded on April 30 (i.e. **not** simultaneous with those earlier recorded at IOA) on both short and long dipole arrays. Although this station is not yet calibrated we might guess that it is precursor of an EQ similar to that of May 13, because its amplitude (10 mV/km) is comparable to those at IOA. The epicenter should be different due to the following facts:

(i) The two long dipoles installed at VOL (with lengths 22 km and 5 km) show comparable ΔV/L-values thus indicating a non nearby source (and hence the neighbouring area of VOL should be excluded).

(ii) As the SES activity was not recorded at IOA, the areas belonging to IOA selectivity map (large hatched area in Fig. 1) should be excluded. As the epicenter of May 13 seems to belong to IOA selectivity map, the regions lying in its immediate vicinity (and especially those in its western side) should also be excluded.

(iii) As the SES activity was not recorded at ASS, the area belonging to ASS selectivity map (Fig. 1) i.e. that surrounded by the regions a, b, and c (and that of Chalkidiki peninsula and the neighbouring sea) should be excluded. (Note however that the VER area still remains as a candidate area as it is not well verified that it belongs to the ASS selectivity map. Unfortunately VER is out of operation).

(iv) As the SES activity was not recorded at KER, the area lying in the vicinity of Athens (and those in Peloponese) should be excluded. GOR is not operating and its immediate vicinity cannot be excluded.

(v) The central Aegean sea (recorded at ASS or at KER) should be excluded. However the area around Skiros and Alonisos islands cannot be excluded (they are of smaller probability).

By summarising: The new EQ might occur in the remaining part [i.e. after deleting the areas due to the points (i) to (v)] of continental Greece. The spectrum of the SES activity at VOL is **strikingly similar** to those recorded at IOA (on April 18, 19) thus indicating that the EQ of May 13 and the expected EQ might belong to the **same** tectonic process which, according to our opinion, is still going on. The time evolution might follow Fig. 22 of Varotsos et al (1993).

a.

Fig. 14. Prediction issued on May 19, 1995 (see Appendix III):
     a. Photocopy of the detailed text (cf. a map showing the sites of the VAN stations as well as the selectivity maps of IOA and ASS was also attached).
     b. Time evolution of the seismic activities (full bars) related with the SES activities (hollow bars): the upper time chart refers to the SES activity recorded on August 31, 1988 while the lower to that recorded on April 30, 1995 (on the basis of which the prediction of May 19, 1995 was issued).

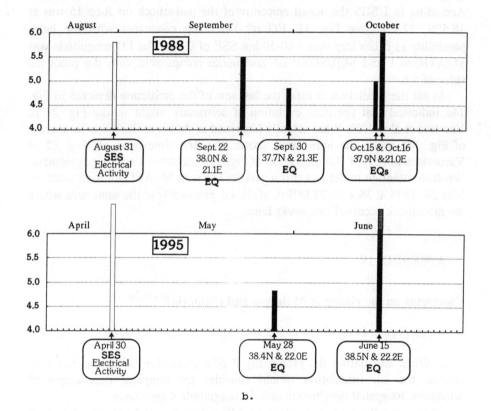

b.

around the EQ of May 13. Of the remaining small part of continental Greece, that lying in the vicinity of GOR was the more probable (cf. the region to the north of GOR was already excluded due to the aforementioned argument that arouse from the same $\Delta V/L$ value collected at the long dipoles of VOL). According to USGS the actual epicentre of the mainshock on June 15 was at 38.5°N, 22.2°E (see Fig. 11, EQ n° 14) being consistent with the latter possibility as it lies lees than ~40-50 km SSE of GOR. The EQ magnitude was $M_s(ATH)=6.1$ and $M_s(USGS)=6.5$ and hence comparable with the predicted value of 6.6.

As for the prediction of time, the last row of the prediction depicted in Fig. 14a indicated that the time evolution of seismicity might follow Fig. 22 of Varotsos et al (1993). What actually happened is shown in the lower time-chart of Fig. 14b while the upper part shows (the first time-chart of) Fig. 22 of Varotsos et al (1993). An inspection of Fig. 14b indicates a striking agreement. We draw attention to the fact that a smaller EQ with $M_s(ATH)=4.8$ occurred on May 28, 1995 at 38.4°N, 22.0°E (USGS), i.e. practically at the same area where the mainshock occurred two weeks later.

APPENDIX IV

Comments on the claims of Mulargia and Gasperini[22,27,28]

Additional remarks on the calculation of Mulargia and Gasperini[22]

a. When calculating the probability P of a prediction being successful by chance, the s.l. calculation should consider the temporal dependence of seismicity, its spatial distribution and the magnitude dependence.

Concerning the *spatial distribution*, MG calculation did not consider at all that each of our predictions had a limited spatial extent. This point is of major importance. Takayama[23], as mentioned, suggested a way of removing this deficiency of MG-calculation in the simplified case of geographically homogeneous seismicity. It should be recalled here that, as mentioned, Hamada's calculation[18] actually considered this point and drew his conclusion mentioned above in paragraph 5 after "... *taking into account a regional factor which includes high seismicity in the predictions area*".

Concerning the *magnitude dependence* it has been disregarded by MG. Obviously, the larger an EQ is, the greater the achievement in successfully predicting it. On the other hand MG used Poisson distributions and hence treated all EQs above a certain magnitude equally.

We finally turn to the *temporal dependence*. If n denotes the number of EQs which occurred during a test period T and Δt is the allowed time indetermination for a predicted EQ, MG claimed that the probability of a prediction being satisfied by chance is given by

$$p = n \ \Delta t/T$$

First of all this formula totally ignores, as mentioned, the limited spatial extent of each prediction as well as the magnitude dependence. Additionally it suffers from the following mistake: obviously when $n>T/\Delta t$ then $p>1$ which is unacceptable because p represents a probability. It is easily verified that MG calculation used in many cases values of $n \ \Delta t/T$ larger than unity and hence their corresponding s.l.-values are wrong. Rhoades and Evison[25] indicated this important point and furthermore suggested a correct formula for the calculation of the probability p (in the time domain).

b. As mentioned, predictions are issued *only* when the expected magnitude M_{pred} is equal to (or larger than) 5.0 units. However MG, by misinterpreting the true meaning of $\Delta M=0.7$-units, demand that all 547 EQs with $M_s(ATH)\geq4.0$ or all 204 EQs with $M_s(ATH)\geq4.3$ *should* have been predicted. The Gutenberg-Richter relation indicates that the number of EQs decreases drastically with increasing M, i.e. the number of EQs with $M_s(ATH)\geq5.0$ is drastically smaller (i.e. by a factor 6 to 7) than that of EQs with $M_s(ATH)\geq4.3$. In other words MG grossly overestimated the number of the EQs that should have been predicted.

c. A correct statistical procedure should necessarily consider the appropriate Δt-value for each prediction. The prediction text clarifies, in each case, whether it is based on GVEF, single SES or SES electrical activity. On the other hand MG considered the *same* Δt for all predictions. This is not correct for the following reason: when they assume $\Delta t\leq11$ days for all predictions they artificially exclude all the successful predictions referring to the SES electrical activities and GVEF. Alternatively, when they consider $\Delta t\leq22$ days for all predictions, the probabilities of those predictions that are based on single SES to be fullfilled by chance are increased by a factor of two (while the predictions based on GVEF or SES electrical activities may be rejected, see below).

d. MG calculation was based, as mentioned, on Poisson distribution. We recall that this distribution holds under the restriction that the events occur independently of each other and the probability does not change with time. However MG included both mainshocks and aftershocks in their calculation, which is not correct.

e. MG claim that in two cases they found a backward time association beyond chance. Varotsos et al[24] showed in detail that this does not hold. We shall briefly discussed here these two cases for the help of the reader:

First case. **MG found** s.l.≈0.03, for the case Δt≤11 days and Δr≤30 km, by considering in their Table 4 that 2 out of the 19 EQs with actual magnitude $M_{EQs} \geq 5.1$ were followed by predictions with (predicted) magnitude $M_{pred} \geq 5.8$. We draw the attention of the reader to the fact that if one of these two "backwards associations" is questionable, it has the following consequence: their s.l. value of 0.03 turns immediately to 0.2 (i.e. larger than 0.05) and hence the MG conclusion is not valid. They considered the following backward association: the prediction of October 21, 1988 with the 6.0 EQ (mainshock) of October 16, 1988. This association however should not have been done by MG because the correct text of the prediction of Oct. 21 clearly stated[11,26] that the expected epicenter is several tens of km *away* from the actual epicenter of the EQ of Oct. 16. It is therefore obvious that the expected displacement of the epicenter was definitely larger than the value Δr<30 km assumed by MG in this case as a condition for the backward association.

Second case. MG found s.l.≈0.04, for the case Δr≤120 km and Δt≤11 days by considering in their Table 5 that three out of the 18 predictions with $M_{pred} \geq 5.1$ are correlated "backwards" with EQs having $M_{EQs} \geq 5.8$.

We again draw the attention that if one of these 3 backwards associations is questionable, their s.l. value of 0.04 turns to s.l.≈0.2 thus invalidating MG conclusion. We focus our attention to the association of the prediction of May 21, 1988 with the 5.8 EQ of May 18, 1988.

What actually happened is the following[11]: the 5.8 mainshock of May 18 was succesfully preceded by a prediction issued on May 15; later a 5.5 aftershock occurred on May 22 which was successfully predicted by a telegram issued on May 21. Therefore MG correlated backwards a successful prediction of an aftershock with the mainshock (which, of course occurred earlier). Such an "association" however had to be definitely excluded from the calculation of the s.l.-value on the basis of Poisson. We recall that, as mentioned, when dealing with Poisson the time-dependent events have to be excluded.

f. *Check of the reliability of the MG-procedure by applying it to an ideally perfect earthquake prediction method:*

A (fictitious) Ideally Perfect Earthquake Prediction method is defined as a method which does not issue any false alarm and successfully predicts *all* the EQs above a certain magnitude threshold with a reasonable accuracy in time (e.g. Δt≤22 days), epicentre (Δr≈0, i.e. Δr≈ within the dimensions of the focal volume) and magnitude ($|\Delta M| \leq 0.7$-units).

In the following we shall follow exactly the MG-procedure by applying it to an IPEPM. In this application we shall intentionally use the same number of EQs (for various magnitude thresholds) as in MG, the same period of almost 3 years, i.e. T≈1000 days, and the same total number of predictions (i.e. 29) as in the VAN-case. Assume now that in a large area (1000 km x 1000 km) 10 remote EQs (mainshocks) with M≥5.5 occurred and an IPEPM successfully predicted *all* of them (i.e. 29 successful predictions with expected magnitude $M_{pred}≥5.0$ were totally issued of which 10 referred to the mainshocks and 19 to some of their aftershocks). For simplicity we assume that the seismicity was concentrated only to these 10 regions.

According to MG (their Table 2) the mean value μ in Poisson is calculated from (Rule of the game : Earthquake range $M_{EQ}≥5.5$, magnitude of earthquakes ±0.7, $\Delta t≤22$days, $\Delta r≈0$): $\mu = 2\dfrac{29}{1000}\dfrac{10}{1000}1000 = 5.8$

which (according to the formula $s.l.= \sum\limits_{x=n}^{\infty}\dfrac{\mu^x}{x!}e^{-\mu}$, n=10 successes) leads to s.l.=0.07. As s.l.>0.05, MG procedure leads to a paradox that all the above successfull predictions can be ascribed to chance. Of course, this result is not physically acceptable.

Remarks on the recent claims of Mulargia and Gasperini[27,28] that there is a backwards time association between predictions and earthquakes.

Mulargia and Gasperini,[22] as mentioned, used in their calculation both mainshocks and aftershocks although this is not allowed when using Poisson. In addition to this deficiency we shall indicate below with simple examples how the recent procedure of Mulargia and Gasperini[27,28] leads to another paradox, i.e. that an IPEPM has a smaller number of forward time associations than that of the backward associations.Figure 15 shows that an IPEPM ($\Delta r≈0$, $\Delta M≤0.7$, $\Delta t≤22$ days) achieved successful predictions for a 6.0 mainshock and for three aftershocks having magnitudes 4.8, 5.2 and 4.7. Regularly, in an experimental method, the predicted magnitudes are randomly slightly either larger or smaller than the actual ones. Let us assume that the predicted magnitudes were 5.9, 5.4, 4.7 and 5.1 respectively. Of course, if we do not put any arbitrary magnitude threshold (both for predictions and earthquakes) the number of forward time associations is always larger than that in backwards time (i.e. in Fig 15 there are 4 associations in forward time while 2 backwards). But let us now put a threshold M≥5.0 both for the EQs and predictions as Mulargia and Gasperini[27] did in their recent calculation: we now find in Fig. 15 one association in forward time (i.e. the 5.9 prediction with the 6.0 mainshock) and two association backwards (i.e. the 5.4 prediction with the 6.0 mainshock, and the 5.1 prediction with the 5.2 aftershock). We are thus led to the erroneous

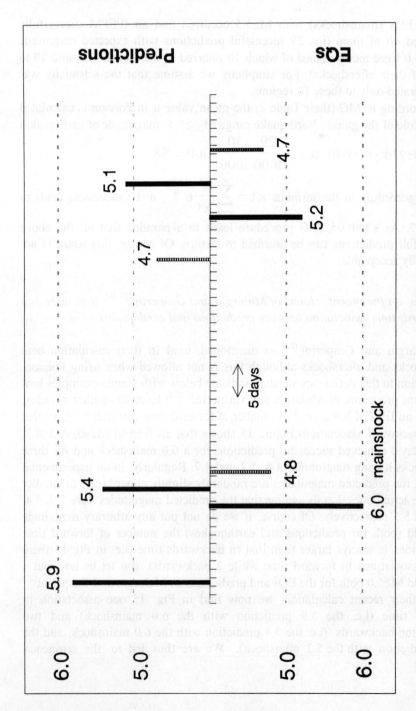

Fig. 15. How a significant association in reverse time can be erroneously obtained in an IPEPM if we put a threshold M≥5.0 both for the EQs and the predictions and also include time-dependent events (i.e. aftershocks): By taking Δt≤22 days we find one association in forward time (mainshock) while two in reverse. Each mark in the horizontal axis corresponds to 1 day.

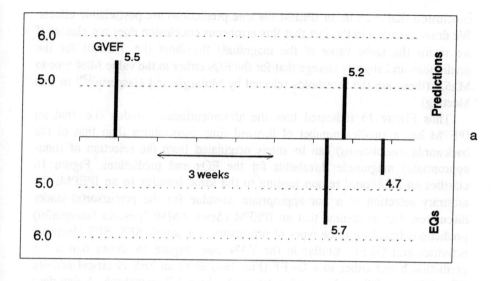

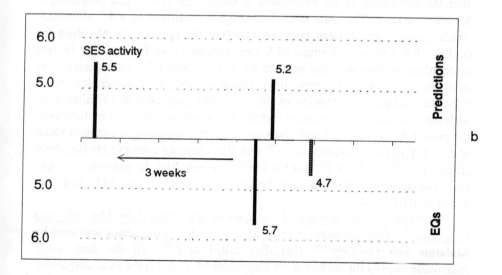

Fig. 16a,b. How a significant association in reverse time can be erroneously obtained between predictions and earthquakes in an IPEPM (Δr≈0, ΔM≤0.7) which issues predictions based on the same type of precursors (i.e. GVEF, SES electrical activity, single SES) as VAN: we find zero associations in forward time and one in reverse time if we put a magnitude threshold M≥4.9, 5.0 or 5.1 and also assume (for all precursors) Δt≤22 days. On the other hand, when (do not put any magnitude threshold and) considering the appropriate Δt-values for each type of precursor (see Section III) we find two associations in forward time (as it should for an IPEPM) and one association in reverse time.

conclusion that even in an IPEPM the true predictions are postseismic effects. We draw attention to the fact that this erroneous conclusion does not change if we retain the same value of the magnitude threshold (i.e. M≥5.0) for the predictions and slightly change that for the EQs either to the value M≥4.9 or to M≥5.1 (these are the thresholds selected by Mulargia and Gasperini[28] in this Meeting).

Thus Figure 15 indicated how the aforementioned paradox (i.e. that an IPEPM has a smaller number of forward time associations than that of the backwards associations) can be solely originated from the selection of (non-appropriate) magnitude thresholds for the EQs and predictions. Figure 16 clarifies an *additional* reason leading to the same paradox in an IPEPM: the arbitrary selection of a non-appropriate Δt-value for the precursor(s) under discussion. Let us assume that an IPEPM (Δr≈0, ΔM≤0.7) issues (successful) predictions based on three types of precursors, e.g. single SES, SES electrical activities and GVEF, similar to the VAN case. Figure 16 shows that a 5.5 prediction based either to a GVEF (Fig. 16a) or to an SES electrical activity (Fig. 16b) was followed, e.g. after 4-6 weeks, by a 5.7 mainshock. A few days after the occurrence of the mainshock, a single SES is recorded predicting a M=5.2 aftershock. The latter prediction is again confirmed by a 4.7 aftershock which occurs, e.g. ~1 week later. By recalling the *appropriate* Δt-values for GVEF, SES activity and single SES (see Section 3) we find in Fig. 16 two associations in forward time between predictions and EQs, as expected, and only one association in reverse time; (obviously the latter association arises from the backwards correlation of the successful prediction of the aftershock with the mainshock that occurred earlier). Let us count again, but by following the procedure of Mulargia and Gasperini, i.e. after assuming a common value of Δt≤22 days for all predictions[27,28] and also adopting a magnitude threshold either M≥5.0[27] or M≥5.1[28] (or M≥4.9[28]): we now find the paradox, i.e. zero (0) associations in forward time between predictions   and EQs and one association backwards.

Comments on the number of the associations counted by Mulargia and Gasperini[28]. The arithmetic average M_{ave} of the following scales was used by Mulargia and Gasperini[28] (hereafter called MG[28]): (i) the body wave magnitude m_b, (ii) the surface wave magnitude M_s and (iii) the local magnitude M_L. MG[28] studied the period from January 1, 1987 to June 30, 1992 by adopting two alternative magnitude thresholds for the EQs: either $M_{ave} ≥ 4.9$ or $M_{ave} ≥ 5.1$ (i.e. 27 EQs for the first case and 11 for the second). They found that in both cases, the number of associations between predictions and EQs in reverse time is larger than that in forward time and then claimed it stands for a prevalent postseismic nature of VAN signals. As it will become clear below this statement is not valid. We shall only discuss in detail the case of the 11 top EQs (i.e. those with $M_{ave} \geq 5.1$) as the other case (i.e. $M_{ave} \geq 4.9$) is clarified (see below) with the same arguments.

Table 4 lists the totality of the 11 EQs with $M_{ave} \geq 5.1$ and explains, for each case, if a successful prediction (succ) was issued well in advance. (We recall that the totality of predictions are given in Table 1). An inspection of this Table indicates that the alarm rate is 6/10 and not 3/11 claimed by MG[28]. This comes from to the following facts: (i) the number of EQs is 10 and not 11 because MG [28] included one EQ (i.e. that on Jan. 9, 1988) which occurred in Albania although it was clearly stated[11,15] that no predictions are issued referring to that area; (ii) MG[28] did not consider the three predictions in bold face: the first one (i.e. that corresponding to the EQ n° 2) was based on a GVEF and was "rejected" by MG[28] because they did not consider the appropriate Δt-value (see Section III) but assumed $\Delta t \leq 22$ days (case similar to Fig. 16a); for the second prediction (i.e. that corresponding to the EQ n° 9) MG found that Δr slightly exceeds 120 km but they should not "reject" this case because the prediction text -submitted to Tectonophysics[11] before the EQ occurrence- clearly stated that a displacement by a few tens of kilometers is expected from the point from which they measured the Δr-value; as for the third prediction this is obvious as it is similar to Fig. 16b, i.e. MG[28] "rejected" the prediction issued on Dec. 27, 1991 -after a long period of quiescence, see Fig. 10b- because it preceded the EQ on Jan 23, 1992 by 27 days while they assummed $\Delta t \leq 22$ days. (Note that in the other case $M_{ave} \geq 4.9$, if we do not disregard the aforementioned three predictions, the number of associations in forward time results larger than that backwards thus invalidating MG claim).

Table 5 lists the six EQs (which $M_{ave} \geq 5.1$) which according to MG[28] are associated with predictions in reverse time. In the last column, we marked with X, three (out of 6) of these associations which referred to a backwards correlation of a successful prediction of an aftershock with the mainshock that occurred earlier: footnotes (i), (ii) and (iii) clarify that they are similar to those depicted in Figs 16a, 15 and 16b respectively which, as already mentioned, explain that arbitrary assumptions lead to the paradox that an IPEPM has a smaller number of forward time associations than that of the backward associations. (Two of the remaining three associations, labelled with an asterisk in the last column, include predictions that were successfully followed[11,12] by EQs with epicenters at different areas than those of the earlier occurred mainshocks; the remaining case, labelled with "unsucc ($\Delta M > 0.7$)," includes a prediction which -as mentioned in Ref(11)- was followed by an EQ that was too small so that $\Delta M > 0.7$).

By comparing the results drawn from the study of the Tables 4 and 5 we can conclude the following: as regards the 10 EQs with $M_{ave} \geq 5.1$, six of them were preceded by successful predictions and also six of them associate with

TABLE 4. The totality of eleven EQs with $M_{ave} \geq 5.1$ treated by Mulargia and Gasperini[28]. For six of these EQs, successful (succ) predictions were issued well in advance but Mulargia and Gasperini disregard three of them labelled in bold face. Period: Jan. 1, 1987 - June 30,1992.

	EARTHQUAKE				PREDICTION	
			EPICENTER[a]			
No.	Y M D	M_{ave}	°N	°E	Y M D	Remarks
1	1987 02 27	5.5	38.4	20.4	1987 02 26	succ
2	**1987 06 10**	**5.2**	**37.2**	**21.5**	**1987 04 27**	**succ (case similar to Fig 16a)**
3	1988 01 09	5.5	41.2	19.6		EQ in Albania and hence it should not be included
4	1988 05 18	5.4	38.4	20.5	1988 05 15	succ
5	1988 10 16	5.6	37.9	21	1988 09 30	succ
					1988 10 03	succ
6	1989 03 19	5.3	39.3	23.6		missed
7	1989 08 20	5.5	37.2	21.1	1989 08 20	unsucc
8	1989 08 24	5.1	37.9	20.1	1989 08 20	unsucc
9	**1990 06 16**	**5.4**	**39.1**	**20.4**	**1990 05 28**	**succ**
10	1990 12 21	5.9	41	22.4		missed
11	**1992 01 23**	**5.1**	**38.3**	**20.4**	**1991 12 27**	**succ (case similar to Fig 16b)**

a. According to SI-NOA monthly catalog

TABLE 5. Earthquakes with $M_{ave} \geq 5.1$ associated with predictions in reverse time according to Mulargia and Gasperini[28]. In the last column we mark with x three cases (out of 6) in which a successful prediction of an aftershock is associated with the mainshock which occurred earlier (that was also successfully predicted). Period: Jan. 1, 1987 - June 30,1992.

PREDICTION	EARTHQUAKE				
			EPICENTER[a]		
Y M D	Y M D	M_{ave}	°N	°E	Remarks
1987 06 13[i]	1987 06 10	5.2	37.2	21.5	x
1988 05 21[ii]	1988 05 18	5.4	38.4	20.5	x
1988 10 21	1988 10 16	5.6	37.9	21.0	unsucc ($\Delta M > 0.7$)
1989 08 24	1989 08 20	5.5	37.2	21.1	*(Ref. 11)
1991 01 02	1990 12 21	5.9	41.0	22.4	*(Ref. 17)
1992 01 02[iii]	1992 01 23	5.1	38.3	20.4	x

a. According to SI-NOA monthly catalog

i. This prediction is successfully correlated with the aftershock occurred at 37.1°N 21.4°E on 1987/08/21 with Ms(ATH)=4.7

ii. This prediction is successfully correlated with the aftershock occurred at 38.4°N 20.5°E on 1988/05/22 with Ms(ATH)=5.5

iii.This prediction is successfully correlated with the aftershock occurred at 38.3°N 20.5°E on 1992/02/01 with Ms(ATH)=4.5

predictions following them. (This invalidates MG[28] claim that the number of the backwards associations is larger than that of the associations in forward time). The equality of these two numbers cannot "stand for a prevalent postseismic nature of VAN signals" because the following fundamental difference exists: the associations in forward time include *solely independent events,* i.e. these six predictions (see Table 4) successfully preceded six mainshocks; on the other hand half of the six associations in reverse time (see Table 5) refer to *dependent events,* i.e. association between successful prediction of aftershocks with mainshocks that occurred earlier (which note that they were also successfully predicted).

References

1. Varotsos P. and Alexopoulos K., *Philos Mag.*, A **42,** 13-18, 1980.
2. Varotsos P., Alexopoulos K. and Nomicos K., *Physica Status Solidi* (b) **111,** 581, 1982.
3. Varotsos P. and Alexopoulos K., Stimulated current emission in the earth (monograph, 474 pages): piezostimulated currents and related geophysical aspects. In S. Amelinckx, R. Gevers and J. Nihoul (Editors), Thermodynamics of Point Defects and their Relation with Bulk Properties. *North Holland, Amsterdam,* p.p. 136-142, 403-406, 410-412, 417-420, 1986.
4. Varotsos P., Alexopoulos K. and Lazaridou M., *Tectonophysics* **224,** 1-37, 1993.
5. Slifkin L., *Tectonophysics* **224,** 149-152, 1993.
6. Varotsos P. and Alexopoulos K., *Tectonophysics* **110,** 73-98, 1984a.
7. Varotsos P. and Alexopoulos K., *Tectonophysics* **110,** 99-125, 1984b.
8. Varotsos P. and Alexopoulos K.,*Practica Athens Academy* **59,** 51-116, 1984.
9. Varotsos P. and Alexopoulos K., *Tectonophysics* **136,** 335-339, 1987.
10. Varotsos P. and Alexopoulos K., *Tectonophysics* **161,** 58, 1989.
11. Varotsos P. and Lazaridou M., *Tectonophysics* **188,** 321-347, 1991.
12. Varotsos P. , Eftaxias K. and Lazaridou M., *Jishin Journal, (Assoc. Develop. Earthq. Pred.)* **17,** (6), 18-26, 1994.
13. Varotsos P., Uyeda S., Alexopoulos K., Nagao T. and Lazaridou M. *in "Electromagnetic Phenomena Related to Earthquake Prediction" (Edited by M. Hayakawa and Y. Fujinawa), TERRAPUB,* pp. 13-24, 1994 (Tokyo).

14. Hadjioannou D., Vallianatos F., Eftaxias K., Hadjicontis V. and Nomicos K., *Tectonophysics* **224**, 113-124, 1993.
15. Varotsos P., *Bullettino di Geodesia a Scieze Affini XLV*, **2**, 191-202, 1986.
16. Varotsos P. *in Convegno Scientifico: "Sono prevedibili, terremoti e le eruzioni vulcaniche (Taormina, 13 e 14 aprile 1989).* Protezione Civile, Publicazione n. **17**, see pages 81-84.
17. Varotsos P., Alexopoulos K., Lazaridou M. and Nagao T., *Tectonophysics* **224**, 269-288, 1993.
18. Hamada K., *Tectonophysics* **224**, 203-210, 1993.
19. Shnirman M., Schneider S. And Dmitrieva O. *Tectonophysics* **224**, 211-221, 1993.
20. Nishizawa O., Lei X. And Nagao T. *in "Electromagnetic Phenomena Related to Earthquake Prediction" (Edited by M. Hayakawa and Y. Fujinawa), TERRAPUB*, pp 459-474, 1994 (Tokyo).
21. Uyeda S., *Zisin* **44**, 391-405, 1991.
22. Mulargia F. and Gasperini P., *Geophys. J. Intern.*, **111**, 32-44, 1992.
23. Takayama H., *Geophys. J. Int.*, **115**, 1197-1198, 1993
24. Varotsos P., Eftaxias K., Vallianatos F. and Lazaridou M., *Geoph. Res. Lett.*, to be published, 1995.
25. Rhoades D.A. and Evison F.F.,*Geoph. Res. Lett.*, to be published, 1995.
26. Dologlou E., *Tectonophysics* **224**, 189-202, 1993.
27. Mulargia F. and Gasperini P., Geoph. Res. Lett., to be published, 1995.
28. Mulargia F and Gasperini P. (*Royal Society Meeting* 1995).

THE TELEMETRIC SYSTEM OF VAN GROUP

K. NOMICOS

Technological Educational Institute of Piraeus
154 Charilaou Tricoupi ,Athens 11472, Greece

J. MAKRIS AND M. KEFALAS

University of Athens ,Solid State Physics Section
Panepistimiopolis ,Zografos 15771 , Athens

A detailed description of the VAN group telemetric system that records variations in the Earth's electrical field from field stations in Greece is presented. Each field station measures potential differences between 8 or 16 pairs of electrodes inserted into the ground in two orthogonal directions N-S and E-W. Two techniques have been applied for the telemetric system. The first one is the real-time telemetry, where the field station takes the information, digitizes it and transmits it via leased telephone line to a central station in Athens, which decodes the information, offsets it automatically and displays on chart recorders the small variations of the electric field. The second technique uses a datalogger, where the information is digitized and stored in the datalogger memory and then transferred to a notebook PC. The central station has commercial personal computer which communicates with the field notebook PC, takes the information via a standard telephone line and plots the recordings.

Introduction

The VAN telemetric system used to have 17 field stations around Greece, and one central station in Athens using real-time telemetry during the period 1983-1989. [1]

From 1989 until 1994 the real-time telemetric system consisted of 4 field stations located at Ioannina (IOA), Assiros (ASS), Pirgos (PIR) and Keratea (KER) using real-time telemetry (fig 1).

Since 1991 we started gradually to install 4 field stations using datalogger telemetry technique, and one Central station which is located in Athens. The locations of these field stations are shown in fig 1.

Configuration of the field station

Each field station measures variation of electrotelluric field using unpolarized electrodes Pb-PbCl$_2$ [2], which are inserted into the ground at a depth of 2m. We call each pair of electrodes an electrical dipole.

Each field station has at least 4 to 6 short electrical dipoles, half of them are deployed on NS direction and the other half to EW. The length of each dipole varies from 50m to 400m. A layout of a typical field station is shown in fig 2. Another two or four long dipoles are installed. The spacing between each pair of these electrodes is in the range of 1km to 10km and the wiring is achieved using leased telephone lines. In this way the reliability of the system is improved since the electric field is simultaneously measured on lines of variable length separations (from 50m to few kilometers). Each pair of electrodes is connected to a sensitive analog device which measures the potential difference. This measured voltage V is the sum of three quantities V_{ch}, V_e and V_n where V_{ch} is the chemical potential of the electrodes, a strong effect, which is reduced by using unpolarized electrodes, V_e is the electric field component which is of the main interest and V_n is the electric noise which may be drastically reduced by measuring far away from sources of electrical noise. Most of the field stations are installed in military bases, because they are situated far away from towns and thus far from electrical sources of noise. Also we have available some military personnel who maintain the stations. Military bases like shooting ranges are preferred since they have the required area necessary for the electrode spacing, and normally they are far away from electrical noise sources.

The analog sensitive device that measures the potential difference in each dipole has the configuration of fig 3. The input of the device is fitted with a spark protection system in order to avoid spikes arising from atmospheric transitions (rise time 10^{-9} sec). The next stages are a low noise differential amplifier, followed by a low-pass active filter with a cut-off frequency of 0.1 or 1 or 10 Hz upon request, and a low noise amplifier having a gain of 25.

We have also performed experiments with the configuration shown in fig 4, where the electrodes are connected to the spark protection system, followed by low pass active filters and a differential amplifier having a gain of 25.

The configuration of fig 4 has the advantage of avoiding the saturation of the direct input amplifier due to spikes that some times come from the earth, and the disadvantage of reducing the common mode rejection ratio (CMRR) to approximately -60 db, while the configuration of fig 4 has a CMRR above -88 db. [3]

The real-time telemetry system

The output of the 8 analog devices is used as input to the real-time telemetric transmitter system which is shown in fig 5a. The telemetric transmitter system consists of a multiplexing system, a buffer and an analog to digital converter (A/D). The A/D converts the analog sampling of each channel to a 12 bit digital information. The multiplexer and the A/D converter are controlled by a single chip microprocessor, which receives the digital data, codes them according to an error detecting and correcting code, and transmits them serially via a driver system to a modem. The modem transforms the digital information to sinusoidal signal of two frequency (FSK technique) in order to feed this signal to the telephone line. The transmission baud rate is 1200bits/sec. The sampling rate for each analog channel is 3 samples per second and per channel.

The receiver in Athens, where the central station is installed, performs the decoding of the digital information. A block diagram is shown in fig 5b. The modem which is connected to the telephone line translates the FSK code to digital data. The microprocessor takes the digital data, performs the demultiplexing, and (most importantly) is that it can recognize any distortion on the digital information due to disturbance in the telephone line . For example if the distortion is one bit per byte, the system can recognize and correct the information. Otherwise the microprocessor rejects the information and warns that a fault exists on the telephone line. The microprocessor in the receiver can also average the sampling every 2, 4, 8, 16 and 32 samples for each channel, upon request. In this way we achieve noise reduction. The D/A converters (8 bits) converts the eight less significant bits into analog form so that the analog information is offseted and amplified by a factor of eight. A gain of 25 from the analog device multiplied by a gain 8 from the receiver gives 200 which is the overall gain of the system.

The analog output of each channel which is connected to a chart recorder corresponds to the equivalent input on the sensor from 0 to ±12.5mV. Absolute values greater than 12.5mV and up to 200mV, which appear across the electrodes, are automatically offseted so that only variations up to ±12.5mV, are displayed on the chart recorder.

The Dataloggers Technique System

From 1986 onwards we had a long experience with datalogger of Campbell Company (21X type), which is the pioneer datalogger in the international market. We used this type of datalogger for magnetotelluric measurements in order to determine the tensor between electric and magnetic field for all field stations located in Greece. Using these tensors we subtracted the magnetic noise from the electric recordings.[4]

Since 1992 we gradually installed four datalodders close to those instruments which are connected with the central station using real-time telemetry as described before. These datalogger's field stations are using separate electrodes, power supply etc., and in some field stations like IOA we have three components coil magnetometers.

The general specifications of the 21X type datalogger of Cambell Company (USA) which we use are the following:

1) A/D Converter of 14 bits

2) Data storage memory circular of 19328 sampling (38kb)

3) 8 analog differential inputs or 16 single-ended

4) The communication software protocol gives reliability 99.998% of the transferred data.

5) Full Scale ranges ± 5000 mV with resolution 333 µV

±500 mV	33 µV
±50 mV	3.3 µV
±15 mV	1.0 µV
±5 mV	0.33 µV

6) Ability to add the date and time to each sampling

7) Extremely low consumption (5mA at 12V) due to CMOS construction.

At the beginning we used the 21X datalogger as a stand alone datalogger, with the following requirements :

1) ± 500 mV full scale range for the electrical dipoles, and ± 5000 mV for the coil magnetometers.

2) Sampling rate 1 measurement/channel·sec

3) In the final memory of the datalogger we save the average value of 20 samples for each channel with low resolution format (4 decimal digits), so the final resolution is 0.1mV.

4) We save also in the final memory date, minutes and seconds.

5) The final memory of the datalogger can store data (8channels) for a period of 8.8 h without overlapping the final memory. The big disadvantage of 21X datalogger is the slow baud rate transfer data via modem (1200 bits/sec). Although the serial datalogger output has 300, 600, 1200, and 9600 transferred data capability, it is impossible to use the 9600 baud rate via smart modem without hardware handshaking which doesn't exist at the 21X datalogger.

Due to the disadvantage of low storage memory of the 21X datalogger and the low speed of transfer of data to the Central station, and the necessity of the experiment to have high sampling rate especially during SES we installed a combination of 21X Campbell datalogger and a notebook PC as shown in Fig 6. The analog electronics of the 16 inputs consists of spark protection system, differential amplifier, and a low pass filter. These 16 analog outputs are fitted to the 21x datalogger, which takes the data one sample per second and for each channel, and stores them in the memory of the datalogger (High sampling rate (HSR)). Every half hour the notebook PC communicates via the serial port com2 of the PC and an opto-insulation interface with the datalogger, takes the data and store them to the hard disk of the PC. After that the program running on the PC, splits the data on files of five hour duration and compresses them. Also the program creates another type of files which is the average of sampling every 20 seconds for each channel. We call these types of files low sampling rate (LSR). The central station which has commercial PC configuration, communicates with the Notebook PC every day and takes the compressed data LSR, via the port Com1 and a smart modem at a baud rate of 9600bits/sec, and plots them on a plotter. When a SES appear on the recording the central station asks the field station to send the appropriate HSR files. The combination of 21X datalogger and the Notebook-PC has the only disadvantage of the delicate hard disk of the PC, which we have to replace every approx. 6 months.

The four datalogger's field stations have been installed within EPOC/CTC91 project.

Recently we designed and constructed another type of datalogger based on an industrial computer having an A/D card of 16 bits resolution as shown in fig 7.

The advantages of this system are:

1) Creation and storage of 3 type of files having 10Hz, 1Hz and 0.1 Hz sampling rate (USR, HSR and LSR files).

2) The central station can communicate with the datalogger on the field station in unattended or attended mode, and can get the compressed LSR files using smart modem 14400 bits/sec.

3) In case the central station would like to have HSR or USR files of a specific time period, this could be done via telemetry if the files are few. For more data USR or HSR, for economical reasons, we may ask a locally authorized person to change the removable hard disk with a spare one, sending the disk with the data to the central station.

References

1. Varotsos P. et al, previous paper in this issue
2. Ptiau G, and Dupis in *A Geophysical Prospecting* **28**, 792 (1980)
3. K. Nomicos and P. Chatzidiakos in *Tectonophysics* **24**, 39-46 (1993).
4. D. Hadjioannou, F. Vallianatos, K. Eftaxias, V. Hadjicontis and K. Nomicos in *Tectonophysics*, **224**, 113-124 (1993)

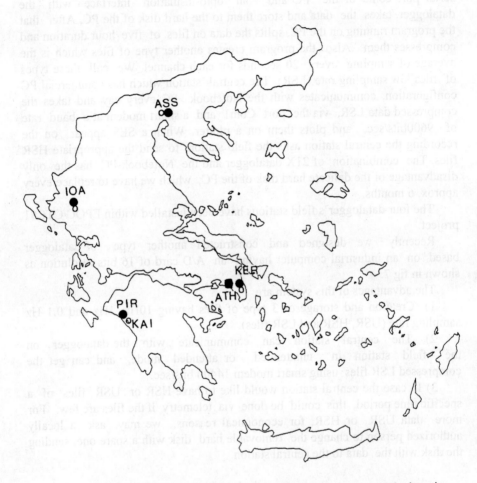

Fig 1 : Location of the field stations using real-time telemetry •, and those using datalogger's telemetry ° .

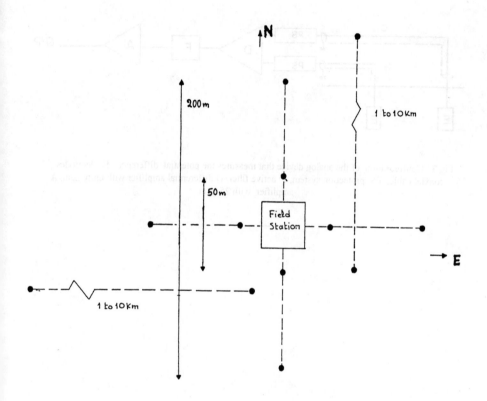

Fig 2 : Arrangement of the electrodes in each field station.

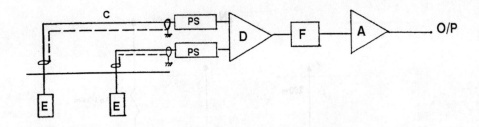

Fig 3 : Configuration of the analog device that measures the potential difference. E: electrodes, C: coaxial cable, PS: protection system, F: active filter, D:differential amplifier with unity gain, A: amplifier with 25 gain.

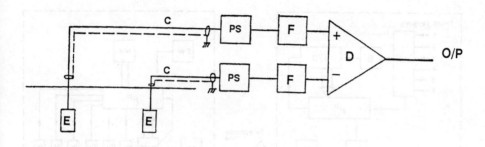

Fig 4 : Another configuration of the analog device that measures the potential difference. E: electrodes, C: coaxial cable, PS: protection system, F: active filter, D: differential amplifier with 25 gain.

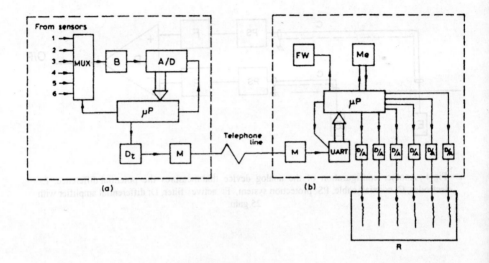

Fig 5 :Arrangement of the telemetry transmitter (a) and receiver (b) MUX: multiplex, B: buffer, A/D: Analog to digital converter, μP: microprocessor, Dr: driver, M: modem, UART: serial to parallel conversion, Me: memory, FW: warning indicator, D/A: digital to analog converter, R: chart recorder.

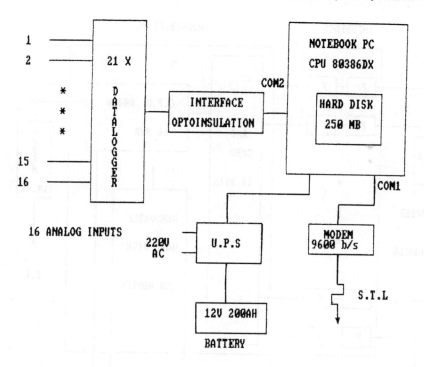

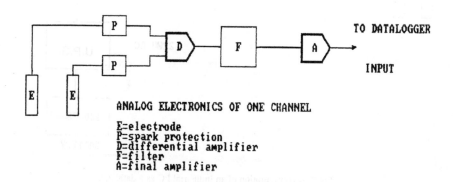

ANALOG ELECTRONICS OF ONE CHANNEL

E=electrode
P=spark protection
D=differential amplifier
F=filter
A=final amplifier

Fig 6 : A combination of 21X datalogger of Campbell and a note-book PC .

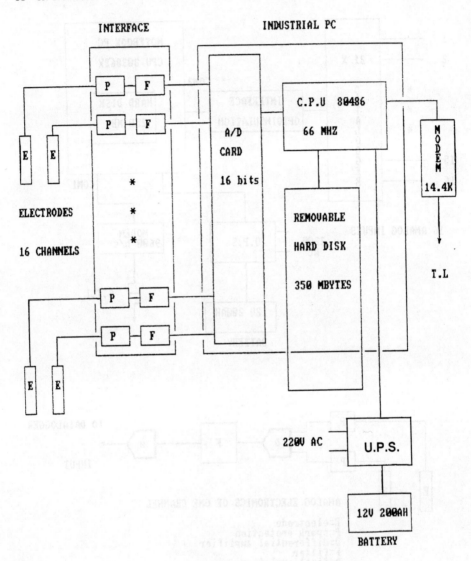

Fig 7 :A configuration of an Industrial PC as a datalogger.

Part II
Possible SES Mechanisms

PHYSICAL MECHANISMS FOR GENERATION AND PROPAGATION OF SEISMIC ELECTRICAL SIGNALS

DAVID LAZARUS

University of Illinois at Urbana-Champaign, Department of Physics,
1110 West Green Street, Urbana, Illinois, 61801-3080, USA

This paper discusses a possible physical model to account for the generation and propagation of seismic electrical signals as precursors to earthquakes, as reported by the VAN program. It is based on a novel three-threshold model for generation of an earthquake. Propagation of the signals from their source at the epicenter to remote detectors is presumed to occur largely along high-dielectric paths in the earth.

1 Introduction

The Seismic Electrical Signals (SES) reported by the VAN group as precursors to earthquakes (EQ's) in Greece have a number of unusual and unique characteristics. It is difficult to posit a mechanism which will provide even a tentative model which can tie together all of the characteristics of SES, as described. It is impossible, at this time, to specify a precise model which can bear the test of detailed scientific scrutiny. The burden of this paper is to describe a plausible scenario which, for all its possible faults; at least provides a coherent framework which seems to tie together the observed phenomenology of SES, without obvious internal inconsistencies and without violating the laws of physics.

It is not the intention of this paper to question the validity of the VAN program nor of the reported nature of their findings. Rather, the only question which will be addressed here is whether or not the reported SES could be generated and propagated by any possible physically supportable process.

This paper will describe the overall framework of such a model, in fairly general terms, indicating how SES might be generated and propagated. The companion paper by Professor Slifkin will concentrate on a possible detailed atomistic mechanism for generation of the electrical signals.

2 Characteristics of Seismic Electrical Signals

As reported by the VAN group, SES appear to have the following specific characteristics which require physical justification:

1. The signals appear to be entirely electrostatic: there are no associated magnetic effects.

2. The signals appear to be transmitted over long distances (hundreds of km) through the earth; no associated electromagnetic waves are measured in the atmosphere above ground far from the epicenter. (If the observations by the Stanford group of signals with above-ground antennas prior to the Loma Prieta EQ are to be equated with the VAN SES, an exception to this generalization must exist very near to the epicenter.)

3. The signals have a duration which varies from perhaps 30 seconds to several hours and appear days to weeks before an EQ, emanating from the same locale as the eventual EQ epicenter. The SES precursor signals have identifiable signatures which are very different from the much larger electrical and magneto-telluric signals often observed a few minutes prior to and during actual EQ's.

4. SES signals are measured only at special "sensitive" detectors at "preferred" sites, not generally, and not even necessarily at sites closest to an epicenter. At any one site, SES are detected only from EQ's occurring in specific regions of coverage. Determination of "sensitive" detectors, "preferred" sites, and regions of coverage is a trial-and-error procedure which requires careful calibration. Signal characteristics and triangulation patterns permit prediction of epicenter locations within fairly narrow limits.

5. The magnitudes of SES correlate well with magnitudes of the subsequent EQ's, permitting prediction of the magnitude of the EQ within fairly narrow limits.

6. The frequency spectrum of SES signals has not been measured directly. The signals are known to be of very low frequency and are at least consistent with a possible $1/f$ spectrum measured by the Stanford group.

3   Models for Earthquakes

The traditional model for an EQ considers it to be the result of a build-up of tangential stress along a pre-existing fault plane until the critical stress for slip is reached, set by dry frictional forces between adjacent rough and incommensurate surfaces at the fault. The stress is then relieved, or at least greatly reduced, by the tangential motion associated with the EQ and possible aftershocks. Some models also presume the possibility of fracture of surfaces and generation of new fault planes during an EQ.

The traditional model of an EQ does not obviously lend itself to the generation of any sort of precursor signals, whether VAN-type or any other, since there is only a single critical threshold to be overcome before the occurrence of an EQ, and there is no significant time delay anticipated between reaching the threshold and the start of the EQ.

The traditional model is based largely on the behavior of terrestrial minerals at the earth's surface, where the ambient temperatures and pressures are low, and the materials are nearly all extremely brittle. It is difficult to understand how such behavior could result in the extremely rapid ground motion during an actual EQ frequently reported by observers close to the surface intersection of a fault plane. It is also by no means obvious that such conventional, dry-frictional limited motion would be characteristic of the behavior of materials at typical sources of EQ's, at the considerable depth of some tens to hundreds of kilometers below the earth's surface, where high ambient pressures and temperatures exist.

As shown by P. W. Bridgman[1] in his classic work over a half-century ago, at hydrostatic pressures of a few kilobars and temperatures of perhaps 1000 degrees Celsius, such as exist at such depths, even very brittle, glassy, materials become ductile and can be readily deformed by very large amounts without fracture. Indeed, these phenomena have been exploited for commercial-scale extrusion of many materials.

In an alternative model, here proposed, an EQ is presumed to arise from a three-step process, still arising from build-up of a uniaxial stress across a pre-existing fault plane.

The stress may be considered decomposed into two components, one orthogonal to the fault plane and the other tangential. The first step in the EQ production occurs when the orthogonal stress component, in combination with the overall hydrostatic stress caused by the overburden of matter, reaches a threshold sufficient to effect a phase transition in the material at the fault plane. The unsupported areas at the misfit region along the fault plane assure a maximum of the stress there, compared to the bulk of the material. The phase transition region, therefore, would be confined to a thin region immediately adjacent to and parallel to the fault plane. Phase transitions are well-known and ubiquitous at high pressures and temperatures, as matter which is not in its closest-packed state transforms to a phase of higher density. The transition would be exothermic and of first order as the lattice is compressed, and would be rapid, as the transition of one region triggers that in an adjacent region. The compression of the lattice at the phase transition would cause large-scale motion of lattice defects. As described in Professor Slifkin's paper, this defect motion would result in a large electrostatic signal, roughly proportional to the length of the region transformed.

A likely candidate for such a phase transition would be from a hydrous to an anhydrous form of a mineral. Nearly all terrestrial minerals are formed with varying amounts of water of crystallization in the lattice. Under sufficient compression, they must transform to anhydrous phases, expelling this water.

The second stage of EQ formation would occur when the water (or any other low-density, low friction substance) released in the phase transition in microscopic amounts diffuses to the interface at the fault plane (where there is excess free volume), to form a coherent layer.

The third, and final, stage for formation of an EQ would occur when the tangential component of the stress reaches a threshold sufficient to cause macroscopic rapid slip along the (possibly now "lubricated") fault plane.

In this alternative model, there is a natural time delay between the first threshold, at which an SES would be generated, and the third, at which the EQ itself is triggered. The time delay would be a function of the angle made by the uniaxial stress with the fault plane; the more nearly orthogonal this stress, the longer would be the time delay between the phase transition and the EQ. The total energy buildup in the field surrounding the fault would also naturally be higher, in this case, suggesting that there should also be a correlation of the time delay between an SES and the magnitude of a subsequent EQ: longer time delays for larger EQ's or, very likely, clusters of EQ's. When the major stress component is not closely tangential to the initial fault plane, other nearby fault planes may also be activated in the SES-generation stage, even before the first EQ.

The magnitude of the SES generated by the phase transition should be proportional to the length of the fault region transformed, which, in turn, should be proportional to the one-third power of the total stored energy in the volume adjacent to the fault. The frequency spectrum anticipated would be a $1/f$ type: rather like "flicker noise" from oxide-coated cathodes.

4   Propagation of SES

As reported by the VAN program, the SES are detected only at "sensitive" locations at "preferred" sites, and coverage is limited to EQ's arising in particular regions. Accepting these findings at face value, it is clear that SES, once generated, must not propagate uniformly in all directions from the source. Thus, the earth cannot be considered a homogeneous and isotropic medium for transmission of SES. Indeed, it is difficult to imagine that the SES signals would not be severely attenuated within a few tens of kilometers from the source if they were transmitted isotropically.

It must be remembered that the SES are of extremely low frequency, 0.1 Hz at most, so their wavelengths are extremely long, far larger than the whole earth in a medium of low dielectric constant. Everything is within the "near field" for such waves, and one should not expect an inverse-square radiation pattern. By the same token, various media in the earth can be expected to have extremely high dielectric constants at such frequencies, factors of a hundred or more above the vacuum. Even ordinary water has a dielectric constant of nearly 100 at a frequency of around 0.1 Hz. Electrical conductivity in a medium can increase the dielectric constant by very substantial amounts, even though the attenuation is also increased.

The SES, therefore, would be channeled into strata of high dielectric constant and possibly of high conductivity, and would propagate with far less loss through these paths, quite analogous to the very long range propagation of light through optical fibers or electrical signals through wires. "Sensitive" detectors and "preferred" sites are those which can couple efficiently to these strata, by virtue of local geological structures and subterranean pathways.

5 Future Research

The current SES signal processing equipment is comprised of a simple low-pass filter and digital amplitude-measuring circuits at the field site, which serve to exclude most industrial and atmospheric noise from the detection circuitry. Coincidence of signals between long and short dipoles is a further critical identification requirement.

The single most important need is for improved instrumentation to permit precise characterization of the SES: the amplitude and frequency spectra and any other special characteristics. Such information is required to determine whether or not these signals are reproducible and arise from a single mechanism or a small set of mechanisms. If it can be determined that the SES have well-characterized "signatures," special sophisticated digital filters can be devised which would permit enormously improved enhancements in signal detection. This, in turn, would permit adaptation of the system to other regions with higher noise background than Greece, as well as improved resolution in the predictive ability of the network.

Such measurements could be most easily effected by design of special apparatus which could be moved close to known active faults. It should include both standard VAN-type "antennas" and large solenoids to detect any direct radiation, similar to that used at Loma Prieta by the Stanford group.

In addition to field research, it would also be valuable to conduct model laboratory studies to investigate possible mechanisms. These might include:

1. Measurements of the generation of signals by ionic solids as a result of plastic deformation, and during phase transitions (e.g., KCl, RbCl under pressure);

2. Study of high-pressure, high-temperature phase transitions in terrestrial minerals;

3. Measurements of diffusion of water through model media; and

4. Model studies of electromagnetic wave propagation through stratified media of varying dielectric constant and conductivity.

Greece is a nearly ideal location for the urgently needed field studies because of its relatively low industrial noise level and high seismicity, compared to many other countries. These would permit careful calibrations of the measuring equipment. It may well be advisable to establish a major international Earthquake Research Center in Greece, since the need for a universal earthquake prediction system is clearly of great international importance, and the cost of conducting such studies may well be beyond the current means available to the VAN program.

Despite questions which have been raised about the VAN program, it is clear that lives have already been saved in Greece by heeding the warnings which VAN has issued. Clearly, a continued and enhanced research program is merited to improve the network and permit its extension to other sites and other countries. The lack of a completely acceptable physical theory must not be used to constrain this effort.

The physical model proposed here is just that: a model which appears to fit most of the VAN findings. It is far from being an established, quantitative theory. It would come as welcome news to this author to learn, eventually, that this model has been supplanted by something far more rigorous and substantial.

Reference

1. P. W. Bridgman, The Physics of High Pressure, (G. Bell and Sons, London, 1931; and, with supplement, 1949).

A DISLOCATION MODEL FOR SEISMIC ELECTRIC SIGNALS

L. M. SLIFKIN

Department of Physics and Astronomy, University of North Carolina,
Chapel Hill, NC 27599-3255, USA

It is suggested that a possible mechanism for production of seismic electric signals involves the motion of charged dislocation loops upon rapid change in a shear component of the stress acting on blocks of rock in the earthquake preparation zone. A calculation which uses likely values of the relevant parameters indicates that such a process could produce significant voltages at the surface of the Earth.

1 Introduction

Detection of telluric voltages at the Earth's surface has been reported by several groups, such as the VAN team in Greece[1]. In order to relate these signals to impending earthquakes, it would be helpful to understand the mechanisms of both their production and also their transmission in the highly anisotropic and heterogeneous media found in the Earth's crust. A feasible mechanism for transmission of the signals over long distances has been proposed by Lazarus[2], and is further discussed in a companion paper in these proceedings; the present article is concerned only with the process of the generation of the voltage.

It has been experimentally demonstrated some years ago, by Hoenig[3], that severe loading of laboratory specimens of rocks can result in the appearance of very large voltages at the surface. Piezoelectric effects and stress-induced orientation of impurity-vacancy electric dipoles do not appear capable of giving rise to large enough signals in the usual polycrystalline materials found in nature. Hoenig proposed that his observations were the result of a sequence of microfractures, and somewhat later it was shown by Gershenzon et al[4] that the production and motion of electrically charged dislocations at the tip of the fracture crack would produce voltage differences. They were able to estimate the magnitudes of electromagnetic emission to be expected from this process during crustal fracturing, and concluded that this phenomenon could produce significantly large signals.

A number of possible mechanisms for production of seismic electric signals have recently been examined by Bernard[5]. He finds that, given the right conditions, one could perhaps understand these effects in terms of electrokinetic phenomena: he suggests that a change of stress may trigger fluid instabilities that would give rise to large streaming potentials. One would still require some "channeling" of the electric signal in order to explain the detection at distances as great as involved in the VAN experiments. The model proposed by Lazarus[2] would seem to be a possible explanation of such preferential channeling.

It has been suggested by the present author[6] that, even in the absence of large amounts of plastic flow or of fracture, an abrupt change in the shear stress -- either increase or decrease -- should cause a voltage difference to appear between opposite surfaces of a specimen of most minerals. In such materials as silicates, oxides, and the like, lattice vacancies and aliovalent impurity ions carry effective charges. Moreover, single-stepped jogs on edge dislocations (essentially, the edges of partial lattice planes which do not extend completely through the crystal) are also electrically charged. Because these jogs must establish simultaneous equilibria with all of the different types of point defects within the crystal, it follows that, in general, the edge dislocations in ionic crystals carry a net electric charge[7].

Now, although the dislocation may not be able to move through large distances in response to the applied shear stress, segments of it can indeed bow out between the points at which the dislocation is pinned, as has been discussed by Granato and Lücke[8]. The pins may be impurity ions, clusters, or points of intersection with other dislocations. The dislocation loops between the pinning points respond to applied shear stresses much as if they were non-Hookean elastic bands. Upon a change in stress, the relaxation rate of the loops is primarily a function of the Peierls stress -- the stress required to cause dislocation displacement in an otherwise perfect lattice. At high temperatures, the dislocation loops can be quite mobile.

The purpose of this paper is to make an estimate of the magnitude of the electric field produced by the motion of these dislocation loops when the stress is rapidly increased or decreased, causing the loops to expand or contract, respectively. The numerical results are quite sensitive to the values chosen for the relevant parameters, some of which are uncertain. At best, then, this discussion is a feasibility test of the question: given a reasonable set of values for these parameters, is it possible for the proposed mechanism to account for signals comparable to those that have been reported?

2　Charged Dislocations

Since the proposed mechanism depends heavily on the net electrical charge of dislocations in non-metallic mineral crystals, it seems worthwhile to review briefly the origin and extent of the phenomenon. We are only concerned with edge dislocations, since screw dislocations are not charged. Moreover, the points of particular interest to us along the dislocation are the "jogs", where the edge of the extra half-plane that produces the dislocation makes an abrupt step from one slip plane to an adjacent, parallel one. In a crystal composed of ions, as in the minerals of interest to us, such a step leaves an exposed ion at the jog. Seitz[9] showed that the jog thereby has an effective charge equal to half of that of the exposed ion. It was earlier pointed out by Frenkel[10] that jogs on surface terraces and also, by implication, on dislocations must act as sources and sinks for each of the individual types of intrinsic point defect, such as cation and anion vacancies. The process of emitting or absorbing a vacancy causes the jog to change electrical sign and to move along the dislocation by one interatomic spacing.

The jogs must therefore establish multiple equilibria -- i.e., with each of the species of point defect. This results in the presence of more jogs of one sign than the other, so that the dislocations carry a net charge. The linear charge density on the dislocations is compensated by an approximately cylindrical space charge around the dislocation, consisting of an excess of point defects of the opposite sign. The electric potential difference that is thereby established between the dislocation and the distant interior just compensates for the differing formation free energies of the different species of intrinsic point defects.

In materials with significant concentrations of impurities, as in the case of geophysically interesting minerals, the space charge consists largely of aliovalent dopant ions. These ions will, in general, be very much less mobile than dislocation segments bowed out between pinning points. Hence, after any abrupt change in stress, the bowed loops will quickly respond, leaving the space charge distribution unrelaxed; the center of the space charge no longer coincides with the line of the dislocation, and an electric dipole has been produced. This is the basis of the mechanism proposed here for the generation of seismic electric signals.

3 The Model

First, we estimate the dipole moment per unit length of dislocation that can be expected to result from a large and sudden change of shear stress. For the purposes of this rough calculation, it is convenient to ignore the various cosine factors that arise from the fact that the slip planes are not necessarily aligned with the major faces of the rock specimen or parallel to the axis of any folding to which the specimen had previously been subjected.

The stress-induced dipole moment lies in the slip plane and is oriented perpendicular to the dislocation line. Its magnitude per unit length of dislocation is given by the product of the dislocation charge density and the change in mean displacement of the bowed segments, caused by the change in applied stress. For a number of alkali and silver halides, experimental values have been deduced for the charge density, as a function of purity, temperature, and dislocation dynamics; these studies have been reviewed by Whitworth[7]. A typical value is of the order of 0.1 e/b, where e is the electronic charge and b the lattice spacing. If we assume comparable numbers for the rocks of interest to the present problem, and take b to be about 5×10^{-10} m, then the net charge density is of the order of 3×10^{-11} C/m.

The second factor determining the dipole moment is the mean displacement of the dislocation loops bowed out between pairs of pins. If the applied stress is a significant fraction of the flow or fracture stress, then the loops will bow out through a distance large compared to b, and we can approximate the mean displacement by one-fourth the spacing between pins (for example, if the loop were nearly semicircular, then its maximum displacement would be half the interpin distance). We now need some reasonable estimate of this interpin spacing, which must depend on the crystal structure, the presence and state of aggregation of impurities, the dislocation distribution, and -- overall -- the thermal and mechanical histories of the specimen. Analyses of various mechanical properties of a range of solids have suggested that the interpin spacing is often of the order 10 - 100 b. If we choose a value in the middle of this range, then our mean dislocation loop displacement under the maximum supportable stress is about $(50/4)$ b $\approx 6 \times 10^{-9}$ m.

If we now consider a sudden change in stress of magnitude comparable to the stress at failure, the dipole moment produced by the motion of the charged loops, per unit length of dislocation, is equal to the product of these two factors: $(3 \times 10^{-11}$ C/m$) \times (6 \times 10^{-9}$ m$) \approx 2 \times 10^{-19}$ C-m/m. The resulting electric polarization, the dipole moment per unit volume, is obtained by multiplying this quantity by the appropriate density of dislocations (i.e., in the simplest case, the total dislocation length per unit volume, or the number intersecting a surface of unit area).

The dislocation density, of course, depends on the mechanical history of the specimen. While it is possible to grow crystals of such semiconductors as very pure silicon with virtually no dislocations, most annealed simple crystals have densities in the range $10^8 - 10^{10}/m^2$, and heavily deformed material contains $10^3 - 10^4$ times this. These values, however, are not appropriate to our problem, because some dislocations have their extra half-planes coming from above, while others are from below (one says that they have opposite mechanical signs). Under an applied stress, the interpin segments of these two types bow out in opposite directions; hence, they make opposite contributions to the polarization. Only the net excess of one type over the other produces a net signal. If the distribution were quite random, this excess would be approximately equal to the square root of the total number.

The distribution of dislocations in most rock strata is not, however, at all random, because of the folding they have undergone in earlier history. Plastic bending of crystalline material is accomplished through introduction of a set of edge dislocations, all of the same mechanical sign, to accommodate the differing lengths of the upper and lower surfaces. Consider, for example, the specimen sketched in Fig. 1, a cube of edge length 1 m. Suppose that it has been bent through an angle of

$1°$ (≈ 0.02 rad). Then the upper edge must be longer than the lower one by 0.02 m. If we again take b = 5×10^{-10} m, then the bending must have resulted from the introduction of $0.02/(5 \times 10^{-10}) = 4 \times 10^7$ new edge dislocations, all of the same mechanical sign, as shown in the diagram. In this case, the excess density of dislocations is then $4 \times 10^7/m^2$. This is a quite modest value, and probably greatly underestimates the dislocation densities in naturally occurring rock.

If, nevertheless, we use this result for our illustrative example, and recall our earlier estimate of the stress-induced dipole moment per unit length of dislocation (i.e., 2×10^{-19} C-m/m), we then obtain a volume polarization of $(4 \times 10^7/m^2) \times (2 \times 10^{-19}$ C-m/m) $= 8 \times 10^{-12}$ C/m^2; this is just the charge density that appears on the surfaces at the ends of the slab (ignoring such factors as the cosine of the angle between the normal of the end surfaces and the direction of displacement of the dislocation loops).

We are now ready to calculate the electric field appearing at the Earth's surface. The numerical result depends sensitively on the choice of such parameters as the size and depth of the block. It will also be greatly perturbed by heterogeneities and anisotropy in the dielectric properties of the surrounding medium. In particular, the decay of electric field with the cube of the distance from the dipole is only applicable for a homogeneous, isotropic medium; this is considered at greater length in the companion paper in these proceedings by D. Lazarus. With these limitations in mind, the following estimate can only be taken as an illustrative example, at best.

For concreteness, consider a horizontal block 1000 m on each side, and 100 m thick. Suppose that it has been folded about a horizontal axis perpendicular to one of the end faces, as shown in Fig. 2. If we take the angle through which the block has been bent to be $1°$, as in the calculation relating the Fig. 1, so that the density of excess dislocations is 4×10^7 /m^2, and if we use the previously estimated stress-induced dipole moment per unit length of dislocation (i.e., 2×10^{-19} C-m/m), then the resulting dipole moment is $(2 \times 10^{-19}$ C-m/m) $\times (4 \times 10^7/m^2) \times (10^5$ m$^2) \times (10^3$ m); the last two factors are the cross-sectional area and the length of the dipole, respectively. The stress change has thus produced an electric dipole moment of magnitude 8×10^{-4} C-m.

To estimate the effect of this dipole moment on observations made at the Earth's surface, we must know the depth of the block and the dielectric structure of the intervening material. Let us first assume that the medium is homogeneous, isotropic, and has a dielectric coefficient of unity. Then, for a dipole of length L, the electric field at a distance d perpendicular to its axis is given by:

$$E = (1/4\pi \ \varepsilon_0)p(d^2 + L^2/4)^{-3/2}, \qquad (1)$$

where p is the dipole moment, estimated above to be 8×10^{-4} C-m, and the depth d is taken, for sake of illustration, to be 10 km = 10^4 m. Ignoring the term $L^2/4$ as compared to d^2, and inserting $\varepsilon_0 = 9 \times 10^{-12}$ C^2/Nm^2, we obtain an electric field at the earth's surface of 7×10^{-6} V/m, or 7 mV/km. This is in the range of signals obtained for large earthquakes by the VAN group, but, of course, their electrodes are not located directly above the active regions. For large distances from the dipole ($d > L$), E falls off as $1/d^3$, and one can understand the experimental observations only in terms of some sort of channeling by dielectric inhomogeneities. This problem is not unique to the present mechanism; virtually any reasonable model would have the same requirement. An attractive solution to this puzzle is proposed by D. Lazarus elsewhere in these proceedings.

Another question relates to the rather long durations of the signals recorded by the VAN group, as compared to the quite short RC time constant for relaxation of electric fields in typical wet minerals. It is possible that the recorded signals consist of unresolved superpositions of many rapid pulses, generated by the propagation of mechanical relaxations in a sequence of neighboring blocks.

It thus appears that the charged dislocation loop mechanism is a likely candidate but certainly not the only possibility, for explaining stress-induced voltages in rocks. For example, if significant microplasticity occurs, then the long-range motion of the charged dislocations would greatly enhance the electric signal. In any case, if a wave-guiding mechanism for long-range transmission, such as that suggested by Lazarus, proves to be an acceptable means of precluding the $1/d^3$ decay, then the voltages estimated here are indeed consistent with the observations that have been made in Greece. It is clear, however, that the predicted signal is quite sensitive to the chemistry, morphology, and mechanical history of the relevant strata. It thus cannot be expected to be a general phenomenon, independent of the local geology.

References

1. P. Varotsos, K. Alexopoulos, and M. Lazaridou, Tectonophys. **224**, 1 (1993).
2. D. Lazarus, ibid. **224**, 265 (1993); also, paper in these proceedings.
3. S. Hoenig, Nat. **279**, 169 (1979).
4. N. Gershenzon, M. Gotchberg, A. Karakin, N. Petviashvili, and A. Rykunov, Phys. Earth Planet. Inter. **57**, 129 (1989).
5. P. Bernard, J. Geophys. Res. **97**, 17 531 (1992).
6. L. Slifkin, Tectonophys. **224**, 149 (1993).
7. R. Whitworth, Adv. Phys. **24**, 203 (1975).
8. A. Granato and K. Lücke, J. Appl. Phys. **27**, 583 (1956).
9. F. Seitz, Rev. Mod. Phys. **23**, 328 (1951).
10. J. Frenkel, "Kinetic Theory of Liquids" (Oxford Univ. Press, 1946); p. 36.

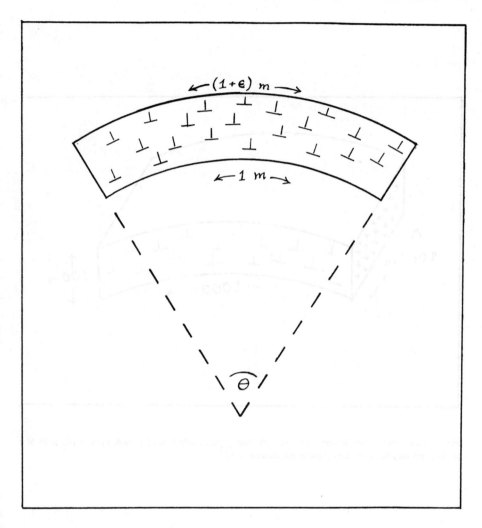

Fig. 1. Introduction of edge dislocations by plastic bending. If the angle $\theta = 0.02$ radians, the upper edge of the block is longer than the lower by 0.02 m. The dislocations are indicated by upside-down "T"s, with the base representing the slip plane and the stem the end of the extra half-plane. (Not to scale)

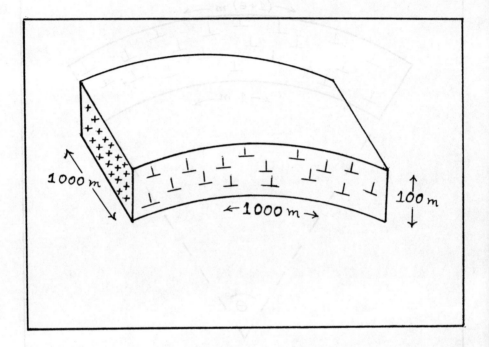

Fig. 2. Production of an electric dipole in a previously plastically bent specimen, upon application of a shear stress parallel to the slip plane of the dislocations.

LABORATORY INVESTIGATION OF THE ELECTRIC SIGNALS PRECEDING EARTHQUAKES

V. HADJICONTIS   C. MAVROMATOU

Section of Solid State Physics,
University of Athens, Panepistimiopolis, 15785 Zografou, Athens Greece.

During the uniaxial compression of dried rock samples two kinds of electric signals were detected, prior to rock's failure:
a) Slow transient electric currents of the order of some nA/cm^2 , which are emitted before the sample gets into the microfracturing stage.
b) Near field electromagnetic disturbances in the frequency range of some kHz to some MHz, which start to be detected when the sample reaches the microfracturing stage.
The intensity of the electromagnetic emission drastically increases as approaching to failure. These laboratory results lead to a simulation pattern according to which a plausible explanation of the origin of the preseismic electric signals (SES) can be suggested.

Introduction

It is well known [1,2] that slow transient variations of the earth's electric field , called Seismic Electric Signals (SES), are detected on the earth's surface, prior to earthquakes. These precursor electric signals are detected by the VAN network in Greece and resulted in successful predictions of imminent earthquakes. Laboratory experiments of uniaxial compression of both rock samples and pure ionic crystals were carried out, in order to investigate the origin of these signals. The main scope of these experiments was to investigate:
1. The existence of electric signals prior to rock's failure and hence, on a larger scale, prior to earthquake.
2. Whether the order of magnitude of these signals is compatible to measurements resulting from field experiments.
3. Which are the physical mechanisms possibly responsible for the emission of such signals.
This experimental investigation was made in the framework of the VAN project.

Experimental apparatus

The experimental apparatus is placed in an electrically shielded room made of copper foil and consists of the following parts:
1)A uniaxial hydraulic loading machine, which is hand operated in order to avoid the motor noise. This machine is supplied by a load cell (Sensotec, Model 53) and an amplifier for recording the external load variations (fig.1).
2)A system for the detection and recording of the slow transient electric currents.This consists of two probing electrodes, an electrometer amplifier (with

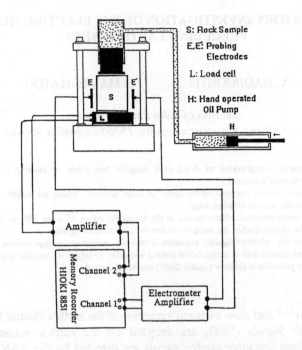

S: Rock Sample

E,E: Probing Electrodes

L: Load cell

H: Hand operated Oil Pump

Amplifier

Memory Recorder HIOKI 8851

Channel 2

Channel 1

Electrometer Amplifier

Fig. 1: The experimental apparatus.

$R_{inp} \sim 10^{13}$ Ω) and a memory recorder. The electric signals are recorded as potential differences between the two electrodes painted on two opposite sides of the sample's surface with a conductive paste, covering an area of about 2 cm². In this way a capacitor is formed having as dielectric the compressed rock. Alternatively two very thin copper plates can also be used (having a surface of about 2 cm²) placed very close to the sample's surface (~0.5 cm) but not touching it. The result, concerning the detection of the signals, was found to be similar to that arising from using the painted electrodes. The output of the electrometer amplifier is connected to the digital memory recorder (Hioki 8851) which has two input channels. In this way both the variations of the externally applied load (the electric output of the load cell) and the potential difference (from the electrometer's output) can be simultaneously recorded. Additionally the time derivative of the externally applied load can be computed and displayed. The equivalent electric circuit of the measuring set-up (fig.2) indicates that the total leakage resistance R_L consists of the following components: a) a bulk resistance b) a leakage resistance through the sample's surface and the air (depending on the humidity) and c) an externally (intentionally) connected resistance. The latter resistance must be lower than the remaining leakage resistances but, in any case it cannot be lower than some tenths of MΩ, (otherwise the signals are too low to be discriminated from noise). The effective leakage resistance determines of course the time constant of the circuit.

3) A system for the detection and recording of the EM and acoustic disturbances the block diagram of which is depicted in fig.3. This system has the following capabilities: a) To set a discrimination level depending on the noise level. b) To record continuously the intensity of the EM and acoustic emission (in a chart recorder). c) To trigger the bi-channel Transient Recorder, so that events of very small duration (like the electric and acoustic pulses corresponding to one crack opening) can be captured and recorded. The Transient Recorder system consists of a PC-99 Amplicon card, with sampling rate up to 20 Msamples/sec and suitable software.

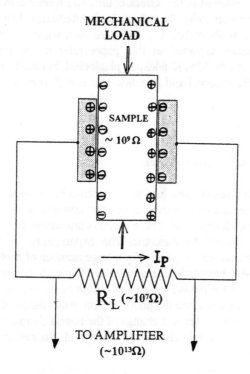

Fig.2: The equivalent circuit of the polarization currents measuring system.

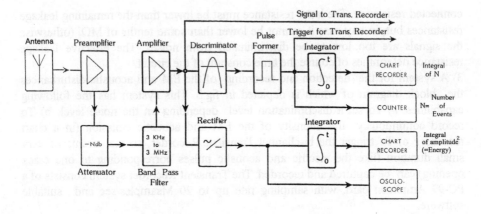

Fig. 3: Blocck diagram of the system for the recording of EM and acoustic emission.

4) The rock samples collected in the Greek countyside were cut in cubic shape and have dimensions, approximatelly, 3cmx3cmx3cm. Afterwards they were dried in an oven for some hours, at about 100° C , and were used without further preparation. The experimental results reported in this paper refer to: a) granite, containing approximatelly 45% quartz, 35% K-feldspar, plagioclast, biotite) b) mineral quartzite, c) limestones from Thessaloniki and Ioannina area and d) pure ionic crystals of LiF and NaCl.

Results

A. Stress induced polarization currents.

Before the compressed sample gets into the microcracking propagation stage, during the compression rate changes, transient electric currents were recorded, in the absence of an external electric field. These currents are attributed to the variations of the electric polarization of the dielectric. The experiments for the detection of transient electric currents were repeated for a large number of rock samples and of ionic crystals as well. Characteristic examples are depicted in figures 4,5 and 6. A careful consideration of the equivalent electric circuit of the measuring set-up (fig.2) could help in better understanding the phenomenon. Any change of the external load results, as will be explained later, in a change of the bound charges surface density of the sample. Concequently this change of the bound charges induces charges of opposite sign on the adjacent conductive plates resulting in the motion of free charges in the external circuit. This is a transient polarization current, with density

$$j = d\Pi/dt, \qquad \text{(where } \Pi \text{ is the polarization of the sample)}$$

which flows <u>only during the variation of the polarization of the specimen</u> from an initial steady state to a final one. The potential difference $U = i_p.R_L$ along the total leakage resistance R_L is proportional to the polarization current. In this case R_L is of the order of 20 MΩ. By taking into account that the measured voltage is of the order of some decades of mV, the current density is some nA/cm^2. This value is compatible to the electric field values measured at the surface of the earth at epicentral distances of ~100 Km by the VAN network [3].

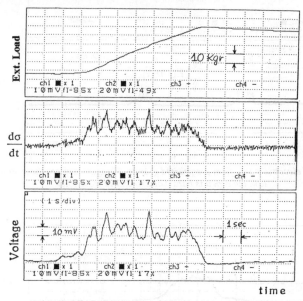

Fig.4: Transient electric signal from compressed marble.

Summarizing, we must point out the following points:

1. The stress induced polarization currents seem to follow the first time derivative of the externally applied stress.

2. After a large number of experiments using many kinds of rocks, we concluded that the stress induced polarization currents are detectable only when the rock is a good insulator (R > of some tens of MΩ) and contains ionic crystalline materials.

3. The stress induced polarization currents are emitted from non piezoelectric materials as well, e.g. non piezoelectric ionic crystals of LiF or NaCl emit such signals, so they are not necessarilly attributed to the usual piezoelectric effect.

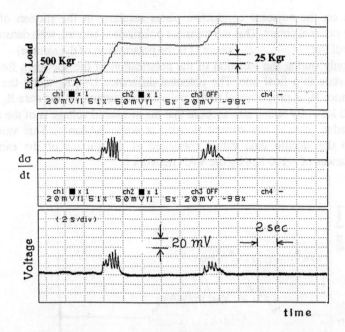

Fig. 5: Transient elecctric signal from compressed granite. In region A where the compression rate is low, the signal is hardly discriminated from noise.

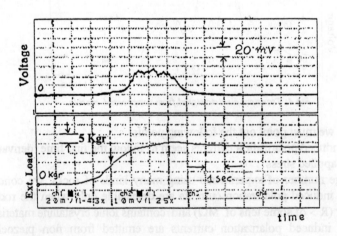

Fig.6: Transient electric signals of pure ionic crystalline LiF: Note the threshold at the stress curve , for the initiation of the emission of the signal.

4. These transient currents are emitted from dry rocks, hence they cannot be attributed to the electrokinetic effect (streaming potentials). On the contrary, no signals were detected from wet rocks.

A tentative interpretation of the stress induced polarization was proposed by L.Slifkin [4]. This explanation is based on the movement of segments of charged dislocations (Whitworth, 1975) in relation to their less mobile Debye-Huckel cloud which consists of oppositely charged point defects [5]. In other words as the stress increases, the pinned dislocations bow forward, as elastic strings between the pinning points[6] while the less mobile cloud of point defects lags behind. In this way a dipole is formed and the sample becomes polarized. The aforementioned procedure is depicted in fig.7.

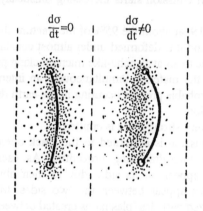

$$\frac{d\sigma}{dt}=0 \qquad \frac{d\sigma}{dt}\neq 0$$

Fig.7: Cartoon type diagram depicting the dipole formation due to the relative displacement between the dislocation core and their Debye-Huckel cloud, during stress variation.

A relaxation time τ characterizes the rate at which the defect concentrations in the charge cloud can adjust to a perturbation caused by the dislocation being displaced. This relaxation time depends strongly on the temperature and increases almost exponentially with T^{-1}. For example, at 200 ^{0}C, τ is approximatelly of the order of a couple of days, while at 500 ^{0}C it is of the order of some milliseconds. So, for relativelly high temperatures, i.e. above 400 ^{0}C no polarization can be practically observed, because the surrounding cloud, (due to the very small relaxation time), follows the movement of the dislocation. On the contrary, at lower temperatures, e.g. below 200 ^{0}C the cloud is quite difficult to move and it is separated from the moving dislocation core thus giving rise to an electric dipole moment. As the sample's stress induced polarization Π is proportional to the externally applied axial load, provided that the relaxation time is high enough, the rate dΠ/dt should also be proportional to the first time derivative of the mechanical load.

B. Electromagnetic and acoustic emission

Figures 8 and 9 depict the appearance of near field electromagnetic disturbances in the frequency band from 3kHz to 3MHz, as the external stress increases till failure. This electromagnetic (EM) emission is detected using a simple monopole antenna placed close to the sample. We draw attention to the following points:
a) There is a stress threshold for the initiation of the EM emission, which is around the 50% of the final fracture stress, depending on the brittleness of the material.
b) The intensity of this EM emission starts increasing drastically as approaching to failure.
c) As depicted in figs.8 and 9, at about the 95% of the fracture stress the sample gets into an unstable situation and it is deformed under almost constant load. During this creep stage the samples due to an avalanche-like microcracking propagation fail after a time lag, depending on the material. Simultaneously an intence EM emission is observed. This is a quite probable process occuring in the earth during the final stage of the earthquake preparation.
d) Just before fracture an very strong pulse is emitted.
The origin of this EM emission is attributed to the crack opening as the material undergoes an irreversible microfracturing procedure. As each crack opens net charges of opposite sign appear on its sides due to their abrupt separation and consequently high voltages appear between the two sides. Moreover conductive material consisting of ionized particles (plasma) is created between the crack walls.

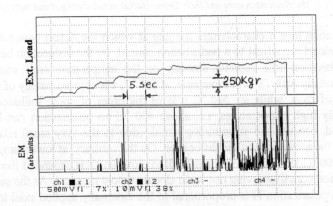

Fig.8: The integral of the EM emission intensity from granodiorite which is compressed until fracture.

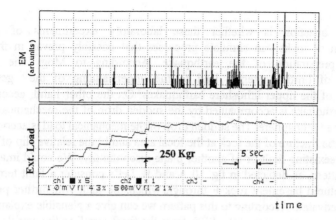

Fig. 9: The integral of the EM emission intensity from granite which is compressed until fracture.

Therefore as the voltage increases, a discharge bursts out resulting in the emission of EM pulses. Each pulse has a very short duration of the order of some microseconds and can be recorded by a transient recorder simultaneously with an acoustic signal (fig.10). The simultaneous recording of the acoustic signal following the crack opening seems to verify the above explanation.

A critical value of the stress is necessary in order to start the process of crack propagation and hence there must be a stress threshold for the initiation of EM activity.

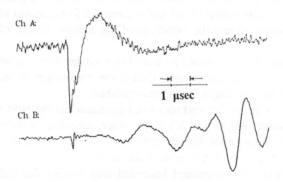

Fig. 10: Simultaneous recording of the electric (channel A) and acoustic (channel B) pulses corresponding to crack oppening. Note the time difference in the arrival of the electric pulse (detected by an antenna) and the acoustic pulse (detected by a transducer).

Discussion

The crucial question is whether these laboratory experiments of uniaxial ccompression of rocks can simulate processes that possibly prevail in the earth during the preparation of the earthquake. According to Mogi[7] the fracture experiments of brittle materials lead to the understanding of the generation mechanisms of the upper lithoshere earthquakes. On the other hand, according to Brace[8] an earthquake is accompanied by the sudden drop of the mechanical load in the compressed lithosphere. The sudden drop of the mechanical load is accompanied by a) the brittle fracture of the rock at the focal area and b) the relative slip of the two sides of a preexisting fault or of another plane discontinuity. We could imagine the following pattern, as depicted in fig. 10 A and B. Let's consider a fault terminating on an unruptured basement rock F through which the fault will further propagate after being activated. According to this pattern we can give a plausible explanation of the propagation mechanism of the SES, from the focal area F to the earth's surface, based on our laboratory experiments. This can be done if we consider the following similarities between figures 2 and 10:

a) The focal area is a block of crystalline dielectric and corresponds to the rock specimen.

b) The active fault provides the driver of the stress field so that a differential uniaxial compression to a preferred direction occurs.

c)The boundary interfaces between the insulating and the conductive blocks form a large capacitor which corresponds to the electrodes capacitor in the experiment.

d) The conductive channel from the focal area to the earth's surface plays the role of the leakage resistance R_L of the experiments.

We empasize that: The existence of a boundary interface between the crystalline dielectic and the conductive medium (and consequently the resulting capacitance) is a crucial factor for the stimulation of induced currents (of free charges) which flow in the conductive medium, during the variations of the surface density of the bound charges.

For the attempted simulation it would be useful to consider figure 11 which depicts the stress variations at an activated fault. During the preparation of an earthquake (associated to faulting), two stages can be distinguished:

a) The shear stress on the unruptured rock mass increases but the fault sides do not move yet (static friction) (fig.11 part A). In this stage, according to our experimental results, suitable conditions exist for the emission of slow transient electric signals similar to SES detected by the VAN network ($d\sigma/dt>0$).

b) When the stress on the unruptured basement rock reaches the yield point the microfracturing process starts. This is a suitable condition for the initiation of the EM emission. Then a relative displacement of the two blocks of the fault starts (dynamic

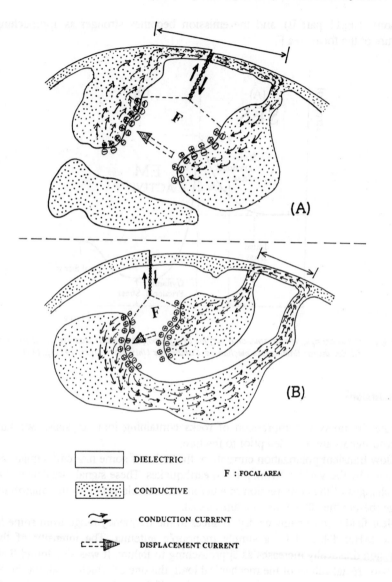

DIELECTRIC

F : FOCAL AREA

CONDUCTIVE

CONDUCTION CURRENT

DISPLACEMENT CURRENT

Fig. 11: An SES propagation pattern.

friction) (fig.11 part B), and the emission becomes stronger as approaching the fracture of the focal area F.

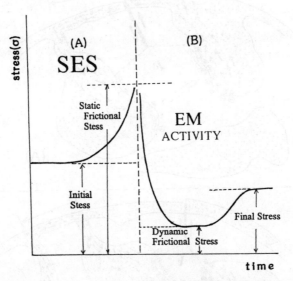

Fig. 12: *Stress change, versus time, on a surface of a propagating shear fracture in the focus area (K. Kasahara: Earthquake mechanics, Cambridge University Press, 1981, p. 139).*

Conclusions

During the uniaxial compression of rocks containing ionic crystals, two kinds of electric signals are recorded prior to fracture:

1. Slow transient polarization currents of the order of some nA/ cm^2 similar to those detected by the VAN network prior to earthquakes. These signals are detected during the changes of the compression rate before the sample gets into the microfracturing stage (before the 50% of the fracture stress).

2. Near field electromagnetic disturbances in the frequency range from some kHz to some MHz, detected by a simple monopole antenna. The intensity of the EM emission drastically increases as approasching to failure. It was also found that there are two crucial values of the mechanical load, the one of which (at about the 50% of the fracture stress) is related to the initiation of the EM emission and the other (at about the 95% of the fracture stress) is related to the initiation of the uncontrolled state of microcracking propagation (avalanche breakdown), which necessarily leads to failure.

References
1. Varotsos P. and Lazaridou M., a, Tectonophysics, 188, 322, (1991).
2. Varotsos P. Alexopoulos K. and Lazaridou M., Tectonophysics, 224, 1, (1993).
3. Varotsos P. and Alexopoulos K., Thermodynamics of point defects and their relation with bulk properties, North-Holland Amsterdam, p.419,(1986).
4. Slifkin L., Tectonophysics, 224, 149, (1993).
5. Robinson W.H. and Birnbaum H.K., Journal of Applied Phys. Vol.37, No 10, (1966).
6. Granato A. and Lucke K., Journal of applied physics Vol.27, No 6, (1956).
7. Mogi K., Bulletin of the EQ Research Institute, Vol.41, p.615, (1963).
8. Brace W.F., J.Phys.Earth 25, suppl.202, (1977).

ON ELECTROTELLURIC SIGNALS

P. BERNARD, J.L. LE MOUEL

Institut de Physique du Globe de Paris, 4, place Jussieu, 75252 PARIS CEDEX 05

In a first part, examples of electric and magnetic anomalous signals are given which have been recorded in different geological settings with different length scales, and whose time constants cover a large band. Most of them are interpreted as generated by a fluid flow in the surrounding rocks through the associated electrokinetic effect. The flow is forced or perturbed by time changes in the local stress/strain field. Such signals with a local origin are expected to be of general occurrence. In a second part, we investigate wether a plausible source mechanism, with sources located in the epicentral area, could generate precursor electric signals that could be recorded up to several hundreds of kilometers away, as advocated by the VAN group. It is argued here that no plausible mechanism can be found which could account for these long distance electric precursors. To conclude, the interest of performing electric and magnetic recordings in many different active areas, in association with other geophysical measurements, is underlined.

1 Introduction

Electric and magnetic signals related to tectonic, seismic or volcanic activity have been reported for a long time, and the present day literature on the subject is abundant (e.g Chen Zhangli [1]; Zlotnicki [2]).

We started our own observations at the end of 1976, after the 1976 seismo-volcanic crisis of La Soufrière volcano (Feuillard et al.[3]). The main purpose of our experiments was to determine wether electric and/or magnetic signals generated in porous or fissured rock by stress variations were to be expected in the field; or, conversely, in the case such signals were observed in the field, to understand the mechanism generating them.

Starting with measurements on active volcanoes, we soon concluded that, despite the highly magnetized rocks these edifices are made of, piezomagnetism was not efficient enough a mechanism to account for the magnitude of the observed magnetic signals. We then focussed on experiments which, however different by the length scale, time scale, geological setting, were all propitious to the manifestation of electrokinetic effects. And we were indeed led to the conclusion that electrokinetics is commonly at work wherever water is made circulate in a porous or fissured rock - by a variation of the local stress field or some kind of external forcing -, and that easily measurable electric or/and magnetic signals are generated.

In this paper we will first present the results of some of these experiments, from four sites: an underground quarry of limestone in the vicinity of Paris, the Piton de la Fournaise volcano in Reunion Island, the Djibouti geodynamically active area, the Ioannina region in Greece. We won't give here any quantitative proof of the electrokinetic origin of these signals. Instead, in a second part of the paper, focussing on the Greek case, we will briefly resume the discussion of the plausibility of long distance electrotelluric precursors to earthquakes (Bernard [4]).

2 The Meriel Quarry

We have now used for ten years the Meriel quarry, close to Paris, as a natural laboratory. This Eocene limestone quarry is dug according to the rooms and pillars technique (Fig. 1). The height of the pillars is approximately 3 m, their section about 15 m^2, the thickness of the roof of the order of 30 m. Porosity of the rock is in the 25-45% range, water saturation (the percentage of pore volume filled with water) in the range 60-85%. Vertical stress σ_{zz} measured within the pillars by the flat jack technique varies from 12 to 40 bars. Quite interestingly for electric measurements, the temperature of the air in the gallery - far from the door - is constant within 1°C.

2.1. First experiment

The West face of pillar 39 was equipped with an array of electrodes implanted at the (7 x7) nodes of a square mesh - 20 cm side - grid (Fig. 1). Each electrode is indexed (i, j) ($i = 1, 2, ..., 7$ for the line; $j = 1, 2, ..., 7$ for the column). Potential differences $\delta V (i, j) = V (i, j) - V (4,4)$ between the different electrodes and the central one (4,4) were measured during a number of 4 weeks periods (Morat et Le Mouël [5]).

OBSERVATIONS. - Coherent variations of $\delta V (i, j)$ - we mean with similar forms for all the (i, j) -, with amplitudes of the order of 10 mV, with time constants from a few hours to some ten days, were observed each time (Fig 2). And it turns out that most of the variations were correlated with the variations of the atmospheric pressure (a pressure sensor had been installed in the gallery). As the changing hydrostatic pressure is uniform, electric signals need rock heterogeneity.

CONCLUSION - An electric potential takes place at the surface of a volume of porous non saturated rock - of metric size in the present case - when submitted to a time-changing hydrostatic stress. The amplitude of this potential is typically 10 mV for a stress change of 10 mbars (for time constants from a few hours to a few days). The time changes of the potential are closely correlated with the time changes of the stress. The typical length scale of the potentials at the surface of the pillars is of the order of a few tens of centimetres and depends on rock heterogeneity. Those observations might recall, although the scales are different, electric signals measured at the surface of a stressed crust, whose amplitude can also change drastically from place to place.

2.2. Second experiment

Two electrodes (0) and (1) were driven 15 cm deep in the western face of pillar 39, and six in the eastern face of pillar 40, 3 cm deep (in fact between 0 and 3 cm, the length of the electrodes), numbered 36 to 41 (Fig. 3). Electrodes are made of tubes of stainless steel (3 cm long) with an inner diameter of 2 mm and an outer one of 4 mm. Potential differences $dV_{no} = V(n) - V(0)$ between the current electrode (n) and electrode (0), generally taken as the reference, have been measured with a high impedance ($>10^8\Omega$) voltmeter. Six thermistances with a sensitivity of 10^{-3}°C have been installed in the gallery, along a vertical line one meter from pillar 40 as well as four relative-humidity and two pressure sensors (Fig. 3). All the measurements are simultaneously recorded with a sampling rate of twelve per minute (Morat et al.[6]).

OBSERVATIONS - Results are illustrated by figures 4 and 5 which display recordings collected on May 19, 1994, from 14h29 to 15h56. On all the curves representing temperature T, relative humidity RH, potential differences dV_{no}, quasi-sinusoidal variations with periods of the order of a minute occur simultaneously (except for a phase shift). More precisely, a succession of trains of oscillations is observed; whereas the period and amplitude of the quasi-sinusoidal oscillations is often remarkably constant over a given train, they change rather suddenly from one train to another. Amplitudes are typically a few hundredths of a degree for T, a few millivolts for the dV_{no} (the humidity sensor is not calibrated).

DISCUSSION AND CONCLUSION - The temperature of the air in the gallery is remarkably constant; its annual variation is of the order of a degree (Morat et al.[7]). We suggested that this constancy of temperature is due to exchanges between the air of the gallery and the surrounding rock, highly porous with a high content of water. We also showed that, due to this same mechanism, the annual variation of the temperature of the embedding rock, a few cm inside, is 50 times smaller than in the air: the rock protects itself against temperature variations through some feedback mechanism. Therefore we consider that the observed sinusoidal variations of T, RH, dV are all due to the reaction of the system consisting of the gallery-and the embedding rock to perturbations (probably thermal). An increase of T is accompanied by an increase of RH and, through evaporation-condensation processes, by an outward motion of the fluid contained in the rock pores, a decrease in T by the fall in RH and inward motion of fluid. This water flow creates the observed electric potentials through electrokinetic effect. The constancy of the amplitude of the sinusoidal variations -for a given period- results from the control (amplitude saturation) by the non-linear terms entering the equations governing the system. In conclusion we reiterate that electrical measurements are a powerful tool to detect and investigate fluid flows in natural porous media; in the present case they shed new light -to our knowledge- on the exchanges between a cavity and its embedding rock, a situation to be expected in other geophysical contexts.

3 The Piton de la Fournaise Volcano

The Piton de la Fournaise is a shield volcano settled in the Southeastern part of Reunion Island in the Mascareigne archipelago (latitude 21°.2 S, longitude 55°.5 E) (Fig. 6). The building of this 2632 m high (above the sea level) 40 km wide volcano started 360 000 years ago (Bachèlery [8]). The last stage of this building was the formation of the Enclos Fouqué caldera (Fig. 6), around 4800 years ago, which is considered as the starting point of the activity which still prevails nowadays.

Two craters cut the summit of the central edifice, the Bory crater (300 m in diameter) and the Brûlant (or Dolomieu) crater (1000 m in diameter). Several fractures cut the volcanic massif. The most prominent one is called the "rift zone" and crosses the Enclos Fouqué caldera from north to south (Fig. 6) spreading out outside the caldera with a N 10° direction north of the caldera and a N170° direction south of it (Bachèlery et Mairine [9]).

Most eruptions, whose frequency according to historical records is higher than 0.5 event per year, take place inside the Enclos Fouqué caldera. Fluid basaltic flows (essentially oceanites and aphyric basalts) are extruded through cracks which open either in the summit craters or on the flanks of the volcano. Most earthquakes are shallow, between less than 1 and 3.5 km under the summit, and of small magnitude (less than 2.5) (Sapin et al.[10]). The swarms of quakes turn into a tremor when the intrusion reaches the surface of the volcano. This tremor disappears with the effusive activity. Observations show that the stress variations related to the volcanic activity are weak (~10 bars).

THE MAGNETIC ARRAY - Six magnetic stations were installed in 1985 and 1986 on the flanks of the volcano; there are now ten of them, equipped with a high sensitivity magnetometer (.25 nT) which measures continuously the intensity B of the geomagnetic field and teletransmits the values to the volcanological observatory 15 km away from the volcano (Fig. 6). One of the stations (CSR, Fig. 6) being chosen as a reference, the differences

$$\Delta B \ (t, \ P_K, \ CSR) = B \ (t, \ P_K) - B \ (t, \ CSR)$$

are continuously computed (t is time, P_K the current station) in order to get rid of the external ionospheric variations of the geomagnetic field.

THE OBSERVATIONS - It comes out that the ΔB differences present a lot of time variations with a wide spectrum of time constants.

First, variations which we call volcano-magnetic are observed every time an eruption occurs (twenty of them from 1986 to 1991). The time constants of these magnetic signals spread from a few minutes to several weeks, their magnitude is of the order of a few nT during the days preceding an eruption and can be larger than 10 nT when the magma is migrating towards the ground surface (Zlotnicki and Le Mouël [11]). Some examples of these volcanomagnetic variations are given on Fig. 7.

DISCUSION AND CONCLUSION - Formerly this kind of anomalous magnetic variations on a volcano was given a piezomagnetic origin (e.g. Pozzi et al.[12]). But we now think they are generated by electrokinetics; the flow of water in the fractured edifice is moved or perturbed by the stress changes taking place before and during an eruption; the resulting electric currents create the observed inhomogeneous varying magnetic field. The role of the big fracture zone or rift zone (see above) is prominent (where major fluid circulation is expected).

We have three reasons to propose this interpretation. First, taking into account that the time changes of the stress field are small (of the order of 10 bars), piezomagnetism does not give the right order of magnitude. Second, electric signals can be shown to be associated with the magnetic ones (Michel [13]). Third, on a larger time scale, rather unexpected variations are observed: an annual wave whose amplitude can reach 15 nT, and a long term drift which can reach 50 nT in the six years time span considered (recall that our magnetic stations are only a few km apart !) (Fig. 8) (Zlotnicki and Le Mouël [14]; Zlotnicki et al.[15]).

We now see the whole body of signals as due to a general water circulation in the massif, perturbed by different causes: variation of the local stress field during the periods of activity, annual external forcing - not yet well understood -, slow modification of the general circulation pattern due to the migration of the global activity of the volcano (Zlotnicki et al.[15]; Michel and Zlotnicki [16]). This migration is also clearly seen on the self potential maps of the summital part of the volcano drawn at different epochs (Michel [13]). Fig. 9 shows that, although the general features of the maps are conserved during the 1981-1994 time span, differences also show up between the maps which can be correlated with the displacement of the activity in the different parts of the volcano.

The time and space characteristics of these electric and magnetic signals are complex, as expected. But the set of data, regularly increasing, buttresses more and more firmly our general interpretation.

4 The Djibouti Experiment

The Djibouti region, in the horn of Africa, is of an extraordinary geodynamical interest. It displays a subaerial ridge (there are only two examples in the world) and plays a key part in the plate tectonics of a large area, being located at the junction of two active oceanic ridges (Red Sea and Gulf of Aden ridges) and the low activity East African rift system (Courtillot et al.[17]). A new volcano (Ardoukoba) showed up on the rift floor between the Gubbet gulf and the Asal lake (Fig. 10) in 1978. A geophysical observatory was installed at Arta in 1972 to monitor - in particular - the seismic activity of the region (Abdallah et al.[18]).

OBSERVATIONS - A small array of electrodes was installed in 1990 across a major fault of the rift floor (PTR) between the Gubbet gulf and the Asal lake (Fig. 10). Differences of potential were measured between the North, South, East and West electrodes and the central one Co (Fig. 11) and continuously recorded through the whole 1991 year. Data from January 1st to March 31 are displayed on Figure 12. At the bottom of the figure one has reported the daily seismic energy, computed from the seismic array of the Arta observatory. A complete presentation of the data will be found in Zlotnicki et al.[19]. Let us just make a few comments. Up to day 40 the signals (North-Co and East-Co on the figure) only display a weak long term drift on which the regular ionospheric daily variation is superimposed - hardly visible at the scale of the figure - with the expected amplitude of 10 mV/km. On day 40 dramatic changes occur, with amplitudes up to 0.5 V/km. Then comes again, from day 50 to day 60, a quiet period with the regular daily variation alone. On day 60 a spectacular enhancement of the daily variation occurs which will last a few months.

DISCUSSION - Such an amplification of the daily variation has been reported a few times before (Meyer and Pirjola [20]; Thanassoulas and Tselenis [21]). One can first think that it results from a sudden change in the local conductivity distribution. But may be the small ionospheric variation before day 60 and the big daily variation after day 60 have different origins. The second one could be due to a periodic water circulation forced by the atmosphere temperature daily variation (same type as the annual variation of the ΔB in the Reunion Island experiment): the stress/strain changes accompanying the seismic activity induce a sudden qualitative change of the transport properties of the rock in the neighbourhood of the fault (at an unknown scale), making possible a flow which was not possible before.

5 The Ioannina Experiment

The longest, most complete and most intriguing series of electric telluric signals ever recorded in a seismic area is undoubtedly the one collected in Greece by a team of physicists (Varotsos et al.[22]). These physicists proposed and currently run a prediction method, the VAN method, which is the main subject of the present book. It is fair to say that the VAN method is highly controversial, all the more since no signal similar to the ones recorded in Greece has ever been observed anywhere else. That is why we have installed a magnetic and an electric station close to Ioannina.

THE ELECTRIC AND MAGNETIC STATIONS - The city of Ioannina is located in a NW-SE basin within the Pindus chain, in the Ionian zone, embedded in a neritic limestone environment extensively and pervasively karstified. A VAN station, 7 km north of Ioaninna, referred to as IOA, has been continuously working for years and has reported the largest number of SES (Seismo Electric Signals) which, according to the VAN prediction method, are related to earthquakes occurring in Western Greece.

We have installed our telluric station, JAN E, near the village of Likotrikion, in a location far from anthropogenic activity, 5.5 km from the IOA station and in a similar geological context (Fig. 13). The electric field sensor system comprises two sets of orthogonal electric lines C_1 and C_2, laid down in a NS-EW configuration at a separation of approximately 50 m (see Fig. 14); non-polarizable Pb/PbCl$_2$ electrodes have been used and driven 60 cm into the clayey ground. The sampling rate is three points per minute.

A three-component, observatory-type fluxgate variometer, with a sensitivity of a tenth of nT, was installed in April 1994 at JAN M, Ioannina airport.

OBSERVATIONS (Gruszow et al.[23]) - In order to be able to detect seismoelectric or tectonoelectric signals, if any, it is first necessary to identify and characterize the regular components of the recorded signal. This analysis allows us to clearly distinguish three components of the signal:

(a) A class of signals which can be attributed to electric fields induced within the earth by primary external ionospheric or magnetospheric sources. The signals exhibit an high correlation with the magnetograms of the Greek observatory of Penteli, 400 km SE of JAN E, as well as of the nearby magnetic station JAN M (installed in April 1994). The polarization of these class (a) signals is linear and frequency independent along the direction N15°-25°W. Induced electric currents are channelled along this direction according to a well-known mechanism (see e.g. Le Mouël and Menvielle [24]) (Fig.15).

(b) A class of signals due to anthropogenic activity. This class (b) comprises powerful transient events with amplitudes varying from five to a few tens of millivolts per kilometer, consistently polarized in the directions N50°E. A significant change of this direction of polarisation from C_1 to C_2 probably indicates nearby source(s) (Fig. 16).

(c) A low amplitude (<3 mV/km), uniform background noise [class (c)].

Are signals of another kind observable in the recording ? Yes; they are polarized in the general direction of class (a) signals, with an amplitude varying between a few and a few tens of millivolts per kilometre on the NS lines, and a duration varying between 2 and 20 min. These transient sporadic events cannot be correlated with any magnetic field variation, in marked contrast to class (a) signals. A few examples are given in Fig. 17 and 18.

DISCUSSION AND CONCLUSION - Our analysis has identified more than 50 sporadic events on the recordings since the installation of the station in August 1994. They may be generated by some industrial source, but it cannot be excluded, at the present state of the analysis, that they might as well be tectonoelectric effects. Some of them were observed on days (e.g. August 22, October 23 and 31, 1993) when SES (Seismo Electric Signals) were announced by the VAN group. Future recordings will hopefully allow us to determine the source of these sporadic signals.

If they are natural -although most of them look like industrial signals -, one would again favour a local electrocinetic source. Water circulation certainly takes place in the karstified limestone, which can give signals when perturbed by a local stress variation.

6 Electrokinetics Signals

As announced in the introduction, we won't give here any quantitative interpretation of the electric and/or magnetic signals we have just briefly described. We interpret all the signals recorded on the pillars of the Meriel quarry (§ 2), on the flanks of Piton de la Fournaise volcano (§ 3), or in the Djibouti active area (§ 4) as generated by electrofiltration - at least we think that electrofiltration is the dominant mechanism in their generation -. This is because some characteristics are shared by all the experiments and observations and since, in most cases, we see of no other interpretation. Let us stress out that in all these experiments we measure the electric or magnetic signals where they are generated. The atmospheric pressure acts on the pillar on the surface of which electrodes are implanted; magnetic differential signals - volcanomagnetic signals - are collected on the flanks of Piton de la Fournaise volcano at - or close to - places where stress changes occur (§3); changes in the rock properties in the Djibouti experiment are generated - according to our views - by a change of the local stress field in the immediate vicinity of the electrodes. And if the sporadic signals of the Ioaninna experiment (§5) are not due to human activity, we would still attribute them to a perturbation of a water flow - the underground is karstified - by local weak changes in the stress field (we don't adress here the question of the correlation with far away earthquakes; see the other papers of this volume).

On the contrary, the interpretations which have been proposed up to now to account for the VAN signals call for sources located in the epicentral area, up to hundreds of kilometers far from the electric recording station. We will now examine the plausibility of long distance electric precursors, giving again special attention to streaming potential signals.

7 Plausibility of Long Distance Electrotelluric Precursors

The claim by the VAN group that some electrotelluric signals with specific characteristics (SES) recorded in Greece are well correlated with the largest regional earthquakes (e.g., Varotsos et al.[22]) has not only to stand the tests of sound statistical analysis - a topic which remains up to now quite controversial- but it has also to be based on a plausible physical mechanism of electrical current generation and diffusion through the conductive earth. This last point is the focus of the following discussion.

The question here is not about the existence of electromagnetic precursors to earthquakes: They do exist, as evidenced by a few reported cases of observations at distances within a few dimensions of the seismic source (e.g., see a review presented by Bernard[25]). The most striking evidence up to now is probably the strong increase of the ULF magnetic noise about 10 days before the Loma Prieta 1989 earthquake (Fraser-Smith et al.[26]). The trouble with the SES in Greece is that they are mostly associated by VAN with earthquakes located at typical distances ranging from

100 km to 200 km, which is 10 to 100 times the seismic source dimensions for magnitudes 4.5 to 6.5 (see e.g. Scholz [27] for scaling source sizes) . At these distances, no signal should normally be observed above the natural noise level, owing to the fact that no significant electrical anomaly has ever been reported in the epicentral area of a future earthquake at the time of an SES observed at long distance.

Let us first quantify the possible sources of electrical currents. First, we want to emphasize that there are *a priori* several good candidates for generating DC or ULF electrical currents in the earth crust in relation with tectonic stress changes (Bernard [4]; Park et al.[28]; Johnston [29]). However, only a few among these models lead to quantitative estimates of the source amplitude, thanks to numerous, well-controlled laboratory experiments on rocks. The three best candidates are electro-kinetic effects, piezo-electric effects, and resistivity changes. But as pointed out by several authors (e.g., Bernard [4]; Fraser-Smith [30]), the electrokinetic effect is the most efficient way to convert mechanical energy into electrical energy at these low frequencies in the earth crust.

The electrokinetic effect - or streaming potential - results from the flow of salty water in porous or fractured rocks, which can generate a typical electrical potential of 100 mV for a pressure drop of 1 MPa by the trapping of ions from the flowing water near the surface of the solid matrix. (e.g., Mizutani et al.[31]; Morgan et al.[32]). With a favourable electrical environment, this potential is not self-cancelled by return electrical currents inside the source volume, so that such a source can be seen from far away as an electrical dipole (or as a distribution of dipoles). For example, an electrokinetic source with a 1 km dimension may generate an electrical current of the order of 100 A, for a 1MPa pore pressure drop. Note however that such an electrical source is necessarily associated with volume changes in the rocks, as will be discussed later.

The attenuation of the electrotelluric field from electrical dipoles located inside the crust decays with the distance R as $1/R^3$ in an homogeneous medium (Fig. 19). In a horizontally layered crust, the decay reduces to $1/R^2$ at large distances when the dipole is inbedded in a strongly conductive layer (Fig. 20). The diffusion of the electrical field brings an exponential attenuation with distance, quantified by the skin depth d related to the frequency f and the conductivity s: $d=(pfms)^{-1/2}$, where m is the magnetic permeability (e.g., Honkura [33]) (Fig. 21). At a distance d, the signal is attenuated by the factor 1/e. For a conductive layer with s=0.1 S/m , d=15 km at 0.01 Hz, and d=45 km at 0.001 Hz. Thus, reducing the geometrical attenuation by some channeling in a more conductive structure is at the expense of a stronger attenuation by Joule effect.

Let us combine these attenuation characteristics with an electrokinetic source. Consider first a volume of 1 km in which such an effect takes place, due to a 1 MPa pore pressure gradient within the source volume. The local electrical field is of the order of 100 mV/km. At a distance of 10 km, in an homogeneous structure, it has decreased by a factor of 1000, and should not be observable. With the most favourable, layered conductive structure, it would decay only by a factor of 100, reaching the typical noise level of 1 mV/km at 10 km. But at 100 km, the signals would be more than 2 order of magnitudes below the noise level, depending on the frequency. In order to reach the 100 mV/km SES level at 100 km, the electrical field should be enhanced by a factor of at least 10000!

We can think of two ways for gaining this factor, by increasing the source amplitude, and by assuming an ad hoc, local site amplification near the electrodes as proposed by VAN. It is possible to increase the power of the electrokinetic source either by increasing its size, or by increasing the available pore pressure gradient. The former possibility is not really suitable for fitting the SES, according to their duration lasting from minutes to hours: At a length scale of 10 km, the water flow in the crustal rocks has typical time constants of days or weeks. The second possibility seems more plausible: two confined aquifers at very different pore pressure may be temporarely connected by some sudden changes in the rock permeability, when the local stress reaches some critical level, allowing rapid fluid flow and a strong electrokinetic effect, as proposed by Bernard [4] based on the models of crustal fluid instabilities (e.g., Nur and Walder [34]) (Fig. 22). The available pore pressure difference may reach a significant fraction of the theoretical limit given by the difference between the lithostatic and the hydrostatic pore pressure; at a depth of 10 km, one may thus expect pressure drops reaching 100 MPa. Hence, with such a mechanism, we have gained a factor of 100 in the source amplitude. An additional, local amplification by a factor of 100 would then be required for observing 100 mV/km at the VAN electrodes; although not impossible, such an amplification seems very unlikely. In particular, this local electrical structure should also strongly amplify other sources of electrical noise, which has not been reported by VAN.

Such a model with a powerful electrical source has two additional drawbacks. Firstly, it produces in the source volume an electrical field of 10 V/km which may generate anomalies of the order of 10 V/km at the ground surface in the source area, if one takes the same favourable criteria for the diffusion of the electrical field as for the SES observations (cylindrical decay, local amplification by a factor of 100); however, such a level has never been reported. Secondly, a pore pressure difference of 100 MPa generates strain changes of the order of 10^{-2} to 10^{-3}, which would increase the total stress well above the elastic strain limit of the crust. Indeed, typical coseismic strain changes are less than 10^{-4}, which means that an increase of strain by an absolute quantity of 10^{-4} is very likely to generate significant earthquakes in a faulted, seismically active area. At the strain level 10^{-3} to 10^{-2}, *a fortiori*, one would expect a noticeable, anomalous seismicity; however, none has ever been reported at the time of the SES anomalies.

Let us describe more specifically the strain associated with the electrical source. The water pressure perturbation forms a dipole of volumic strain sources - pore pressure drop in one fluid reservoir, pore pressure increase in the other -, whose field decays as $1/R^3$, just as the electrical field in an homogeneous medium.

However, the typical relative spatial variations of elastic parameters inside the crustal rocks is of the order of 10, and does not allow a strong channeling of the strain field, contrary to what may occur for the electrical field, for which resistivity contrasts may easily reach 3 orders of magnitude. The strain is thus expected to decay more rapidly than the electrical field with the distance. However, near the epicentral area, the source of strain associated with an electrical source should be large enough for being detected at the surface. Indeed, taking our first example with a reasonable 10 MPa pore pressure difference over 1 km, at 10 km in depth, one gets a source strain of the order of 10^{-4}, which decays to 10^{-7} at the surface above the source (Fig. 23). Such strain levels are of the order of the earth tidal strain, and are detectable with sensitive strain- or tiltmeter.

The discussion above for the electrokinetic effect and the associated strain may apply in fact to any kind of electrical sources inside the earth crust: the latter can all be represented as electrical dipoles, and are all associated with sources of dipole strains. This last point is usually missed or disgarded in the published attempts to generate strong electrical signals observable at the surface: to our knowledge, all the proposed models require large strain sources, greater or equal to 10^{-3}, which are very unlikely (e.g., Dobrovolsky et al.[35]; Draganov et al.[36], Fenoglio et al.[37]).

In conclusion, there are presently no plausible, quantitative physical model for supporting the claim that the SES reported by VAN can be generated at one or two hundreds of kilometers from the recording stations, owing to the unlikeliness of the existence of natural electrical channels connecting the source area to the VAN electrodes, and to the absence of reported electrical anomalies in the epicentral area at the time of the SES, where the effect should be maximal. In addition, any model of crustal electrotelluric source would have to face a severe difficulty: electrical sources in the rocks imply necessarily some crustal strain, and increasing the former to reach the SES levels is expected to rise the strain perturbation orders of magnitude above the coseismic changes - which is difficult to admit, would it be only because no geophysical anomalies have ever been reported at the time of the SES. The SES are thus probably generated within a few tens of kilometers from the measurement site, whatever their correlation may be with the preparation of earthquakes at longer distances.

However, the cases where anomalous electrotelluric signals - including a few SES - are reported at short distances from impending earthquake sources - say at a few tens of kilometers - may well reveal an electrotelluric phenomenon truly associated with the earthquake preparation process.

8 General Conclusion

Electrical signals with time constants from a minute to hours, days, weeks and even years are currently and easily observed in various geological settings, as shown by the few examples given here (and many others). As said above, electrokinetic effect is the most efficient way to convert mechanical energy into electrical energy

(Bernard [4]; Fraser-Smith [30]). Wherever some forcing mechanism - most often stress/strain variations, but also outside temperature variations - generate or perturbate a water flow in the surrounding rocks, electric and, in favourables cases, magnetic related signals are to be expected. Such can be the case for seismic regions. Co-seismic signals are of course the most expected.

On the contrary, long distance precursors advocated by the VAN group lack, at the present time, according to us, any plausible source mechanism. All the proposed mechanisms manage to build up some electric dipole in some volume around the focus of the impending earthquake; but afterwards the electric field has to "propagate" to large distances in a conductiviting medium. It was shown in section 7 - the discussion was focused on electrokinetics which rises, by far, less serious energy problems than other source mechanisms derived from materials physics; see the paper by Slifkin in this volume - that the unavoidable strong attenuation of the signal with distance requires unrealistically powerful electrical machines at the source or/and unplausible amplification effects at the recording stations. It has been called for rather mysterious channels linking the focal region to the "sensitive" stations. But, let alone the geological unplausibility of such channels, they would not actually help; furthermore, it now appears that VAN signals are not so localised in small sensitive regions than it was formerly believed (see Varotsos paper in this volume).

Let us say, to conclude, that electrical measurements constitute a potentially powerful tool to monitor and study the earthquake preparation process. To learn more, electrical and magnetic recording stations must be installed, in dense arrays if possible, in a number of active areas, in association with other geophysical equipements measuring stress and strain variations.

References

1. Chen Zangli ed. : Collected papers of the methods of earthquake prediction and estimation of strong seismic risk, *Seismological Press*, Beijing (1990)
2. J. Zlotnicki, Monitoring active volcanoes. Strategies, Procedures and Techniques, Geomagnetic surveying methods ed. by *Univ. Press College*, London (1995)
3. M. Feuillard et al., *Journal of Volcanology and Geothermal Research, 16*, 317 (1983)
4. P. Bernard, *J. Geophys. Res 97*, 17531 (1992)
5. P. Morat et J.L. Le Mouël, *C.R. Acad. Sci. Paris, t. 315, série II*, 955 (1992a)
6. P. Morat et al., *C.R. Acad. Sci. Paris, t. 320, série IIa,*173 (1995)
7. P. Morat et al., *C.R. Acad. Sci. Paris, t. 315, série II*, 1083 (1992b)
8. P. Bachèlery, Thèse de doctorat de spécialité, *Univ. Clermont Ferrand* (1981)
9. P. Bachèlery and P. Mairine, in *Le volcanisme de la Réunion : Monographie*, ed. J.F. Lénat, 213 (1983)
10. M. Sapin et al., *Journal of Volcanology and Geothermal Research*, in press (1995)
11. J. Zlotnicki and J.L. Le Mouël, *J. Geophys. Res., vol. 93*, 88, 9157 (1988)
12. J.P. Pozzi et al., *Journal of Volcanology and Geothermal Research* (1979)
13 . S. Michel, Thèse de doctorat, *Univ. Paris 7* (1995)
14. J. Zlotnicki et J.L. Le Mouël, *Nature, 343*, 633 (1990)
15. J. Zlotnicki et al., *C.R. Acad. Sci. Paris, t. 314, série II*, 661 (1992)
16. S. Michel and J. Zlotnicki, *C. R. Acad. Sci. Paris, t. 321, série IIa*, (1995)
17. V. Courtillot et al., from Coward, M.P., Dewey, J.F. & Hancock, P.L. (eds) *Geological Society Special Publications, n°28*, 559 (1987)
18. A. Abdallah et al., *Nature, 282*, 17 (1979)
19. J. Zlotnicki et al, *submitted*
20. K. Meyer and R. Pirjola, *Tectonophysics, 125*, 371 (1986)
21. C. Thanassoulas and G. Tselenis, *Tectonophysics, 224*, 1 (1993)
22. P. Varotsos et al., *Tectonophysics, 224*, 1 (1993a), *Tectonophysics, 224*, 269 (1993b)
23. S. Gruszow et al., *C.R. Acad. Sci. Paris, t. 320, série IIa*, 547 (1995)
24. J.L. Le Mouël and M.Menvielle, *Geophys. J. R. astr. Soc., 68*, 575 (1982)
25. P. Bernard, *IUGG XXI General Assembly*, Boulder, Colorado, July 2-14. (1995)
26. A.C. Fraser-Smith et al.,*Geophys. Res. Letters, 17*, 1465 (1990)
27. C.H. Scholz, The mechanics of earthquakes and faulting, Cambridge University Press (1990)
28. S.K. Park et al., *Final report from Lake Arrowhead Workshop*, June 14-17 (1992)
29. M.J.S. Jonhston, *IUGG XXI General Assembly*, Boulder, Colorado, July 2-14 (1995)

30. A.C. Fraser-Smith, *IUGG XXI General Assembly*, Boulder, Colorado, July 2-14 (1995)
31. H. Mizutani et al., *Geophys. Res. Lett., 3*, 365 (1976)
32. F.D. Morgan et al., *J. Geophys. Res., 94*, 12449 (1989)
33. Y. Honkura, *Proceeding of theWorkshop*, Lake Arrowhead, June 14-17 (1992)
34. A. Nur and J. Walder, in *Crustal Processes*, 113, (National Academy Press, Washington, DC, 1990)
35. I.P., Dobrovolsky et al., *Phys. Earth Planet. Inter., 57*, 144 (1989)
36. A.B. Draganov et al., *Geophys.Res. Letters, 18*, 1127 (1990)
37. Fenoglio et al., *IUGG XXI General Assembly*, Boulder, Colorado, July 2-14. (1995)

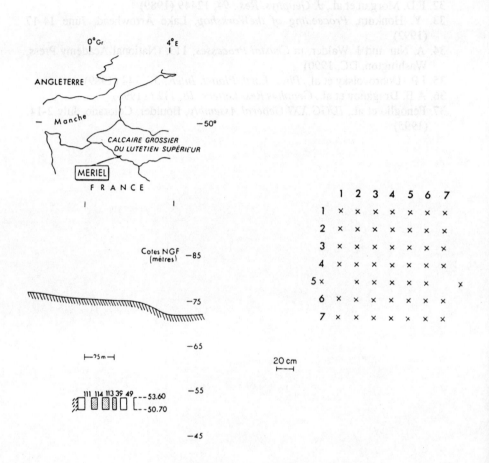

Figure 1: Left top: location of the Meriel quarry. Left bottom: schematic vertical North-South section of the quarry. Right: array of electrodes on the West face of pilar 39.

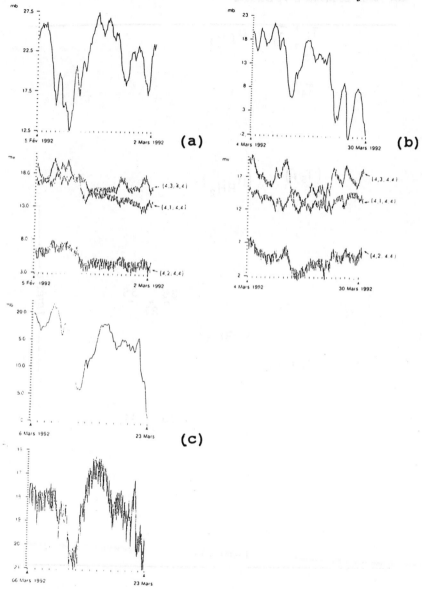

Figure 2: Top: recording of the atmospheric pressure in the gallery; abscissae are in days, ordinates in millibars (approximately), reckoned from an arbitrary origin. Bottom: differences V(4, 1)-V(4, 4); V(4, 2)-V(4, 4); V(5, 3)-V(4, 4). 4a: February 5, 1992 to March 2, 1992; 4b: March 4, to March 30, 1992; 4c: March 6, to March 23, 1992 : V(5, 3)-V(4,4).

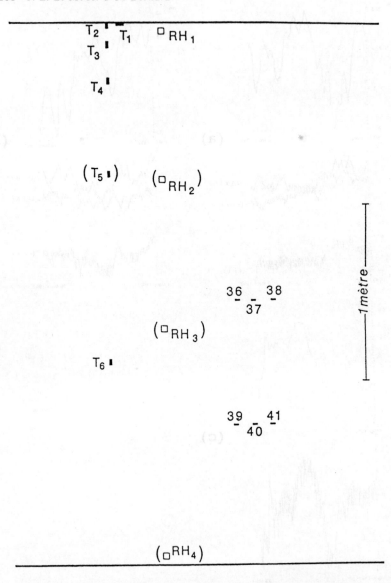

Figure 3: Location of the electrodes on the face of pillar 40. Locations of temperature and humidity sensors are also indicated. T: temperature. RH: humidity. 36-41: electrodes.

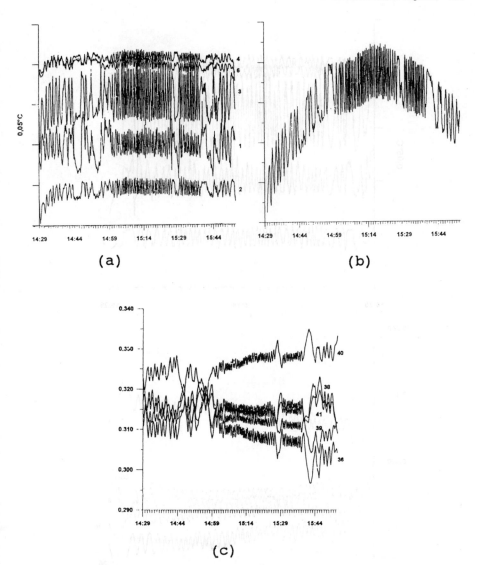

Figure 4: Signals recorded on 19 May, 1994 from 14h 29 to 15h 56.
a) temperatures at points 1, 2, 3, 4, 6: ordinates in hundredths of degrees Celsius.
b) relative humidity at RH; arbitrary units. c) potential differences $\delta V_{n,\,o}$ (n= 36, 37, 38, 39, 40, 41) in volts.

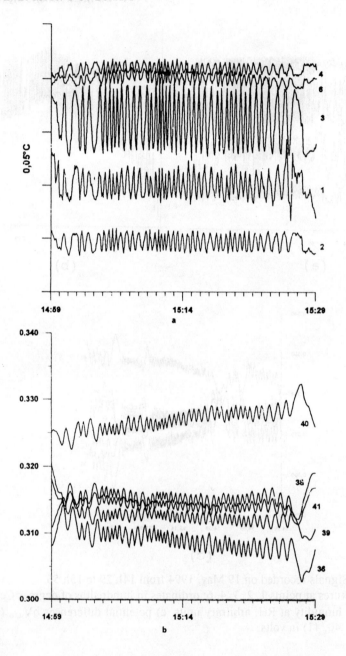

Figure 5: Enlargement of figure 4 for the time interval 14h 59 - 15h 29.

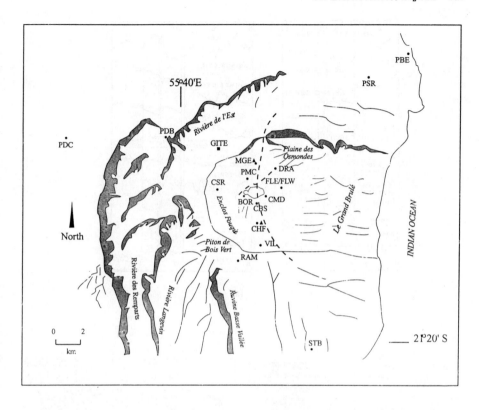

Figure 6: La Fournaise volcano. Black circles and triangles represent magnetic stations. Black squares represent climatological stations. The interrupted line represents the main fracture zone.

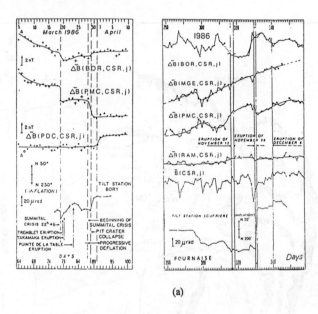

(a)

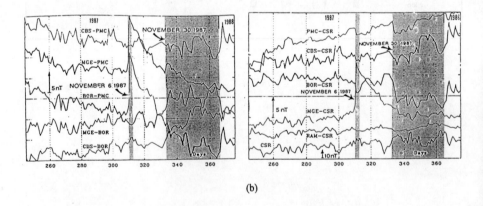

(b)

Figure 7: a) Examples of volcanomagnetic signals; $\overline{\Delta B}$ are daily means of the diferences of the field intensity between the current station and the reference CSR. At the bottom the tilt data of Bory station (BOR on fig. 6) are reported. b) Other examples of the $\overline{\Delta B}$ differences between different stations. Eruptions occurred on November 6 and 30, 1987.

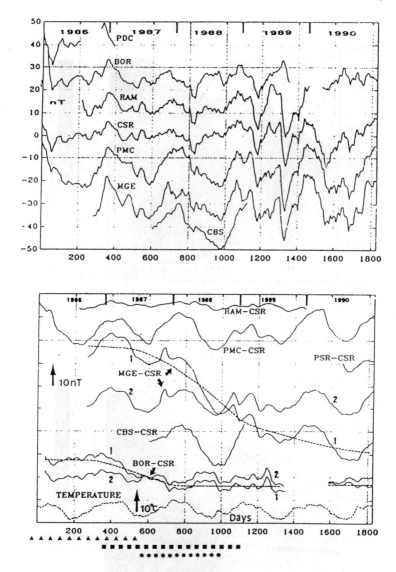

Figure 8: Top: Variations of the field intensity at the different stations from 1985 to 1990. Bottom: Variations of some differences $\overline{\Delta B}$ and of the temperature.

PS 1981

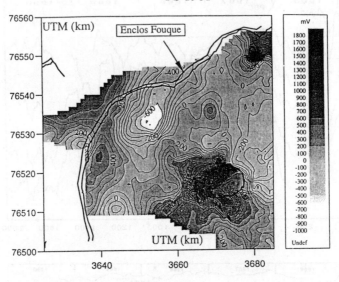

PS 1992

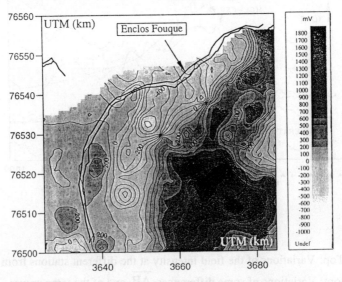

Figure 9: Cartography of the spontaneous potential on Piton de la Fournaise volcano in 1981 and 1992.

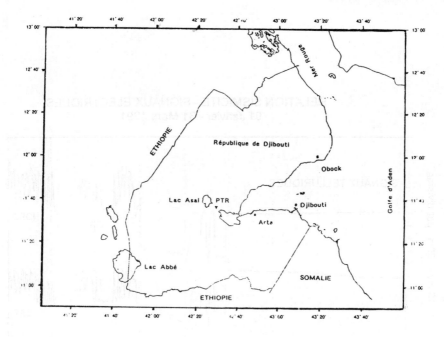

Figure 10: The Djibouti area. PTR is the telluric station; ART the geophysical observatory.

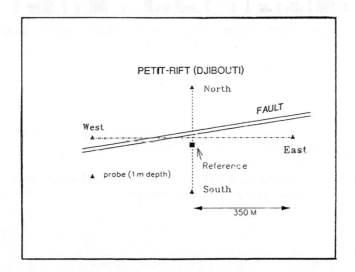

Figure 11: Configuration of the electrotelluric PTR station.

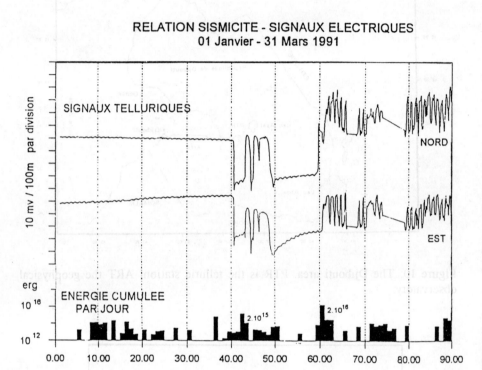

Figure 12: Electric recordings at PTR from January 1th to March 30, 1991. At the bottom the daily seismic energy (in ergs) is reported.

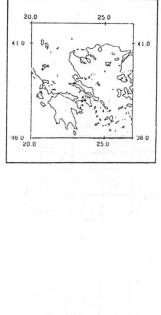

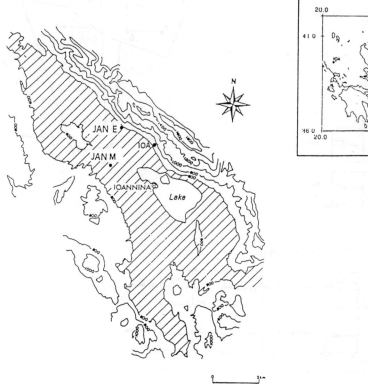

Figure 13: Simplified map of Ioannina basin and location of stations IOA, JAN E and JAN M

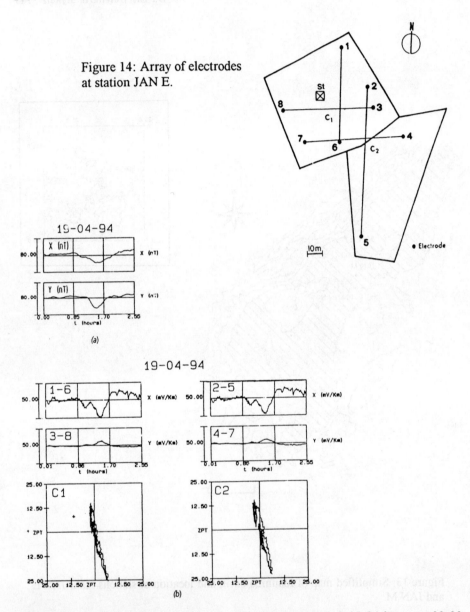

Figure 14: Array of electrodes at station JAN E.

Figure 15: a) Magnetic recordings at JAN M station on 19-04-94 between 00 00 a.m. and 02:30 a.m. (UT). b) Telluric recording at station JAN E on 19-04-94 between 00:00 a.m. and 02:30 a.m. (UT) (NS: 1-6 and 2-5; EW: 3-8 and 4-7). Polarization diagrams for C_1 and C_2 crosses (cat. a).

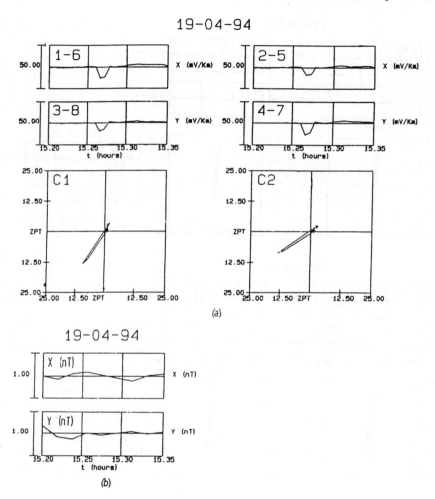

(a)

(b)

Figure 16: a) Example of a high amplitude category (b) telluric signal detected at JAN E (on 19-04-94 between 03:12 p.m. and 03:21 p.m. (UT). Hours and hundredths of hour on the drawings. Bottom: magnetic recording at Penteli.

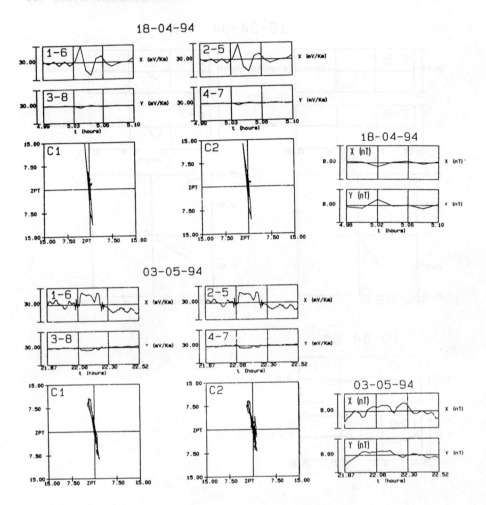

Figure 17: Example of electric signals of (d) category. Top left: recording on day 18-04-94 at JAN E between 05:00 a.m. and 05:06 a.m. UT and polarization of the signal on crosses C_1 and C_2 Right: magnetic recordings. Bottom left: recordings on day 03-05-94 at station JAN E between 09:52 p.m. and 09:31 p.m. UT and polarization of the signal on crosses C_1 and C_2. Right: magnetic recordings at JAN M on 03-05-94 between 09:52 p.m. and 09:31 p.m. UT (hours and hundreths of hours on the drawings).

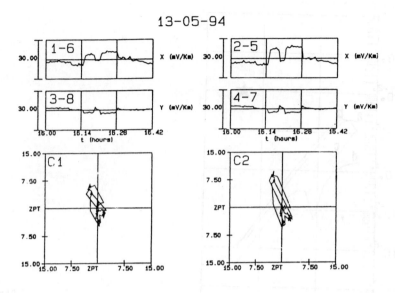

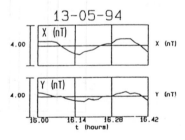

Figure 18: Left: recordings on day 13-05-94 at station JAN E between 04:00 p.m. and 04:25 p.m. UT and polarization of signal on crosses C_1 and C_2. Right: magnetic recordings at JAN M on 13-05-94 between 04:00 p.m. and 04:25 p.m. (hours and hundreths of hours on the drawings).

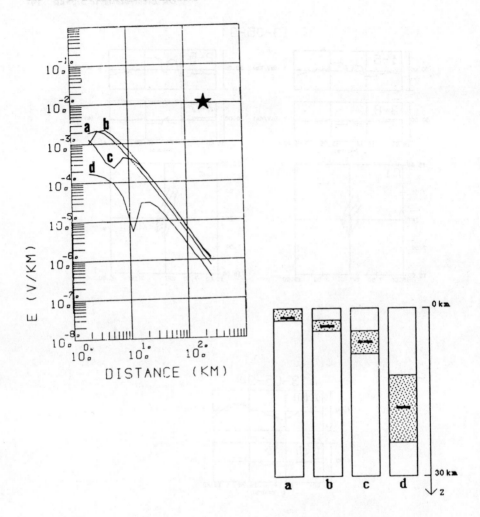

Figure 19: Attenuation of the D.C. electrical field from a buried dipole in a homogeneous half-space. Top: amplitude of the horizontal electrical field at the surface of a half-space with homogeneous conductivity, due to a 1 km long buried dipole with a 100 A current , at various depths. The star represents the typical observation range for SES in Greece. The long distance decay of the amplitude is in $1/R^3$. Bottom: Depth and orientation of the dipole. (from Bernard,1992)

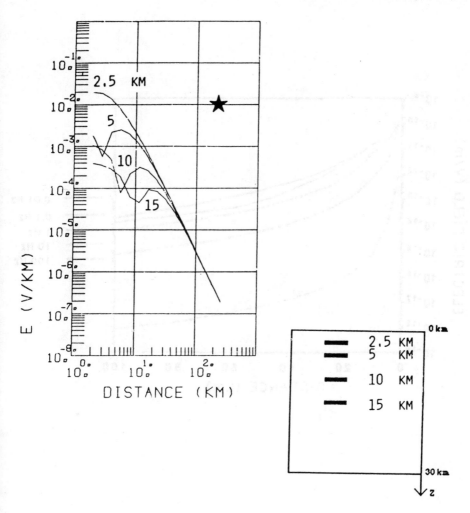

Figure 20: Attenuation the D.C. electrical field from a buried dipole in a layered half-space. Same representation as in figure 1, for a horizontal dipole (100 A, 1 km), in a highly conductive layer (s = 0.1 S/m) between 0 and 2 km (a), 2 and 4 km (b), 4 and 8 km (c), 12 and 24 km (d), inbedded in a more resistive medium (s=0.01 S/m). This is equivalent, in the case of an electrokinetic effect, to considering a pore pressure difference of 1 MPa with sC=10 mV S m/atm., where C is the streaming coefficient. The long distance decay of the amplitude is in $1/R^2$. (from Bernard, 1992)

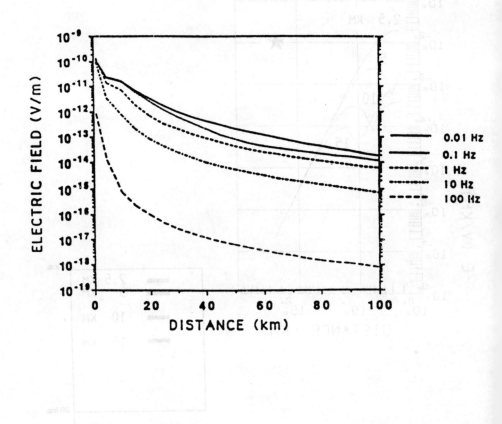

Figure 21: Frequency dependence of the attenuation of an electrical field from a buried dipole. The electric dipole is horizontal, 5 km long, and has an amplitude of 1 A m. (from Honkura, 1992). Note the strong attenuation with frequency at a given distance.

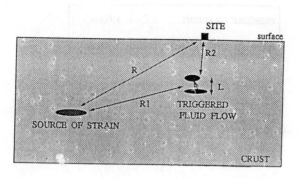

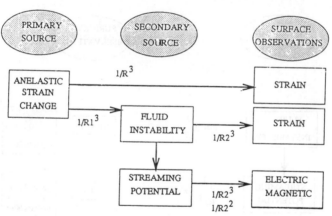

Figure 22: Sketch of the fluid instability model. The triggering of fluid instabilities in the crust may be a powerful source of pore pressure variations, and hence of electrokinetic effects. The amplitude of the electrical anomaly may be much larger to what could be expected by a direct strain effect from the primary source of strain. At the surface, the strain source may have observable effects as well.

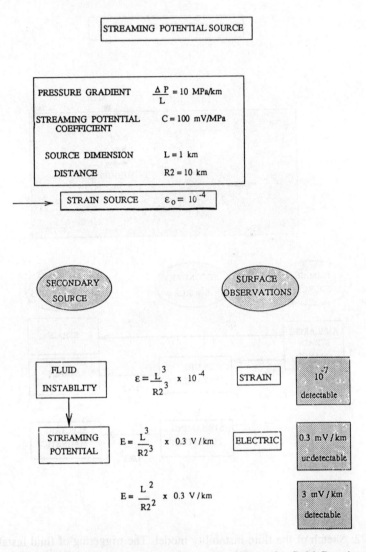

Figure 23: Comparison between strain and electric effect of a fluid flow instability. Top: source characterisitics. Such a source is associated with a source of strain of the order of 10^{-4}, equivalent to a coseismic strain change of a magnitude 4 (typical dimension 1 km). Bottom: estimate of electric and strain field amplitudes. The coefficient 0.3 V/km has been deduced from typical streaming coefficients (see Bernard, 1992). The surface observations are made 10 km above the buried source. Note that strain should be measurable, whereas the electrical field may or may not be detectable, depending on the electrical structure.

Part III
Counterarguments Against the VAN Approach

Part III

VAN: A CRITICAL EVALUATION

ROBERT J. GELLER

Department of Earth and Planetary Physics, Faculty of Science, Tokyo University,
Yayoi 2-11-16, Bunkyo-ku, Tokyo 113, Japan

A detailed examination of statements by VAN (Varotsos, Alexopoulos, and Nomicos) and their supporters reveals many inconsistencies and ambiguities. VAN's telegrams and faxes should not be recognized as "earthquake predictions," because they are insufficiently precise and because they fail to specify the windows (range of values) within which the "predicted" epicenter, origin time, and magnitude should fall. VAN or their supporters have in many cases claimed "success" even when earthquakes have clearly fallen outside VAN's own retrospectively stated time (e.g., 11 d) or space (e.g., 100 km) windows, or where their "predictions" have failed altogether to specify one of the key parameters, e.g., the magnitude. There also are instances in which VAN have claimed "success" for "predictions" issued during swarms of microearthquake activity, or for "predictions" of aftershocks of a large earthquake, but such "successes" should not be regarded as significant. In summary, VAN's claims to be making "successful predictions" of earthquakes are not valid. It is also shown, based on analyses of newspaper and magazine articles and comic books, that VAN and their supporters in Japan have systematically used the mass media to disseminate groundless claims regarding VAN's work. Research on earthquake prediction over the last 110 years is reviewed. From time to time highly optimistic claims have been made, but all have ultimately proven groundless. The failure of these efforts is consistent with recent work in non-linear dynamics that suggests earthquakes are inherently unpredictable.

" ...the professional duty of a scientist confronted with a new and exciting theory is to try to prove it wrong. That is the way science works. That is the way science stays honest. Every new theory has to fight for its existence against intense and often bitter criticism. Most new theories turn out to be wrong, and the criticism is absolutely necessary to clear them away and make room for better theories. The rare theory which survives the criticism is strengthened and improved by it, and then becomes gradually incorporated into the growing body of scientific knowledge."
—Freeman Dyson, *Infinite in All Directions*, Penguin, London, 1988, p.258-259.

1   Introduction

Beginning in the the early 1980's, Varotsos, Alexopoulos, and Nomicos (eponymously referred to as VAN) and their coworkers and supporters have been claiming to be making "successful earthquake predictions" on the basis of electrical observations. [1-4] Both these workers and their methods are now commonly referred to as "VAN." Nine examples of claims by VAN and their supporters are given below.

" ... a transient change of the electric field of the earth (seismic electric signal), hereafter called SES, appears many hours before an earthquake (EQ). ... The uncertainty of [the magnitude] M is around 0.5 units. Following Sobolev (1975) and for the statistics to be beyond any doubt, predictions were officially documented before the EQ-occurrence. For 23 earthquakes with a magnitude equal [sic] or greater than $M_S = 5.0$ two events were missed." (Varotsos and Alexopoulos,[3] 1984, p.99)

"In summary, our present experience indicates that *every sizable EQ is preceded by an SES*, and inversely [sic] *every SES is always followed by an EQ* the magnitude and the epicenter of which can be reliably predicted." (*Ibid.*, p.120; italics follow published ms.)

"The present results, based on officially documented predictions, demonstrate that the determination of the three parameters, origin time, epicenter and magnitude, of an impending earthquake is today possible, provided that the epicentral area is surrounded by a number of, at least, four ellectrotelluric stations. The accuracy of epicenter location and the magnitude is mainly governed by the mean distance between these stations." (Meyer, Varotsos, Alexopoulos, and Nomicos,[4] 1985, p.161)

" ... the VAN system has shown that, under certain conditions in precalibrated areas, electrical signals can be detected up to several days in advance of major earthquakes. During the past few years, several events of magnitude greater than 5 have been predicted with reasonable precision as to magnitude, time, and epicentre location."[5] (Recommendation adopted by the conference on "Measurements and Theoretical Models of the Earth's Electric Field Variations Related to Earthquakes," Feb. 6-8, 1990, Athens.)

"Greek scientists Friday claimed a 75 percent success rate for an earthquake-forecasting technique being tested in eight quake-prone countries. ... Conference delegates were told VAN can predict quakes up to 10 days ahead with a 75 percent success rate."[6] (Reuters story on the above conference.)

"What matters in this case is not being right or wrong in some umpteenth little scientific squabble: what matters are human lives. How can we reject a method when almost all of its predictions have been borne out? A method whose only and infrequent failures, which are moreover becoming more and more infrequent as experimental data accumulate, consist in not having predicted 10% of the tremors that have occurred in more than half a dozen years? Such failures are the lot of any new technique, and they are minor when compared to the 90% success rate, especially since they can be explained essentially by the fact that new seismic zone/recording station pairs have not yet been calibrated, and sometimes by a background noise level that drowns out the SES." (Tazieff,[7] 1992, p.131)

"Contrary to the consensus that short term prediction is impossible, there is already an outstanding case of success: i.e., the so called VAN method in Greece." [8] (Uyeda and Nagao, Symposium on Earthquake Prediction Research, June 29-30, 1994, Tokyo.)

"[Varotsos] claims to have successfully predicted both the approximate timing and the location of the [M6.1] earthquake [in Greece on June 15, 1995]. He has now called on Greek seismologists to 'drop their misguided opposition' to [VAN]. ... Varotsos says that, using the VAN method, he has correctly predicted the timing, magnitude and location of the three most recent earthquakes in Greece. ... [Varotsos asks:] 'How many more people will have to die before this method is recognized as being correct.' " [9] (News story in *Nature*, June 22, 1995.)

"[Prof. Varotsos] is the developer of the VAN method, which predicts close to 70% of the earthquakes in Greece, a seismically active country. ... His record since 1987 is nine successes and four failures." [10] (Newspaper interview of Prof. Varotsos on a visit to Japan in June 1995; my translation)

The meeting, "A Critical Review of VAN" which was held in London on May 11-12, 1995, was sponsored jointly by the International Council of Scientific Unions (ICSU) and the Royal Society. According to the Editor's preface to this volume, the London meeting was intended to address, "in an appropriately critical spirit," three questions:

(a) What is the VAN method?
(b) Do claims for its value rest on scientifically sound bases?
(c) Are there scientifically credible mechanisms for the VAN method?

Note that a detailed discussion of (c) is necessary only if the answer to (b) is positive. On the other hand, if VAN's claims to be able to "predict earthquakes" are without merit, (c) is of little or no interest.

The purpose of this paper is to evaluate VAN's claims to be making "successful predictions" of earthquakes. If these claims were valid, VAN would have made a major scientific contribution. As scientists we therefore are obligated, as noted above by Dyson, to ruthlessly seek out all possible weaknesses in VAN's work. VAN's claims should be recognized if and only if they can withstand rigorous and relentless scrutiny.

On the basis of the analysis presented below, the conclusion of this paper is that VAN's claims to be making "successful predictions" of earthquakes are not valid. In short, the answer to question (b), above, is "no."

2 Past efforts at earthquake prediction

Research on earthquake prediction has been conducted intermittently over the past 110 years. Highly optimistic reports have been presented from time to time, but none has withstood detailed examination.

Before discussing VAN in particular it seems worthwhile to review previous work on earthquake prediction. Some of the key questions to bear in mind are: Did past efforts at earthquake prediction fail simply because they weren't looking for the right precursory phenomena, and have VAN found some genuinely new approach? What past efforts have been made at earthquake prediction based on observations of electromagnetic phenomena? What does the physics of the earthquake process suggest about whether or not any approach to prediction can be successful?

As VAN is claimed to be a "short-term" earthquake prediction method, I will not present a detailed review of work on so-called "long term prediction" on time scales of the order of years. (Many seismologists, including me, think that "long term probabilistic forecasting" is a more appropriate term for such studies, as the public and media seem to think that "prediction" refers only to warnings of an imminent earthquake.) In this paper "prediction" refers only to so-called "short term prediction."

Work on long-term prediction is also controversial. For example, Kagan and Jackson evaluated some long term forecasts based on the "seismic gap hypothesis;" their statistical analysis showed these long term forecasts to have been failures. [11] The U.S. Geological Survey claimed that the 1989 Loma Prieta, California, earthquake was "an anticipated event," but these claims have been refuted by Savage. [12] The U.S. Geological Survey carried out an "earthquake prediction experiment" at Parkfield, California, where it was claimed on the basis of allegedly periodic seismicity that an M6 earthquake would occur between 1985 and 1993. In any case, the predicted earthquake failed to occur. But Savage showed that even if an M6 earthquake had taken place within the time window, it would have been more reasonable, from a statistical point of view, to regard it as a random event rather than as a validation of the "Parkfield Earthquake hypothesis." [13]

2.1   Research prior to 1960

From the time seismology became a recognized discipline in the latter half of the 19th century, earthquake prediction has been a topic of discussion. John Milne, a British engineer and seismologist at Imperial University (now Tokyo University), noted in 1880 that "Ever since seismology has been studied one of the chief aims of its students has been to discover some means which would enable them to foretell the coming of an earthquake ...." [14] Milne discussed possible earthquake precursors, including weather conditions, animal behavior, electrical effects, earthquake lights, earth tides, changes in the temperature of hot springs, and microearthquake activity.

After the Nobi earthquake of 1891, the Imperial Earthquake Investigation Committee was founded by the Japanese government. Its mission was "...to investigate whether there are any means of predicting earthquakes; and ...to investigate what can be done to reduce the disastrous effects of earthquake shocks to a minimum .... " [15] Its members "attacked with every resource at their command the various problems bearing on earthquake prediction, such as earth tiltings and earth pulsations, variation in the elements of terrestrial magnetism, variation in underground temperatures, variation in latitude, secular variation in topography, etc., but satisfactory results were not obtained." [16]

Milne, who had retired to England, noted in 1911 that earthquake prediction could serve to secure public support, and hence funding, for seismology:

> "What the public imagine they would like to know about an earthquake is the time at which it might occur. If this could be stated, and at the same time something about the character of the expected disturbance in earthquake districts, seismology would be liberally supported." [17]

In an extensive report on the 1906 San Francisco earthquake, H. F. Reid presented prospects for earthquake prediction in a rosy light:

> "As strains always precede the rupture and as the strains are sufficiently great to be easily detected before the rupture occurs, in order to foresee tectonic earthquakes it is merely necessary to devise a method of determining the existence of the strains .... To measure the growth of strains, we should build a line of piers, say a kilometer apart, at right angles to the ...fault ...... if the surface becomes strained through an angle of about 1/2000, we should expect a strong shock ..." [18]

Several points stand out. First, there was a naive optimism that precursory phenomena must exist, and that all one had to do was go out and observe them. Second, it was assumed that predictions could be made merely by observing precursors; questions such as how to discriminate signals from noise, or whether, how, and how reliably the time, location, and size of an earthquake could be predicted on the basis of the hypothesized precursors, were not discussed. Third, it was assumed that there was an upper limit on the strain, and that when this limit was reached an earthquake would occur. Finally, nobody seemed to be worried by the lack of a physical theory for earthquake prediction; prediction proponents seemed confident that a purely empirical approach would suffice. All of the above assumptions are questionable in view of what is now known about the earthquake source process.

No organized prediction research programs existed during 1910-1960, but a number of observations of alleged precursory phenomena were reported. A search of *Nature* and *Trans. Am. Geophys. Un.* found five reports of various luminous phenomena alleged to be associated with earthquakes,[19] ten mentions of prediction based on geodetic precursors,[20] and two reports of alleged anomalous animal behavior (one on birds, and one on fish).[21] However, the seismological community in general was quite pessimistic regarding earthquake prediction.

In 1935 H. Landsberg discussed an apparent temporal correlation between deep and shallow earthquakes at the Spring Meeting of the American Geophysical Union, but a news report in *Science*[22] began "Late July should see, somewhere in the world, a severe earthquake with its focus ... relatively close to the surface of the earth." A subsequent discussion by Landsberg began by clarifying the misinterpretation of his work by the news report, saying that "our present knowledge ... does not permit any prediction of location and time of occurrence of a major earthquake with scientific precision." However, after discussing the low level of public funding for seismology and noting that "many [seismologists] believe it is futile to make an attempt to attack the question of forecasting earthquakes," he ended by asserting that " ... the only conclusion which we have to draw is that more research is needed to attack the problem of earthquake prediction successfully."[23]

Similarly, Imamura stated in 1937 that

"... there are some who declare that the prediction of earthquakes is impossible, but [I] do not share in such an idea. Comparing the state of our present knowledge with that of, say, 30 or 40 years ago, it cannot be denied that we are nearer to making practical predictions than we were then. We think that it can at least be said that, though yet very remote, we are steadily advancing toward that desired end."[16]

These two themes — that further research (i.e., funding) is needed and that progress towards a very difficult goal is being made — tend to be widely used, even today, by the officials of earthquake prediction programs.

On the other hand, J. B. Macelwane, S.J., then First Vice President of the Seismological Society of America (SSA), spoke as follows on a national radio broadcast in the U.S. in 1946:

"An earthquake forecast must be *specific*. This means that the forecaster must predict three things about the earthquake. In the first place, the forecast must announce the *time* of the earthquake, at least within a few hours. Secondly the forecast must specify the *intensity* of the coming earthquake, or the extent of the damage to be expected from it. Thirdly, the forecast must state in just what *place* or *places* the destruction of property is going to occur.

"Besides being specific, the forecast must also be *reliable*. That is, a predicted disaster must be sufficiently probable to justify public authorities in removing the threatened populations, cutting off the gas supply and the electricity, and otherwise disrupting the normal life of the people.

"Is it possible, in the present state of scientific knowledge, to predict earthquakes in this definite and positive sense which alone deserves the name of forecasting? Unfortunately, no! All reputable seismologists agree that we have no means at the present time of arriving at a reliable forecast of any earthquake anywhere.

"The problem of earthquake forecasting has been under intensive investigation in California and elsewhere for some forty years; and we seem to be no nearer a solution of the problem than we were in the beginning. In fact the outlook is much less hopeful." [24] (Italics follow original ms.)

Earthquake prediction research remained in disrepute in the 1950's. On p.385-387 of his authoritative textbook, *Elementary Seismology*, which was published in 1958, C. F. Richter commented as follows:

"At present there is no possibility of earthquake prediction in the popular sense; that is, no one can justifiably say that an earthquake of consequence will affect a named locality on a specified future date. It is uncertain whether any such prediction will be possible in the foreseeable future; the conditions of the problem are highly complex. One may compare it to the situation of a man who is bending a board across his knee and attempts to determine in advance just where and when the cracks will appear.

"Amateur predictors are legion, and will continue to be, so long as claiming to predict earthquakes is an easy way to get one's name into the newspapers. Many of them are honestly self-deceived; ....

" ... The student should be aware of the little significance attempts at prediction have had in relation to the actual development of our knowledge and understanding of earthquakes." [25]

Richter was a consistent and caustic critic of earthquake prediction. In a popular article in 1964 he commented as follows:

"Don't worry about predictions. ... Claims to predict usually come from cranks, publicity seekers, or people who pretend to foresee the future in general." [26]

In 1977, in his acceptance of the Medal of the Seismological Society of America, Richter again commented in a similar vein:

"... [Prediction] provides a happy hunting ground for amateurs, cranks, and outright publicity-seeking fakers. The vaporings of such people are from time to time seized upon by the news media, who then encroach on the time of men who are occupied in serious research." [27]

Few people seem to be aware of the prediction efforts before 1960. I was unaware of much of this work when I published an article in *Nature* in 1991 that criticized Japan's earthquake prediction program:

> "The empirical approach [to earthquake prediction] depends on the existence of reliably measurable and unambiguously identifiable precursors. There might arguably have been reasons for supposing, in 1962, that such precursors existed, but there no longer is room for such a belief." [28]

However, now that I have surveyed much of the literature from 1880 to 1960, I think that even in 1962 there were reasonable grounds for being skeptical about empirical earthquake prediction.

2.2   Research after 1960

Earthquake prediction research suddenly became a major priority in Japan, the U.S., and several other countries starting in the early 1960's. It is almost impossible to tell how much of the enthusiasm was sincere, and how much was motivated by using the pretense of working on earthquake prediction as a tool for extracting large-scale funding. Three facts stand out clearly, however.

First, almost no discussion was devoted to the theoretical question of whether or not earthquakes are predictable. There was a naive belief, exactly like that of the 1906 San Francisco earthquake report, [18] that instruments could be set out, precursory phenomena observed, and earthquakes successfully predicted, all on a purely empirical basis. The 1962 "Blueprint" for Japan's earthquake prediction program, which still remains in effect today, said:

> "...it seems highly probable that we would be able to find some significant correlation between earthquake occurrence and observed phenomena merely by accumulating data for several years." [29]

The American prediction program, as outlined by Press and Brace in 1966, followed basically similar empirical lines:

> " ...it seems obvious that a major feature of such a program would be the monitoring, with the greatest achievable sensitivity, of all possible indicators foretelling the occurrence of earthquakes. Networks of instruments would be deployed in seismic belts and would be operated continuously over long periods of time in such a way as to provide the greatest possible likelihood that many earthquakes would be 'trapped' within the arrays. Although this is essentially an empirical and somewhat wasteful approach, the absence of a confirmed theory for the earthquake mechanism justifies it." [30]

Second, there was naive optimism about the time required for the prediction programs to succeed. The Japanese Blueprint, which was published in 1962, promised to answer the question of whether or not earthquakes are predictable within ten years of the onset of the prediction program:

> "Now, when will earthquake prediction become possible and an efficient fore-warning service be available? This question cannot be answered now. But if we start the project presented here we should be able to answer the question with sufficient certainty within ten years." [29]

Japan's earthquake prediction program started in 1965, and the 7th five-year plan began in April 1994. In contrast to the initial claims that an answer would be forthcoming in ten years, K. Mogi, the Chairman of the Coordinating Committee for Earthquake Prediction, commented as follows in 1994:

> "Earthquake prediction is a 100-year national project. In a small and seismically active country like Japan, there is no place to escape from earthquakes. Even if there are no immediate results, the earthquake prediction program must not be discontinued." [31] (My translation)

Third, both the American and Japanese prediction proposals in the 1960's were notable for the complete absence of discussion of where the previous efforts at prediction (during 1880-1960) had gone wrong. One might have expected a detailed literature survey of earlier efforts, together with an analysis of where the earlier programs had failed, so that the mistakes of the past wouldn't be repeated. However, neither the Japanese Blueprint nor the study of Press and Brace made any mention of previous prediction programs (although both mentioned various historic examples of alleged earthquake precursors). Such ignorance regarding past research efforts led to a repetition of the earlier un-successful projects on a larger scale.

It ordinarily is not necessary to study the literature of 50 or 100 years ago, as progress is successively summarized in reviews, monographs, and then textbooks. But in the case of earthquake prediction we are dealing with a field where, despite great efforts, there has not been significant progress. Study of papers published 50 or even 100 years ago might therefore prove useful, as learning from past failures is less painful than repeating them.

2.2.1 The crest of optimism

By 1973 many geophysicists were claiming that prediction would soon become possible. Scholz *et al.* wrote in *Science* that:

> "Earthquake prediction, an old and elusive goal of seismologists and astrologers alike, appears to be on the verge of practical reality as a result of recent advances in the earth and materials sciences." [32]

F. Press, writing in *Scientific American* in 1975, declared that:

"Recent technical advances have brought [the] long-sought goal [of earthquake prediction] within reach. With adequate funding several countries, including the U.S., could achieve reliable long-term and short-term forecasts in a decade." [33]

A 1975 cover story in *Time* entitled "Forecast: Earthquake" touted the allegedly successful prediction in Nov. 1974 of a M5.2 earthquake near Hollister, California, as a major breakthrough:

"The accurate forecast of the Hollister temblor was a dramatic demonstration that scientists are on the verge of being able to predict the time, place and even the size of earthquakes.

"Recently, in fact, U. S. and Russian seismologists have quietly–and correctly– forecast several other earthquakes. In China, where the understanding of earthquakes has become an important national goal, ten quakes are said to have been accurately predicted in the past ten years. Before two large recent quakes, the government confidently issued public warnings and evacuated vulnerable areas. Buoyed by their rapid progress in forecasting, scientists are already talking about an even more exciting possibility: actually taming the more destructive convulsions of the earth." [34]

2.2.2   The bubble bursts

It is now 20 years since Scholz *et al.*'s paper, *Time*'s cover story and Press's highly optimistic forecast that prediction would become a reality in a decade. The widespread optimism typified by these reports has now almost completely disappeared. What went wrong?

Discussing this question might appear to be a waste of time, since the focus of the London meeting and this volume is on VAN. Actually, however, the problems that led to the collapse of the 1970's prediction boom were of the same type as those that appear to be afflicting the work by VAN: uncritical acceptance of data at the noise level as precursory signals, followed by the proposal of *ad hoc* qualitative theories to explain the "observations."

In 1973, at the height of the prediction boom, Scholz *et al.* claimed that:

"A variety of effects premonitory to earthquakes such as crustal movements and anomalous changes in such phenomena as tilt, fluid pressure, electrical and magnetic fields, radon emission, the frequency of occurrence of small local earthquakes, and the ratio of the number of small to large shocks have been observed before various earthquakes. ...Vigorous programs to monitor premonitory effects, particularly those in Japan and the U.S.S.R. during the last 5 to 10 years, leave little doubt that such effects are real." [32]

Several workers, including Scholz *et al.*,[32] quickly proposed a mechanism to explain the alleged precursors: dilatency. Dilatency is in fact a real phenomenon observed in laboratory studies of fracture in rocks: prior to failure rocks undergo an inelastic increase in volume. Although there never were quantitative theories to show that dilatency could account for the alleged precursors, a variety of *ad hoc* qualitative explanations were proposed. In the end, however, the alleged precursors turned out to be noise rather than signals.

One of the best publicized of the reported precursors were alleged 10-20% changes in the *in situ* seismic wave velocities or V_P/V_S, the ratio of P-wave velocity to S-wave velocity. It was claimed that the seismic wave velocities (or V_P/V_S) decreased 10%-20% and that its recovery to its original value was the sign of an impending earthquake.

Unfortunately, however, the reported *in situ* velocity changes were artefacts of poor data analysis. What was really being reported were changes in the "apparent velocity," which is the distance between two seismic observatories divided by the difference of the arrival time of the same wave, say a P-wave, at the two observatories. The apparent velocities can change greatly due to changes in hypocentral locations, without any need to invoke temporal changes of the *in situ* velocities.

Following the initial reports of 10-20% *in situ* velocity changes, a number of more careful studies were performed. Kanamori and Fuis searched for possible temporal variations in *in situ* seismic velocities using travel time data observed for quarry explosions, for which the epicenter is known with extremely high precision. They showed in 1975 that temporal variations in *in situ* velocities, if they existed at all, had to be on the order of less than 1% rather than the alleged 20% changes that had been reported earlier.[35] Other recent carefully controlled studies[36,37] have placed even smaller bounds on *in situ* temporal variations; the reports of 10-20% *situ* changes made in the 1970's are now generally regarded as having been discredited.

The prediction bubble burst quickly. A Research News article in *Science* in 1978, "Earthquakes: Prediction Proving Elusive," said:

> "A few years ago, the scientific community was optimistic, even euphoric, in the wake of several successful predictions of earthquakes around the world. Today, that euphoria is gone. The optimism has not given way to pessimism, but, as one administrator puts it, 'There's a long, hard road ahead.' "[38]

The bubble had collapsed completely by 1982, as evidenced by a special session on earthquake prediction at the Annual Meeting of the SSA. C. R. Allen's report conveys the highly pessimistic mood:

> "In my opinion, we must face up to the fact that our progress during the past 5 yr in short-term earthquake prediction has not been as rapid as we

had envisaged when the program started. ......no plethora of precursors has been claimed. Indeed, some of the results have been downright discouraging. ......But let us continue to be honest with our funding agencies, Congress, and the public. To some degree, we in the seismological community have been guilty of allowing the public to conclude that short-term earthquake prediction is more imminent than most of us really believe." [39]

The 1982 special session at the SSA meeting essentially marked the end of earthquake prediction as a mainstream research topic. Remarkably, however, the Japanese government continues to provide large scale funding for a national program on earthquake prediction, despite the lack of any demonstrable progress towards this goal. [40] Turcotte's 1991 assessment closely matches that of Allen in 1982:

"The empirical approach to earthquake prediction has failed to produce reliable measurements of precursory phenomenon [sic]." [41]

2.3 Prediction efforts based on electromagnetic phenomena

Milne's address to the Seismological Society of Japan in 1880 mentioned electrical signals as one possible precursor. [14] Studies of possible geomagnetic and geoelectric precursors from the late 1800's to the mid 1970's are summarized in Chapter 10 of Rikitake's monograph. [42] Some of the work summarized by Rikitake involved searches for possible precursory changes in earth currents.

In 1976 USGS scientists were optimistic that magnetic observations could be used to predict earthquakes. (It is now clear that these hopes have not been realized.) Smith and Johnston of the USGS reported as follows:

"The most significant changes in local magnetic field observed in central California during 1974 occurred in the 2 months preceding, though at a site 11 km away from, the largest earthquake in this region. Although no definite relation has been established between the magnetic changes and this earthquake, the most probable source of these changes is a tectomagnetic effect in the complex fault system around the earthquake epicenter. If this is the case, these results indicate that magnetic observations can be used to monitor crustal stress changes in areas of sufficient magnetization." [43]

As the focus of the London meeting was VAN, let us consider work using geoelectrical observations. Suzuki's 1982 review commented:

"Geoelectric precursors have been reported many times. Most of the old reports are based on the observation of natural earth-current or self-potential. More recently ground resistivity has been measured, sometimes by sending artificial current into ground to improve the signal/noise ratio." [44]

In 1989 a group from the Japan Meteorological Agency pointed out several problems in reports of "anomalous changes" in electrical potential:

> "Many authors have claimed that "anomalous changes" in earth-potential prior to earthquakes were precursors (e.g., ... Varotsos and Alexopoulos 1984a,b ...). However, we think it is very debatable whether or not those anomalous changes were truly precursors, for the following reasons. (i) The electric potential between the electrode and soil is unstable (especially due to precipitation). (ii) The data can be contaminated by noise due to industrial activities, electrical railways, etc. (iii) Aside from the above two effects, most fluctuations in electric potential are caused by electromagnetic induction within the earth; i.e., by geomagnetic variations of primarily external origin." [45]

Turcotte's 1991 review paper comments as follows:

> "It is relatively easy to measure electrical currents and magnetic fields in the vicinity of a fault. Precursory variations in these measurements have been reported in California (...), Japan (...), and Greece (Varotsos & Alexopoulos 1984a,b, Dologlou 1987, Varotsos *et al.* 1988). Again there are many problems associated with noisy signals and with the statistical significance of the observables. Also, no quantitative theory for piezo-electric or other mechanisms predicts an observable signal." [41]

A recent review by Park *et al.* comments as follows:

> "Anomalous electromagnetic (em) fields and changes of electrical properties of the Earth have been reported before earthquakes by many researchers over the years. The observations cover the entire electromagnetic spectrum from visible light to quasi-dc, and viewed collectively, they present a bewildering, and to some, improbable array of cause and effect relationships." [46]

Apparent electromagnetic precursors of earthquakes — mainly in the form of ultralow frequency (ULF) electromagnetic wave signals — were discussed in a specialized gathering at the URSI (Union Radio Scientifique Internationale) General Assembly in Kyoto, Japan, in August, 1993. At the 7th Meeting of SC-IDNDR (the ICSU Special Committee for the International Decade for Natural Disaster Reduction) in Santiago, Chile, in October 1993:

> "the URSI Secretary-General J. van Bladel informed SC-IDNDR about work within URSI concerned (i) with the detection of electromagnetic (EM) precursors of earthquakes and volcanic eruptions and (ii) with the identification of mechanisms for such EM precursor phenomena. ... Speculations on mechanisms included the hypothesis that a long thin region of high conductivity (possibly associated with a subsurface fluid flow phenomenon — or with triboelectric effects), forming along a fault just before a major earthquake, modifies incident EM waves of external origin. The Special Committee accepted that

the time may now have come for the wider IDNDR community to take seriously the possibility that further study of EM precursors may show them to have true predictive value." [47]

The presentation by Dr. van Bladel in Santiago may have been the first time that most members of SC-IDNDR had been exposed to a discussion of prediction of earthquakes based on electromagnetic observations. They might therefore have thought that this was a new and promising approach. However, the above summary makes it clear that attempts at prediction based on electromagnetic observations have as long a history of failure as other approaches. Claims of success have been made from time to time, but none has withstood a detailed critical examination or proven repeatable.

2.4   Empirical search for precursors

Most empirical research on earthquake prediction consists of a retrospective search for precursors after large earthquakes have already occurred. The difficulty is that any phenomena that happened to occur before an earthquake can be called precursors whether or not they have a causal relation to the earthquake. This is a dangerous area, as there are no known quantitative theories to guide us. Suzuki's 1982 review commented as follows:

> " ... the present state of the art is very chaotic, or at least seems more confusing than in the past. A remarkable variety of earthquake precursors have been reported so far. Some reported precursors seem very strange and open to doubt. Even excluding these ambiguous cases, no general and definite way to successful earthquake prediction is clear." [44]

Proponents of empirical earthquake prediction invariably claim that more "case studies" of precursors have to be obtained, and that eventually it will be possible to sift out the noise and find reliable precursors. However, this may simply be wishful thinking. A recent review by Kagan (p.161 and 187) comments as follows:

> "The intrinsic randomness of seismicity means that most standard techniques would not produce the expected results when applied to earthquake data. For example, case studies of particular earthquakes or earthquake sequences present an [sic] insufficient evidence, since if the phenomenon is stochastic, in a certain sense, any earthquake sequence is possible. Thus unless it can be shown that such cases are proper representatives of the general earthquake population, the evidence based on case studies alone cannot be taken as a final word: by using a biased selection rule in a largely random earthquake population, one can obtain almost any kind of results."

"...anyone who [has] tried to force open a jammed door knows that a minuscule (of the order 10^{-3} or 10^{-4}) change of the system geometry can modify its mechanical behavior in a substantial way. During each earthquake the strain drop is on the order of 10^{-4} (Scholz, 1990), thus the mechanical properties of the fault system should change after each earthquake: new barriers or asperities (disclinations) form as [the] result of an earthquake deformation, and these new complexities modify the fault behavior in a way that is unpredictable by simple models of earthquake quasi-periodicity." [48]

2.4.1 IASPEI's Guidelines for Precursor Candidates

In 1991 IASPEI's (International Association of Seismology and Physics of the Earth's Interior) Sub-commission on Earthquake Prediction proposed a set of "Guidelines for submission of earthquake precursor candidates." [49] The main items of these guidelines are as follows:

Validation Criteria: (a) The observed anomaly should have a relation to stress, strain, or some mechanism leading to earthquakes. (b) The anomaly should be simultaneously observed on more than one instrument, or at more than one site. (c) The amplitude of the observed anomaly should bear a relation to the distance from the eventual mainshock. If negative observations exist closer to the mainshock hypocenter than the positive observations, some independent evidence of the sensitivity of the observation sites should be provided. (d) The ratio of the size of the dangerous zone to the total region monitored shall be discussed to evaluate the usefulness of the method.

Data: There should be a persuasive demonstration that the calibration of the instrument is known, and that the instrument is measuring a tectonic signal.

Anomaly Detection: Anomaly definitions should be precisely stated so that any other suitable data can be evaluated for such an anomaly. The difference between anomalous and normal values shall be expressed quantitatively, with an explicit discussion of noise sources and signal-to-noise ratio. Negative evidence (such as failure to observe the anomaly at other sites nearer the earthquake hypocenter) shall be reported and discussed.

Association of anomalies with subsequent earthquakes: The rules and reasons for associating a given anomaly with a given earthquake shall be stated precisely. The definition of an anomaly and the association rule should be derived from a data set other than the one for which a precursory anomaly is claimed. The probability of the "predicted" earthquake to occur by chance and to match up with the precursory anomaly shall be evaluated. The frequency of false alarms (similar anomalies not followed by an earthquake) and surprises (similar size mainshocks not preceded by an anomaly) should also be discussed.

No precursors satisfying the above guidelines have ever been observed. Although I will not present a detailed discussion in this paper, it appears to me that VAN's claims to have observed Seismic Electrical Signals (SES) fall far short of meeting the criteria of the above guidelines.

2.5   Improved observations — worsening precursors

It is generally considered that if earthquake precursors exist at all, microseismicity and geodetic observations are the most promising possibilities. Turcotte's 1991 review commented as follows:

> "Studies of local seismic activity and geodetic strains should receive the highest priority, since they are directly associated with the earthquake cycle." [41]

2.5.1   Regional seismicity

About 25% of all large earthquakes have foreshocks, but it seems to be impossible to determine whether a particular earthquake is a foreshock until after the mainshock has occurred. [50] Claims of changes in seismicity patterns or seismic quiescence before major earthquakes have often been made but these are always based on retrospective analyses. Turcotte notes that the results of observations of seismicity have been deeply disappointing:

> "It is actually quite surprising that regional seismicity is not strongly correlated with the occurrence of large earthquakes. If a large earthquake significantly reduces the regional stress field, it would be expected that the level of seismicity would also be reduced. As the stress and strain accumulate for the next event, it would then be expected that a systematic increase in the regional seismicity would occur. This pattern is not observed, however. Except for aftershocks, regional seismicity does not seem to be significantly correlated with major earthquakes." [41]

2.5.2   Geodetic observations

The earthquake prediction program in Japan has long viewed geodetic precursors as the best hope. Great emphasis has been placed on the observation of an allegedly clear precursory uplift of a few mm recorded at Kakegawa in the days before the 1944 Tonankai earthquake. [51] Officials of the Japanese earthquake prediction program repeatedly cite this report when defending the earthquake prediction program to the mass media. For example, Prof. H. Aoki of Nagoya University, appearing on a special NHK-TV (Japan Broadcasting Corporation) program on Sept. 1, 1994, cited the 1944 observation as proof that the so-called Tokai earthquake could be predicted. He said (my translation), "If such a clear precursor were to be observed we could unquestionably predict the Tokai earthquake. Make no mistake; we could predict it." [52]

The 1944 observation was based on observations from an expert military surveying party under wartime conditions. It is not clear whether or not the alleged precursory uplift was above the noise level. Also, as noted by Thatcher, [53] the alleged precursor occurred 200 km from the epicenter, and no quantitative model links it to the subsequent earthquake. Thus, even if the 1944 observation was not simply noise, there is no known way it could have been used to predict quantitatively the time, location, and size of the earthquake.

Many efforts have been made in both Japan, the U.S., and elsewhere to detect geodetic precursors, but these have proved unsuccessful or inconclusive. For example, Castle *et al.* [54] analyzed surveying data and reported an extensive uplift (the so-called "Palmdale Bulge") of as much as 450 mm between 1959 and 1974, but Jackson *et al.* [55] showed that this was probably an artefact due to systematic errors in leveling.

Recently GPS data, which have much greater accuracy and precision than previous surveying techniques, have become routinely available. There were great hopes in Japan and elsewhere that GPS data would finally allow geodetic precursors to be detected. However, results from GPS data have been uniformly negative. For example, Bock *et al.* [56] reported there were no significant preseismic motions for the 1992 Landers, California, earthquake. Similarly, GPS observations showed no significant precursors for the 1994 Hokkaido Toho-oki earthquake or 1995 Kobe earthquake.

2.5.3   Implications for VAN

Both seismic networks and geodetic observations have improved greatly in the past 30 years. But these improved observational facilities have had the ironic effect of placing much tighter bounds on the maximum extent of whatever geodetic and seismic precursors might exist. In other words, the standard deviations of the measured data have drastically decreased, but the precursors being sought are still below the new and reduced noise level. Proponents of earthquake prediction are undeterred, claiming that the old reports of precursors are real, and that the new data just prove that some earthquakes have precursors but others don't. However, a much more reasonable interpretation is that the precursors simply don't exist, and that the earlier reports involved various kinds of noise. What does this imply for VAN?

First, improvements in the instrumentation and techniques used by VAN or other groups making similar observations are highly desirable. The work by Gruszow *et al.* [57] is a welcome step in this direction. It is important to understand, identify, and quantitatively model the effects of all possible sources of electrical signals, not just earthquakes. However, based on all other experiences with earthquake prediction, I predict that improving the accuracy of the measurements, rather than revealing clear precursors, will simply lower the amplitude of whatever alleged precursors are observed in the future.

3   Are earthquakes predictable?

As discussed above, a large number of scientists have made massive efforts towards the goal of earthquake prediction for over 110 years, with no positive results to show for these efforts. I am unaware of any other active research field in physical science with a comparable history of sustained failure. Before making further efforts in this field it seems reasonable to consider whether earthquake prediction is, even in principle, a feasible goal.

We understand the occurrence of earthquakes in a general way. Convection in the mantle is dragging the overlying lithosphere along with it at an essentially steady rate. The brittle portion of the lithosphere cannot deform freely, so strain energy builds up. From time to time some of the strain energy is released by earthquakes of various sizes. The amount of energy released by earthquakes depends on the the level of "tectonic stress," which is unknown. However, minimum estimates of the amount of energy released by earthquakes can be made. Kanamori[58] summarizes recent work on this topic.

Kagan's[48] review summarizes the difficulties that confront research on the earthquake source process: (1) There is no comprehensive theory of earthquake rupture; even basic relations cannot logically be derived from fundamental physical laws. (2) Earthquakes occur in the deep interior of the earth; i.e., in places which are largely inaccessible to direct observations and measurements. (3) Experiments with earthquakes are not possible. (4) The modeling of fracture either in a laboratory or by computer simulation is still in a very rudimentary stage; it is not clear whether the results of such modeling are applicable to natural earthquakes. (5) Earthquake occurrence is characterized by extreme randomness; the stochastic nature of seismicity is not reducible by more numerous or more accurate measurements. (6) The degree of seismicity randomness is much greater than that of atmospheric turbulence; whereas it is possible to predict weather with increasing accuracy as the time lag between an observation and the forecast decreases, an earthquake usually strikes without any warning.

Earthquake occurrence on a particular fault is extremely sensitive not only to fine details of the initial stress distribution and the constitutive properties of the medium near that fault, but also to the stress distribution and fine structure of the earth in remote regions. [59,60] Earthquake initiation and rupture propagation are examples of what Bak and Tang [61] call "self-organized criticality." Although this term does not appear to have a rigorous quantitative definition, it means roughly a critical state where there is no characteristic time, space, or energy scale, and all spatial and temporal correlation functions are power laws. Bak and Tang studied the dynamic evolution of a schematic faulting model based on cellular automata, and concluded that:

> "If a mechanism of the type discussed here is indeed responsible for earthquakes, there is virtually no hope for ever making specific predictions." [61]

Bak and Tang's cellular automata model replicates many of the key features of real earthquakes, but is not a physical model of earthquakes. More realistic models of the earthquake failure process that include the effects of fault complexity, interaction between various faults, etc., should be and are being developed. While of great interest from a purely scientific point of view, it seems far more likely that such work will confirm rather than refute the conclusions of Otsuka, [59] Brune, [60] and Bak and Tang. [61]

Except for the fact that the maximum size of earthquakes is limited by the thickness of the seismogenic zone, self-organized criticality seems to be a good model for the actual state of the earth. A spectacular example was provided by the 1992 Landers, California, earthquake. This earthquake triggered a remarkably sudden and widespread increase in earthquake activity across the western United States, at distances up to 1250 km. [62] This has highly negative implications for earthquake prediction studies, as it suggests that the lithosphere is indeed in a state of self-organized criticality and that any small and distant perturbation could be the "straw that breaks the camel's back."

Naive hopes for prediction are based on the belief (hope) that there exist precursory phenomena whose spatial and temporal extent or amplitude will be proportional to the size of the impending earthquake. From time to time scenarios postulating the existence of such precursors are still proposed, but there is little or no evidence to support them.

Both the above recent theoretical work on non-linear dynamics and observational studies of individual earthquakes cast doubt on the existence of precursors that will allow the time, location and size of large earthquakes to be predicted reliably, and within narrow limits. Small earthquakes are continually occurring, and any one of them seems to have the potential to trigger a runaway avalanche of successively larger earthquakes within a few seconds or tens of seconds.

Studies of the rupture mechanisms of large shallow earthquakes have revealed that they are highly complex multiple shocks. Detailed analyses of waveforms of many earthquakes have shown that several spatially separated faults slipped in turn, with each apparently being triggered by the previous segment. [63–66] The extent of fault complexity, and the sensitivity of rupture evolution to the interaction between different faults suggest that earthquake prediction is effectively impossible.

We can summarize the present situation as follows. Empirical efforts at earthquake prediction have failed to find any phenomena that can be reliably identified in advance as precursors of a particular future earthquake with a well-constrained location, origin time, and magnitude. Furthermore, recent results in non-linear dynamics suggest that earthquakes are effectively unpredictable. The continued failure of empirical prediction efforts is entirely consistent with the hypothesis that the precursors being searched for simply do not exist.

3.1   How should we evaluate VAN?

VAN's claims to be successfully "predicting" earthquakes [1–4] stand out against the above gloomy background. The level of success claimed by VAN is remarkable in view of the long and sustained history of failure of other earthquake prediction efforts. Furthermore, the equipment used by VAN is essentially just a few electrodes and voltmeters. If VAN's claims are correct, VAN's work merits approbation. However, as noted by Dyson, we are obligated as scientists to relentlessly search out all possible weak points in VAN's work. VAN's claims should be accepted if and only if they can withstand such scrutiny.

In several respects the discussion at the London meeting fell short of the desired level of critical scrutiny. First, many of the participants in the London meeting seemed willing to accept VAN's claims to have been "successfully predicting" earthquakes without subjecting them to critical scrutiny. The talk by Prof. Varotsos was mainly a qualitative discussion of possible mechanisms for VAN's observations, and the talk by Prof. Uyeda was basically the same talk he regularly gives to audiences of non-specialists. In neither case did they discuss the details of VAN's claims to have made "successful predictions" of earthquakes.

Second, VAN's work was discussed in isolation, rather than in relation to other work on earthquake prediction. For example, as noted above, recent geodetic observations using GPS have consistently found there were no observable precursors. Thus if VAN's so-called SES signals are really earthquake precursors we need a physical mechanism that produces electrical precursors but that does not generate geodetic precursors. However, the discussion at the London meeting failed to consider this vital point. If VAN's signals really were precursors, one could reasonably expect even larger signals at the time of the earthquake. But no convincing arguments were given for the absence of coseismic signals in VAN's electrical data.

Third, many of the participants in the London workshop seemed to accept without question that VAN's observations were earthquake precursors, and immediately began to propose *ad hoc* and qualitative physical mechanisms. This is at variance with one of the fundamental principles of science, Occam's Razor, which states that we should test the simplest hypotheses first. Thus before accepting the claim that VAN's observations are related to earthquakes we should first rule out simpler explanations such as instability of the electrodes due to precipitation, noise due to industrial or other human activity, and electromagnetic induction due to external geomagnetic effects.

3.2 Lessons from "pathological science"

Recently there have been several notable episodes in which apparently revolutionary discoveries in physics have been announced with great fanfare, but later proved to have been completely false. Two representative examples are the claim that a "fifth force" was required to explain gravity[67] and the claim by Pons and Fleischmann to have developed an apparatus for "cold fusion."[68]

The Nobel laureate P. W. Anderson, a theorist of condensed matter physics, pointed out in a 1992 article in *Physics Today* that these time-wasting episodes could have been avoided if Bayesian statistical methods had been applied.[69] Both the fifth force and cold fusion episodes are the scientific equivalent of searching for a needle in a haystack. We should therefore assign a high *a priori* probability, say 50% or more, to the null hypothesis. On the other hand, the non-null hypothesis (that, say, a fifth force exists) is not a well defined hypothesis, as there are several free parameters (the exponent of the decay of the force with distance, r^{-n}, and its amplitude). Thus the non-null hypothesis for each possible value of the free parameters can be assigned an infinitesimal *a priori* probability, with the integral of the infinitesimal *a priori* probabilities over all parameters being equal to the remaining 50%. As pointed out by Anderson, such a Bayesian analysis would have led to the requirement for large amounts of convincing data before we rejected the null hypothesis, and the fifth force or cold fusion fiascoes could thus have been avoided.

Anderson's analysis could profitably be applied to VAN's work as well. VAN has never proposed a well defined hypothesis with a testable quantitative relation between the electrical signals they are observing and earthquakes. Based on VAN's claims, I infer that the hypothesis they implicitly are proposing is: "there is some relation between electrical signals and earthquakes, but the parameters of this relation are as yet unknown."

Due to the failure of all previous earthquake prediction efforts I would have assessed the *a priori* probability of any given earthquake prediction effort, including VAN's, as at most 1% (and probably much less) in 1981. Other workers might argue for a somewhat larger value, but the consensus of the geophysical community would probably be in this general range. I will not present a Bayesian statistical analysis here, but I think that such an analysis would show that the *a posteriori* probability for VAN's "hypothesis" as of 1995 is lower than the *a priori* probability in 1981 for all plausible values of the *a priori* probability in 1981. Thus the conclusion we should draw from VAN's efforts is that once again yet another effort at prediction has proven unsuccessful.

A distinguished senior scientist advised me to delete the above paragraph, as he said it could be cited as evidence of prejudice on my part. However, any good scientist will always be making estimates of what is more and less likely; it's only prejudice if we refuse to change our minds no matter how much supporting evidence is amassed. On the other hand, it's perfectly reasonable to demand a higher standard of proof for a seemingly implausible proposition. Anderson's Bayesian analysis simply quantifies this commonsense approach.

The Nobel laureate chemist, I. Langmuir, collected a number of examples of what he called "pathological science," the science of "things that aren't so." Langmuir gave a delightful lecture on this subject in 1953; this lecture was revived and published in *Physics Today*[70] in Oct. 1989, partly as an oblique commentary on the cold fusion fiasco. The fifth force and cold fusion qualify as pathological science under Langmuir's criteria, as does polywater.[71]

Langmuir summarized the qualities of pathological science as follows: (1) The maximum effect that is observed is produced by a causative agent of barely discernible intensity, and the magnitude of the effect is substantially independent of the intensity of the cause. (2) The effect is of a magnitude that remains close to the limit of detectability, or, many measurements are necessary because of the very low statistical significance of the results. (3) There are claims of great accuracy. (4) Fantastic theories contrary to experience are suggested. (5) Criticisms are met by *ad hoc* excuses thought up on the spur of the moment. (6) The ratio of supporters to critics rises up to somewhere near 50%, and then gradually falls to oblivion.

Langmuir's description of pathological science seems to apply well to all previous efforts at earthquake prediction, including VAN's. The reader is invited to make a detailed comparison. One of the most apt points matching Langmuir's criteria seems to be VAN's claims regarding "selectivity," the effect VAN invoke to explain why allegedly precursory anomalies are not recorded at stations close to the epicenter, while they are observed at more distant stations. This appears to apply particularly to VAN's argument that "selectivity is not reversible." Varotsos and Lazaridou [72] (p.329) claim:

> "If selectivity were only governed by the conditions along the main travel path between A and B, this would imply that when a station S_A is sensitive to an area B, a station at B should also be sensitive to the seismic area A. However, this is not always observed." [72]

It seems difficult to reconcile this claim with electromagnetic theory, as reciprocity should be satisfied.

An obvious way to test VAN's work would be to set up a dense network of seismographs and dense array of VAN stations in the same region and study whether or not there really was a correlation between "SES" signals and the earthquakes in that region. However, VAN (p.60 of reference no. 73) invoke the so-called selectivity effect as the reason such a test would be impossible:

> "[The impossibility of such a test] would become apparent if electrical instruments were installed at a particular region (e.g., close to Patras, i.e. PAT), with a 'dense' seismological network, and [one] tried to establish a correlation between SES and earthquakes (having epicenters in the area of PAT), then the proposed method would certainly fail, because, as explained in [reference no. 2 of this paper], the electrical station of PAT is not sufficiently sensitive to the EQ of that area ..." [73]

Each reader can make his own judgment, but it seems to me that the above statement matches most if not all of Langmuir's above six criteria.

The key point underlying all of Langmuir's examples of pathological science is that they involved observations at or below the noise level; it was very easy for the scientists involved to delude themselves that they were seeing real effects, rather than noise. Once they truly believed what they were seeing, it was easy for the protagonists of pathological science to fudge data, etc., so they obtained the results they expected. This kind of self-delusion seems to have afflicted many of the workers who have claimed positive results for earthquake prediction research, including VAN.

4 SC-IDNDR's study of VAN

At the 8th Meeting of the ICSU Special Committee for the International Decade for Natural Disaster Reduction (SC-IDNDR) in Paris in Jan. 1994 Prof. S. Uyeda, a new member of SC-IDNDR representing IUGS (International Union of Geological Sciences) expressed his agreement with the views of the 7th Meeting of SC-IDNDR, quoted above in Section 2.3, that further study of EM precursors might show them to have true predictive value. Prof. Uyeda drew "attention moreover to important studies in Greece and in China that are showing further promise in this regard." As a result, "It was agreed that persons with detailed knowledge of those studies would be invited to make presentations to the 9th Meeting of SC-IDNDR (Yokohama, 25 May 1994) in order that the Special Committee as a whole can come closer to forming a judgement on this potentially important issue." [47]

The Yokohama meeting of SC-IDNDR was mainly a business meeting, but two talks on earthquake prediction efforts based on EM observations were squeezed in at the end. Prof. P. Varotsos was scheduled to give one of these talks, but he did not attend the Yokohama meeting. The reason given for Prof. Varotsos's absence was that he detected anomalous electrical signals in Greece starting on May 3, 1994 and issued a prediction on May 14. It was reported to the SC-IDNDR meeting (which took place on 25 May) that the occurrence of two magnitude-6 earthquakes in Greece on 23 May (at 06:46 GMT) and on 24 May (at 02:05 GMT) was a sensational yet not isolated proof of the predictive power of VAN's methods. The 23 May and 24 May earthquakes are discussed below in Section 8.4.2. In Prof. Varotsos's absence, a report on VAN was presented jointly by Prof. Uyeda and Dr. T. Nagao of Kanazawa University, Japan. No memos, papers, reports, or other supporting documentation regarding VAN were presented to the Yokohama meeting of SC-IDNDR (Prof. J. Lighthill, personal communication).

The 1994 annual report of SC-IDNDR summarizes the talks on EM methods at Yokohama and SC-IDNDR's subsequent actions as follows:

"The talks along these lines given at Yokohama tended to suggest that one currently existing approach which may offer a realistic prospect of short-term prediction is the method based on passive measurements of electrical field strengths; this approach, known as VAN, calls for the processing of observed data on electrical potential differences between various points of a network, in order to derive components of electrical field strength at the earth's surface from which (as far as possible) contributions of ionospheric origin have been removed; forecasts being based upon temporal changes (including, especially, sudden changes) in those components. Some reasons why the Special Committee tended to take this approach seriously were outlined in its report on the Yokohama meeting.

"After circulation of that report, on the other hand, some weighty counterarguments against these views were brought to the Chairman's attention. He therefore took action — in consultation with several members of the Committee — to organise the international meeting A CRITICAL REVIEW OF VAN (London, 11-12 May 1995) with the Royal Society generously participating and offering its premises as a neutral venue for this review." [74]

4.1 Discussion of VAN at the Yokohama meeting

As the time available for the presentations on EM methods was severely limited, it is understandable that the organization of the Yokohama meeting of SC-IDNDR fell short of what would ideally have been desirable. Several steps that ideally ought to have been taken in advance, but which were, in one form or another, taken after the Yokohama meeting are discussed below.

First, normal procedures of peer review ought to have been followed. VAN and their supporters should have been asked to submit a report summarizing VAN's claims. Such a report, which should have included VAN's publications, copies of their faxes and telegrams, etc., as appendices should have been sent to expert referees for written comments before the Yokohama meeting.

Second, to the extent possible, a wide variety of experts should have been invited to attend the talks on EM methods in Yokohama. Earthquake prediction has long been a topic of great interest in Japan; many scientists in Japan have published extensively on the topic of earthquake prediction; some publications by Japanese scientists have discussed VAN, with a wide range of both positive and negative views. As the SC-IDNDR meeting was held in Yokohama, why were invitations not generally extended to these experts?

The International Association for Seismology and Physics of the Earth's Interior (IASPEI) is in turn a member of IUGG, which is affiliated with ICSU. Although in general SC-IDNDR and the IASPEI cooperate closely (especially in regard to the Global Seismic Hazard Assessment Program), a third less than ideal aspect of the talks on EM methods at the Yokohama business meeting of SC-IDNDR is that the Subcommission on Earthquake Prediction of IASPEI was not informed of these talks. As a result, Prof. M. Wyss, the Chairman of this Subcommission, had no opportunity to make a presentation or submit a written statement on this subject. Fortunately Prof. Wyss attended the London meeting and published his paper in this volume.

Fourth, before the Yokohama meeting, SC-IDNDR apparently made no attempt to ascertain what views, both pro and con, had already been published. Had they done so, they would have found both strongly positive and strongly negative views. Of course the many published criticisms of VAN's work should not necessarily have been regarded by SC-IDNDR as a definitive refutation of this work. However, SC-IDNDR should have been, but evidently was not, aware that VAN's work is highly controversial.

Turcotte's views on VAN, quoted above in Section 2.3, might have been of interest to SC-IDNDR. Another notable criticism of VAN was made by Mulargia and Gasperini, who claimed that " ...the apparent success of VAN predictions can be confidently ascribed to chance; conversely, we find that the occurrence of earthquakes with $M_S \geq 5.8$ is followed by VAN predictions (with identical epicentre and magnitude) with a probability too large to be ascribed to chance." [75] VAN has presented counterarguments against Mulargia and Gasperini. [76] Both sides' views will be presented in the Debate on VAN in *Geophysical Research Letters (GRL)*. Several other strongly critical reports on VAN had been published as of 1994, three of which are cited here. [77–79] Replies to these criticisms were published by VAN. [73,80,81]

4.2   London meeting

Due to a variety of circumstances more VAN supporters than critics attended the London meeting. However, with two important exceptions, all important points of view were represented in London. First, no seismologists from Greece attended the London meeting. This was unfortunate, as one of the key issues is whether some of VAN's "predictions" were issued after medium size earthquakes or microearthquake swarms had already occurred. Another key issue is the actual value of earthquake source parameters such as hypocenter, magnitude, and origin time. Finally, the question of exactly what statements were made by VAN before and after particular Greek earthquakes is also important. Seismologists from Greece could have contributed greatly to the discussion of these questions had they been present in London. Fortunately, a group of seismologists from Greece has been invited to submit a paper to these Proceedings. Nevertheless, the views of the Greek seismologists are underrepresented in this volume, and completely absent in the discussion following each of the talks.

Second, no officials of the Greek government attended the London meeting. VAN is presented by its proponents as an operational method for earthquake prediction which can reduce earthquake damage and casualties. The region in which VAN claim to be "successfully predicting" earthquakes is Greece. It therefore would have been highly desirable for officials of OASP (the Greek government's Office of Anti-Seismic Planning) to have attended the London meeting. As discussed below in Section 8.5.3, OASP does not necessarily regard VAN's "predictions" as being useful for practical purposes.

5   Costs and Benefits of Operational Earthquake Prediction

Let us consider, using Japan as an example, the costs and benefits of earthquake prediction. The property damage due to the Kobe earthquake of Jan. 17, 1995 has been estimated at 10^{13} Yen. Even if a prediction had been issued, it would not have stopped any structures from collapsing. Perhaps a prediction would have allowed fire fighters and other emergency forces to have been put on alert, and other measures to be taken to prepare for disaster relief and reconstruction. A rough guess is that a successful prediction might have reduced the property damage by 20%, or about 2×10^{12} Yen.

The cost of issuing a prediction depends on the measures to be enforced. The Japanese government is empowered to halt virtually all activity if an earthquake is predicted in the Tokai region, but does not have such powers for other regions. The costs (in terms of lost income, etc.) of a mandatory cessation of activity in the Tokai district in the event of a prediction have been estimated by think-tank studies to be about 5×10^{11} Yen per day. Let's hypothetically say that the cost of a similar cessation of activity in the Kinki area (which includes Kobe) might be 2×10^{11} Yen per day. Thus even if an earthquake of the devastating extent of the Kobe earthquake could have been predicted with 100% certainty, implementation of such a prediction would have broken even only with a time window of ten days or less.

On the downside, announcing an earthquake prediction could cause panic. Roads could become clogged as residents of the prediction area tried to flee, hoarding might denude stores of goods, and looting or other types of criminal behavior might break out. Also, in the event of an unsuccessful prediction there might be widespread demands for compensation of economic losses by outraged citizens. (Japanese law exempts the government from responsibility for losses due to an unsuccessful prediction in the Tokai district, but not elsewhere.) There might also be demands for the resignation of the politicians and government officials responsible for issuing the prediction.

The relative cost of shutting down the economy may be much lower in rural areas of less highly developed countries. Thus the criteria that have to be met in order for an operational earthquake prediction system to be worth developing may vary greatly from one country to another; each country's governmental authorities should make the final decision on what criteria are appropriate. However, at least for industrialized areas, earthquake prediction is unlikely to be useful unless the time window of the predictions is on the order of at most a few days, and the reliability is at least 50%.

5.1 *Possible reductions (or increases) in deaths and injuries*

Prediction proponents frequently resort to emotional arguments about the potential for saving human lives. This is a complicated subject that cannot be reduced to purely economic terms, even though after a fatal auto accident or plane crash insurance companies compensate the heirs of the deceased victims by making cash payments. (In the case of Japan such payments usually are in the range from 3×10^7 to 10^8 Yen.)

It is easy to imagine cases where a scientifically successful prediction could increase, rather than reduce, the number of deaths. Many km of the expressways collapsed or overturned during the Kobe earthquake. As the earthquake occurred at 5:46AM (local time), there was almost no traffic on the expressways, so casualties were light. But suppose that a 100% reliable prediction had been issued 6 hr before the Kobe earthquake. Due to large scale attempts to flee the area the expressways would presumably have been packed with cars. Massive fires could easily have broken out among overturned or piled up vehicles at the time of the earthquake. Such a conflagration might have caused a much larger number of deaths than the 6,000 that actually occurred.

An unreliable prediction capability can increase the number of deaths and casualties by lulling people into a sense of false security. Repeated publicity in Japan about the government's program for predicting the "Tokai earthquake" lulled people in the Kobe area into thinking they were not at risk. Many commonsense precautions, such as tough building codes, stringent safety inspections for critical structures, disaster mobilization planning by local authorities, provision of backup water supplies for firefighting, etc., were neglected. Also, most of Japan's seismic networks are operated as part of earthquake prediction research programs rather than for purposes of disaster mitigation. Thus appropriate measures for informing government authorities of the earthquake source parameters and their implications were not in effect for the Jan. 17, 1995 Kobe earthquake. [82] This caused a delay of six hours before the national government even began to mobilize disaster relief forces, which in turn led to the deaths of many people trapped in collapsed structures who might have been saved by a more rapid response.

The situation may be similar in other countries. It is estimated that 50% of the multiple-story buildings in Almaty, the capital of Kazakhstan, would not withstand a major earthquake. But rather than taking measures to improve building safety, the chairman of the Kazakh State Committee on Emergency Situations instead is relying on "experts, to predict the time of a quake to enable people to protect themselves." [83] If these "experts" are making irresponsible and unfounded claims for their ability to predict earthquakes, far from saving lives they instead are placing the population at greater risk.

In summary, claims, including VAN's, that earthquake prediction can save lives should not be made unless a well-documented, tested, and reliable prediction capability actually exists. Furthermore, even if such a capability did exist, it could easily prove counterproductive unless its use had been carefully considered by a wide range of leaders from all sectors of society.

6   Are VAN's statements earthquake predictions?

Before we can evaluate VAN's claims (Section 1) to be "70% successful" in "predicting earthquakes," we must confirm the definitions of "prediction," "earthquake prediction," and "successful prediction." Much of the confusion and controversy regarding VAN is due to a lack of clarity regarding these terms.

6.1   "Prediction"

I will define a *prediction* as a statement about future events that is sufficiently precise so that its correctness or falseness can be objectively and unambiguously determined at some particular future time. Statements that are too vague, or whose truth or falseness cannot be determined for some other reason, are not predictions.

Testable propositions regarding future sporting events are classic examples of predictions. "Brazil will win the World Cup (of soccer) in 1998" is a prediction. On the other hand "Brazil will do well in the 1998 World Cup" is not a prediction, as "do well" is too vague for truth or falseness to be unequivocally established. Similarly, "Brazil will win the World Cup" is not a prediction, as there is no time cutoff.

The truth or falseness of a betting proposition may involve extremely fine differences. For example, minute examination of photographs may be required to determine the winner of a horse race. However, the result is always unambiguous: one horse is announced as the winner. The holder of a betting ticket on the loser will receive scant sympathy, and no money, no matter how loudly he argues with his bookmaker or with a parimutuel ("tote" in the U.K.) clerk that his bet (prediction) was "almost right."

6.2   "Earthquake prediction"

Statements about future seismicity must be as unambiguous as bets on horse races to qualify as "earthquake predictions." The time, space, and magnitude windows must be unambiguously stated when a prediction is issued. A prediction can be considered successful only if the epicenter, origin time, and magnitude all fall strictly within the announced windows.

In horse racing, a loss is a loss, whether by a nose or by 10 lengths. Similarly, even if only one of the three actual earthquake parameters (origin time, epicenter, and magnitude) lies only slightly outside the window, the prediction must be considered a failure. Such rigor is absolutely essential. Once fudging is allowed the tolerance limits will gradually grow until claims for "success" of the predictions become meaningless.

It is essential to look at each of VAN's allegedly successful "predictions" one at a time to see exactly what was stated in advance, and what the parameters of the actual earthquakes were. Such an examination reveals five problems. First, VAN's "predictions" never explicitly state the windows for any of the parameters of the earthquakes. VAN sometimes give the uncertainty ranges of their "predictions" in separate publications, but these are sometimes vague and also have changed greatly with time. Second, VAN sometimes fudge their results, claiming success when one or more of the earthquake source parameters fall outside the nominal windows. Third, in discussing the "success" or "failure" of their "predictions," VAN use earthquake source parameters taken from the "Ten Days Preliminary Seismogram Readings at Athens," rather than the more accurate data from the authoritative final bulletin of the Athen observatory. Also, VAN have on occasion cited parameters published by other organizations when the Athens values are unfavorable. Fourth, VAN's statements are frequently worded in such a way that the "predicted" earthquake parameters themselves aren't stated quantitatively. Fifth, some of the events claimed by VAN as having been "successfully predicted" are aftershocks or events in an ongoing swarm, which could have been "predicted" on the basis of seismicity data alone.

In the discussion following my talk at the London meeting Prof. Varotsos said (my transcription from tape) "Even if I have a delay of a few days or a week [relative to the end of the time window for VAN's predictions] I wouldn't call it [the prediction] a failure." This lax attitude to distinguishing between success and failure makes VAN's claims of "successful predictions" almost meaningless. Furthermore, it makes it impossible to conduct rigorous statistical tests of the significance of VAN's "predictions."

If the spatial window for a prediction is, say, a circle with a radius of 100 km, there is no physical difference whether an earthquake is 101 km or 99 km from the predicted epicenter. The uncertainty of the hypocenter is probably at least 5-10 km, so calling the former a failure and the latter a success seems arbitrary.

However, both public discussion of VAN and debate about VAN within the scientific community is being carried out on terms dictated by VAN and their supporters. VAN have attracted widespread attention by claiming to have made large numbers of "successful predictions." As this is the criterion VAN have chosen for their claims, we seem to be stuck with it. But we must adhere to an ironclad rule of calling any earthquake that falls even an iota outside the prediction window a failure if the terms "success" and "failure" are to have any meaning at all.

6.3 Definition of Earthquake Prediction

Over the years many seismologists have attempted to define what distinguishes an earthquake prediction from a vague general statement about future seismicity. Macelwane's [24] views in 1946 were quoted above in Section 2.1. In 1935 H. Wood and B. Gutenberg said:

> "...the prediction of an earthquake must indicate accurately, *within narrow limits*, the region or district where and the time when it will occur—and, unless otherwise specified, it must refer to a shock of important size and strength, since small shocks are very frequent in all seismic regions." [84] (Italics follow original ms.)

In his 1976 Presidential Address to the SSA, C. R. Allen said that an earthquake prediction should have the following six attributes. (1) It must specify a time window. (2) It must specify a space window. (3) It must specify a magnitude window. (4) It must give some sort of indication of the author's confidence in the reliability of the prediction. (5) It must give some sort of indication of the chances of the earthquake occurring anyway, as a random event. (6) It must be written, and presented in some accessible form so that data on failures are as easily obtained as data on successes. [85]

The Panel on Earthquake Prediction of the Committee on Seismology of the U.S. National Research Council (NRC) proposed the following definition in 1976:

> "An earthquake prediction must specify the expected magnitude range, the geographical area within which it will occur, and the time interval within which it will happen with sufficient precision so that the ultimate success or failure of the prediction can readily be judged." [86]

Guidelines promulgated by the SSA in 1983 state:

> "An 'earthquake prediction' is taken to mean the specification of the time, place, magnitude, and probability of occurrence of an anticipated earthquake with sufficient precision that actions to minimize loss of life and reduce damage to property are possible (*Earthquake Prediction Evaluation Guidelines*, California Office of Emergency Services, 1977, p.1)." [87]

An international working group convened by UNESCO and IASPEI recommended in 1984 that predictions be expressed in probabilistic terms:

> "...it is recommended that predictions should be formulated in terms of probability, i.e. the expectation, in the space–time–magnitude domain, of the occurrence of an earthquake. The strength of the predictive statement is then represented by the increase in the expectation, compared with that prevailing before the prediction was made." [88]

6.4   VAN's "predictions"

VAN's "predictions" fail to meet the above criteria. In contrast to the criteria of Wood and Gutenberg, many of VAN's recent "predictions" have involved small shocks (around M=5) of the type that occur frequently and cause little or no damage or loss of life. Some of VAN's earlier "predictions" (e.g., Varotsos and Alexopoulos,[2] p.74) involved events as small as M=3.2. Rather than the few hours of uncertainty called for by Macelwane, many of VAN's "predictions" involve time windows of a month or more. Allen and the NRC panel said that the time, space, and magnitude windows must be clearly specified, but VAN's "predictions" fail to do this.

As noted above, VAN's statements about future seismicity appear too vague to contribute to mitigating the damage due to future earthquakes, so they don't qualify as predictions under the NRC definition or SSA Guidelines. And, in contrast to the recommendations of Allen and the UNESCO–IASPEI working group, VAN's "predictions" neither state the probability of the earthquake's happening anyway by random chance, nor do they give any indication of the increase in the probability of earthquake occurrence as compared to the state before the prediction was issued.

6.4.1   Time

VAN's telegrams or faxes never specify the time, space, or magnitude windows of their "predictions." This makes their "predictions" essentially useless from a practical point of view.

To see why this is so, suppose you, the reader, were a government official responsible for public safety in some particular area, say, Southern California. A scientist says that an earthquake will occur in Los Angeles. What would you do? First, you would presumably determine the grounds on which the scientist based the statement and obtain information on his general reputation and past behavior and track record. If you decided that his statements needed to be taken seriously you would presumably take action.

You might place the police and fire departments on full alert, close schools and government offices, ask industry to shut down, close roads to non-essential traffic, etc. However, all of the above actions can be continued for at most a few days, perhaps only one day or less in some cases. Whether or not you decided to take these actions would therefore depend critically on the time window specified by the predictor. If the time window was a month or longer you probably wouldn't take any of these actions, even if you had full confidence in the predictor. As VAN's "predictions" apparently cover time periods of one or two months it is hard to see how they could be greatly useful in reducing earthquake damage even if they were 100% reliable.

Your actions would also presumably depend greatly on whether or not the prediction clearly specified an expiration date. If no expiration date was specified you would be in a terrible position, as there would be no clear time at which the prediction could be ruled a failure and the alert called off. This is the situation facing any official who proposes to use VAN's "predictions," which do not specify the expiration of the time window, in operational earthquake prediction.

6.4.2 Epicenter

The spatial window of VAN's "predictions" has a radius of 100 km or more; this too is not clearly specified in VAN's statements. How useful are "predictions" with such a large spatial window? Let's consider the Jan. 17, 1995 Kobe earthquake (Mw=6.9) as an example. Damage in the city of Kobe was severe, but damage in Osaka, which is only 30 km from Kobe, was relatively light. Damage from smaller earthquakes, such as the M5 or M6 earthquakes commonly "predicted" by VAN, can be expected to be concentrated in an even smaller area with a radius of only a few km around the epicenter. VAN's radius of 100km seems unacceptably large if their "predictions" are being claimed as having a practical benefit. Furthermore, as discussed below, in several cases VAN have claimed to have made "successful predictions" even when the distance between the "predicted" epicenter and the actual epicenter exceeded 100km.

6.4.3 Magnitude

VAN's "predictions" are generally considered to have a magnitude range of ± 0.7, although this is not explicitly stated in their announcements. Thus if VAN "predict" an earthquake of $M = 5.0$, any earthquake in the range $5.7 \geq M \geq 4.3$ is regarded by VAN as a "success." This wide magnitude window makes it extremely difficult, if not impossible, for government authorities to use VAN's "predictions" in operational disaster mitigation efforts.

Suppose the VAN "method" were used to "predict" an M6.0 earthquake. If you, the reader, were a government official who had to decide whether to act on the basis of this "prediction," you would be in a terrible quandary. Not all of VAN's "predictions," even by their own reckoning, are "successful." Both an M6.7 event, i.e. an event with the energy release of the 1994 Northridge, California, earthquake, or an M5.3 event that would cause little or no damage would both be cited as "successes" by the "predictors." If you took action, there would be a high likelihood that no damaging earthquake would occur, and you would be blamed for disrupting society. If, as would probably be the case, you took no action but a damaging earthquake did occur, the "predictors" could accuse you of negligence. This might seem like a far-fetched scenario but, as discussed below in Section 8.5.3, VAN apparently filed such a complaint after the Greek earthquake of June 15, 1995.

The Gutenberg-Richter (G-R) relation, an empirical relation between the number of earthquakes N of a given magnitude M within a fixed time period and area, is

$$\log N = a - bM, \tag{1}$$

where b is usually empirically found to be about 1.0. This means, all other things being equal, that there will be ten times $(10^{1.0})$ as many earthquakes of magnitude $M = 4.0$ as of magnitude $M = 5.0$.

VAN or their supporters sometimes point out that about half their "successful predictions" have magnitudes greater than their "predicted" magnitude. They argue that this means their "predictions" are meaningful, not merely randomly successful, but it appears to me that the opposite is the case. The G-R relation tells us that any given earthquake is more likely to be in the lower half of the magnitude window. But as as the number of earthquakes in the magnitude window increases, the expectation value of the magnitude of the largest of the "successfully predicted" earthquakes increases. Thus if over half the "successful predictions" have $0 \leq M_{eq} - M_{pred} \leq 0.7$, far from proving that the "predictions" are meaningful, this tends to prove that the "predictions" had a high chance of randomly being correct, as the number of earthquakes falling within the window was presumably significantly greater than one.

6.5   *Hits, misses and false alarms*

A prediction is successful ("a hit") if an earthquake occurs within the predicted time, space, and magnitude windows, and is unsuccessful ("a false alarm") if no earthquake occurs within the predicted time, space, and magnitude windows. However, there is also a second category of failure, "misses." A "miss" (sometimes called a "surprise") means that an earthquake occurred, but no prediction was issued. The question of how misses should be dealt with when evaluating predictions is non-trivial, and various participants in the *GRL* Debate have differing opinions.

As discussed above, if VAN announce that a magnitude 5.0 earthquake will occur, and a magnitude 4.3 earthquake occurs, VAN will claim this as a "successful prediction." On the other hand, if VAN can claim an M4.3 earthquake as a "successful prediction," does it not follow that if an M4.3 earthquake occurs, but no "prediction" was issued by VAN, this should count as a failure ("a miss")? This seems to be a self-evident proposition, but VAN and their supporters do not recognize its validity.

By neglecting misses VAN and their supporters are able to claim fantastic "success rates" of 60%–70% for their "predictions." But this is achieved largely by (i) counting "hits" as successes, even though many of the hits involve small earthquakes of $M \leq 5$; (ii) fudging their statistics, so that some earthquakes that fall outside the (retrospectively chosen) prediction window are counted as successes, rather than failures; and (iii) not counting as failures earthquakes of $M \leq 5$ that were not "predicted" ("misses"), even though they would have been claimed as "hits" had a "prediction" been in effect. Even if VAN and their supporters do not believe that misses should be counted as failures, they should announce statistics separately for hits, false alarms, and misses so that each scientist can make his own evaluation.

6.6 Outperforming random chance

Even a groundless random prediction method, such as throwing darts at a map, a calendar, and a magnitude chart, will sometimes be successful. As the randomly chosen "predicted magnitudes" become smaller, the proportion of "successful predictions" will greatly increase. The mere fact of making "successful predictions" is thus insufficient to warrant recognition of a proposed prediction method; the method must outperform random chance.

Although there seems to be general agreement regarding this point, there are some subtle issues involved in deciding what constitutes "random chance." Going back to the example of throwing darts at a map, we can dramatically improve our success rate by throwing darts at a seismicity map, and rejecting "predictions" that fall into relatively aseismic areas. Also, it is well known that large shallow earthquakes are invariably followed by aftershocks. One easy way to make "successful predictions" would be to wait for a large event to occur and then issue "predictions" for smaller earthquakes in the same area. Such "predictions" would be meaningless, as the probability of aftershocks occurring by random chance is close to 100%.

The above paragraph only scratches the surface of the issues involved in determining whether or not VAN's "predictions" have significantly outperformed "random chance." These issues are discussed in detail in the *GRL* "Debate," so I refrain from further comment.

7   Parameters of VAN's "predictions"

7.1   VAN's time varying time window

Since 1981 VAN have been issuing statements regarding future seismic activity which are allegedly based on VAN's observations of electrical signals. VAN's claims, especially those regarding the time window of their "predictions," have varied greatly. The duration of the time window claimed by VAN has steadily lengthened since 1981. The abstract of a 1981 paper by Varotsos *et al.* stated:

> "In a former paper (Practika of Academy of Athens **56**, 277, 1981) precursor changes of the telluric currents were described; they occurred a few minutes before each earthquake. Subsequently electric signals of another type have also been found; they occur about seven hours before the impending earthquake." [1]

However, the following statement regarding precursors of the above type was made by VAN on p.22 of a paper published in 1993:

> "During the aftershock period of the 6.5 event that occurred on February 24th, 1981, close to station KOR a 3 month monitoring was carried out using a Carry 401 vibrating rod electrometer. ... Hundreds of cases were observed in which, 1–4 min before the initiation of the seismic recording, an 'electric pulse' was observed (Varotsos *et al.*, 1981). A preliminary study showed that this 'pulse' was up to several milliseconds in length, with an amplitude of a few Volts (for $L = 50$m). The study of this effect was not further continued because later on we had to use filters, which removed all frequencies higher than a few Hertz in order to achieve the low level of electric noise appropriate for SES detection. ... Due to the restricted amount of data, no attempt was made to examine whether the parameters of this short duration signal are correlated with the EQ magnitude and/or the epicentral distance." [89]

If signals that occur reliably a few minutes before earthquakes but not at other times exist, they are a far more important target than SES signals, which (see below), only have a claimed accuracy of 11 days to on the order of one to two months. Thus, if the 5 minute warning signals are real, VAN made a grave mistake in filtering them out. However, the above statement may just be a rationalization for the fact that the 5 min precursors proved not to be reproducible. Such non-reproducibility of apparently promising precursors is a ubiquitous feature of empirical earthquake prediction research.

The abstract of a 1984 paper by Varotsos and Alexopoulos stated:

> "These precursor seismic electric signals (SES) occur 6-115 h before the earthquake (EQ) and have a duration of 1 min to $1\frac{1}{2}$ h. The duration and lead-time in contrast to other precursors, do not depend on EQ-magnitude(M)." [2]

On p.88 of the same paper a more detailed discussion regarding the lead time, Δt, was given:

"The lead times can be classified mainly into two groups: *group I* with values from 6 to $13\frac{1}{2}$h with a strong maximum around 7h. The lead times of about 60% of all events fall into this group. *group II* with lead times between 43 and 60h with a flat maximum between 45 and 54h. Around 25% of all Δt-observations fall into this second group. Lastly there are two intermediate groups with lead times of between 24-36h and 60-115h which are however rarely observed.[*] ...

"In spite of the fact that the Δt values vary roughly by one order of magnitude (6-115h) there is also no correlation between Δt and the corresponding earthquake magnitude." [2] (Italics follow original ms.)

[*]In a few percent of isolated cases there is some evidence that Δt can reach values up to one week.

Meyer, Varotsos, Alexopoulos, and Nomicos made the following claim on p.156 of a paper published in 1985:

"Since the time window of occurrence of the EQ is known, i.e. within the time limits of 6 hrs and 1 week after the observation of the SES, the event time is always predetermined so that it is sufficient to give the arrival times of SES in the telegram." [4]

Varotsos and Lazaridou (p.322) said, in a paper published in 1991, that there were several different types of anomalies, with different time windows:

"For *isolated events* (i.e. when a single SES and a single earthquake allow a one-to-one correlation), the time lag Δt lies between 7 hours and 11 days. No correlation between Δt and M has been observed.

"For cases of prolonged *electrical activity* (i.e. when a number of SES, detected within a time period comparable with the time lag Δt, is followed by a number of earthquakes) it has occasionally been observed that, although the time lag between the onsets of the electrical and seismic activity does not usually exceed 11 days, the time lag between the largest SES and the strongest earthquake may, however, be much longer, e.g., around 22 days." [72] (Italics follow original ms.)

These authors (p.340) introduced another type of electrical anomaly:

"... a 'gradual variation of the electric field' (GVEF) [is] ... [an] unusual type of variation [that] sometimes precedes the detection of an SES at the same station. It always has the *same polarity* and is detected by the *same dipole array(s)* as the subsequent SES (Varotsos and Alexopoulos, 1986). As mentioned in the same publication, a GVEF usually appears a few weeks before the occurrence of a strong earthquake ($M_S \geq 5.5$) and it has an amplitude one order of magnitude larger than that of the corresponding SES." [72] (Italics follow original ms.)

Dologlou (p.196) also said, in a paper published in 1993, that there were several different types of electrical anomalies. However, the parameters of the windows differed somewhat from those given above by Varotsos and Dologlou:

"(1) *In the case of a single SES:* within the usual expected time window of 11d and within a circle of 100km around the predicted epicentre (the current experimental accuracy of the epicentral determination) we investigated whether an earthquake with magnitude M_S compatible with the predicted one had occurred (the accuracy in the prediction of the M_S is ±0.7 units). In a few telegrams (d, e, and h) this time window was violated by only a few days.

"(2) *In the case of electrical activity:* the initiation of SES activity and the initiation of seismic activity usually does not exceed the time lag of 11d. However, the time lag between the largest SES and the strongest earthquake might be of the order of 1 month (Fig. 9 and appendix I of Varotsos and Lazaridou, 1990)." [90] (Italics follow original ms.)

Dologlou (p.194-195) also discussed the so-called GVEF anomaly:

" ... a 'gradual variational in the electric field' (GVEF) [is an] unusual type of variation [that] sometimes precedes the detection of an SES at the same station and usually appears a few weeks before the occurrence of significant seismic activity (Varotsos and Alexopoulos, 1986)." [90]

As demonstrated by the quotations in Section 1, VAN's claims to be making "successful predictions" have scarcely changed since the early 1980's. On the other hand, the accuracy claimed by VAN for their "predictions" has been steadily worsening. They claimed in 1981 to have detected signals that occurred "a few minutes before each earthquake." [1] They next claimed in 1981 to have detected signals that occurred "seven hours before the impending earthquake." By 1984 VAN's window had increased significantly, to "6 to 115hr" (about $\frac{1}{4}$ to 5 days). [2] In 1985 VAN's window had widened to "within the time limits of 6 hrs and 1 week." [4]

By 1990 (the submission date of Dologlou's paper) the accuracy claimed for VAN's "predictions" had further degraded to 11 days. [90] Furthermore, the time window of 11 days was violated for three of the "successful predictions" in Dologlou's [90] Table 2–telegrams d, e, and f–for which the actual time delays were 17, 16, and 17 days respectively. Also, an entirely new category, "electrical activity" was introduced. This category was not mentioned in VAN's 1981 or 1984 papers. In the event of "electrical activity" a time lag of "around 22 days" (Varotsos and Lazaridou [72]) or "on the order of one month" (Dologlou [90]) might be allowed. It is not clear whether this time lag should be measured from the onset of the activity or its termination.

Prof. Varotsos stated at the London meeting, as quoted above in Section 6.2, that he wouldn't consider it a failure if the delay between a prediction and an earthquake was a few days more than the nominal duration of VAN's window. Several newspaper or magazine articles based on information supplied by VAN, cited below, claim that cases with a time delay of 40-50 days between the "prediction" and the earthquake were "successful predictions." The criteria for the window for "GVEF" anomalies is even vaguer ("usually appears a few weeks before the occurrence of significant seismic activity." [90]) The steady degradation of the accuracy claimed by VAN should be viewed as a highly negative factor when evaluating VAN's claims.

7.2 VAN's magnitudes

In the abstract of their 1984 paper Varotsos and Alexopoulos stated that:

"The uncertainty of M is around 0.5 units." [3]

Dologlou (p.196) gave the uncertainty of the magnitude for a single SES:

"The accuracy in the prediction of the M_S is ±0.7 units." [90]

A simple calculation based on the G-R relation (eq. 1) using a b-value of 1.0 shows that the above increase in the width of the magnitude window from ±0.5 in 1984 to ±0.7 in 1990 increases the number of earthquakes that would randomly fall into the window by a factor of about 1.5 ($10^{0.2}$).

There has been much confusion over the magnitudes used by VAN in their predictions. Varotsos and Alexopoulos (p.74) stated in 1984 that:

"Magnitudes throughout this paper equal to M_S taken from the officially certified edition of the preliminary seismological bulletin of the National Observatory of Athens; if M_S is not given, we estimate it from $M_S = M_L + 0.4$." [2]

Dologlou's 1993 paper (p.196) states:

"An attempt has been made to correlate the predictions with earthquakes taken from the Preliminary Bulletin of Observatory of Athens (PBOA).... The magnitude, M_S, is obtained from the formula $M_S = M_L + 0.5$ where M_L denotes the local magnitude." [90]

Based on the G-R relation for a b-value of 1.0, the increase of 0.1 (from 0.4 in 1984 to 0.5 in 1990) in VAN's additive constant for converting M_L to M_S from 1984 to 1990 would increase the number of eligible earthquakes by a factor of about 1.2 ($10^{0.1}$) for those cases where VAN estimated the value of M_S from M_L.

Although VAN use the term "Preliminary Bulletin," they are referring to reports whose correct title is "Ten Days Preliminary Seismogram Readings at Athens."[a] These reports are issued as a source of raw data for other seismological organizations on the basis of data available to the Athens observatory by telemetry. As many of the stations operated by the Athens observatory are not telemetered, the source parameters in the final Bulletin of the Athens observatory, which are based on all of the available data, are more accurate than the interim values given by "Ten Days Preliminary Seismogram Readings at Athens." Recently the differences between the source parameters given by "Ten Days Readings" and those given by the final bulletins have tended to decrease, as a larger proportion of the network operated by the Athens observatory is telemetered. (Dr. G. Stavrakakis, personal communication.) However, it seems indefensible for VAN to use the "Ten Days Preliminary Seismogram Readings" rather than the final bulletin to determine the "success" or "failure" of their "predictions."

At one time the Athens Observatory apparently followed the practice of announcing a magnitude value called "M_S" which was obtained by adding 0.5 to the local magnitude M_L. We henceforth denote the "Athens M_S" as "M_{SA}." VAN's publications appear to have mainly used the values M_{SA} as given by the "Ten Days Preliminary Seismogram Readings at Athens" as the magnitude values for determining the "success" or "failure" of their "predictions," although this is not stated in VAN's telegrams themselves.

Wyss[91] obtained, by least squares, an empirical relation between the global surface wave magnitudes M_S and the Athens local magnitudes M_L:

$$M_S = 1.5M_L - 2.62 \tag{2}$$

If we use $M_{SA} = M_L + 0.5$, we obtain the following empirical relation between M_{SA} and the standard global M_S:

$$M_S = 1.5M_{SA} - 3.37. \tag{3}$$

Eq. (3) shows that an earthquake with $M_{SA} = 5.0$ will have an average global M_S of 4.13, and an earthquake with $M_{SA} = 6.0$ will have an average global M_S of 5.63. In general an M_S value of 6.0 on the global scale can be considered the lower limit for an earthquake that can cause significant damage. $M_{SA} = 6.25$ is required to reach $M_S = 6.0$.

---

[a] See the discussion of "preliminary readings" on p.324-331 of Richter's textbook (reference no. 25).

7.2.1 Physical implications of $\Delta M = \pm 0.7$

Roughly speaking, the energy released by an earthquake is proportional to $10^{1.5M_W}$, where M_W is the moment magnitude.[58] For the size of events we are considering here ($M_{SA} < 6.5$), the surface wave magnitude M_S is not saturated (i.e., $M_S \approx M_W$). Using eq. (3), the earthquake energy is thus roughly proportional to to $10^{2.25M_{SA}}$.

This means that if M_{SA} varies by ± 0.7, the earthquake energy will increase or decrease by a factor of about 30 ($10^{2.25 \cdot 0.7} \approx 30$). If VAN's so-called SES signals are precursors which are physically related to earthquakes, one would expect their energy to be proportional to the energy of the impending earthquake. Physically it seems hard to understand why an SES of some given energy could be a precursor of earthquakes with energy ranging from, say, 0.03 to 30, which is a dynamic range of about 1000.

7.3 Spatial windows

It is unclear what Δr should be used to decide whether or not VAN's "predictions" are "successful." In 1984 Varotsos and Alexopoulos (p.120) said:

> "By evaluating the 18 predictions given in Table 4 we see that in only one case (June 14, 1983) the inequality $|\Delta M| \le 0.6$ was violated; furthermore the deviations Δr between the predicted and the real epicenters were at most 90km except the one case of April 25, 1983 [which had $\Delta r = 130$ km]."[3]

Dologlou (p.196) stated, as quoted above, that for a single SES the epicenter was expected to be within 100 km of the predicted epicenter, but:

> "In cases where no earthquake of compatible magnitude to the predicted one occurred within the aforementioned circle, we extend the radius by a few tens of kilometres (telegrams c, d, 7, 13, 15, 18 and 21)."[90]

The radius for "predictions" based on "electrical activity" or "GVEF" anomalies was not clearly stated by Dologlou. Furthermore, if the radius for some of the "predictions" can be arbitrarily "extended," but this fact is not clearly stated in the telegram or fax announcing the "prediction," how can we decide whether or not a given "prediction" was successful? Dologlou's[90] Table 2 lists the actual distance between the "predicted" and actual epicenters for telegrams c, d, 7, 13, 15, 18 and 21 as 150, (150, 160), >120, 120, (170, 160), and 150 km respectively. The parentheses indicate the two events listed by Dologlou as corresponding to telegrams d and 18 respectively.

7.4   VAN's telegrams and faxes

In 1988 Drakopoulos *et al.* (p.57) criticized VAN as follows:

> "Why do the authors (VAN group) systematically refuse to send their tele-
> grams either to the Government, or to the Seismological Institutes, before the
> occurrence of a particular earthquake? They mainly just exchange telegrams
> between themselves (e.g., Mrs. Varotsos-Lazaridou to Prof. Alexopoulos)." [78]

VAN's reply (p.60) implicitly conceded that some of the telegrams were
not publicly disclosed:

> "The statement by [Drakopoulos *et al.*] 'at this stage, we also want to draw
> the attention of the scientific community ... that the authors (VAN group)
> systematically refuse to send their telegrams ... but *mainly* they just exchange
> telegrams' gives the false impression to the reader that our group does not
> issue official predictions: ..." [73] (Italics follows original ms.)

As some of VAN's telegrams (e.g., some of the telegrams listed in Dologlou's[90]
Table 1) were exchanged among members of the VAN group, it is impossible
to verify independently whether or not a complete list of VAN's "predictions"
has been published.

At some point after 1990 VAN switched from telegrams to faxes as the
means of announcing their "predictions." These faxes are sent to many researchers around the world. However, VAN's faxes (some of which are introduced below in Sections 8.4 and 8.5) are unfortunately not numbered, so there
is no way to determine whether or not one has a complete set. Furthermore,
despite repeated criticism regarding this point, VAN's statements about future seismic activity still do not clearly indicate the windows for the epicenter,
origin time, and magnitude.

8   Case studies of VAN's "predictions"

Some of VAN's "predictions" are re-analyzed below. In the "neutral" analysis
VAN's claims to have sent particular telegrams, etc., are accepted, but the time,
space, and magnitude windows of the "predictions" are rigorously enforced. A
maximum of only one "success" is allowed per "prediction." Any event clearly
falling outside the space, time or magnitude windows by even an iota is scored
as a failure. One telegram that failed to state the magnitude of the "predicted"
earthquake is scored as a failure. All events falling within the nominal windows
are automatically scored as "successes."

The "critical" analysis is designed to provide the rigorous challenge to
VAN's claims demanded by Dyson's above remarks. The critical analysis scores

ambiguous, questionable or insufficiently documented cases as failures. Also, aftershocks or events that occurred as part of a swarm of earthquake activity are considered ineligible for having been "predicted," because such "predictions" could have been made solely on the basis of seismological data.

The results of the various analyses presented below differ greatly. Dologlou's [90] Table 2 listed 37 earthquakes with $M_{SA} \geq 5.0$ in Greece in 1987-89, and claimed that 21 of these had a "corresponding prediction," i.e., were "successfully predicted." The neutral analysis concludes that there were only 7 "successful predictions," while the critical analysis concludes there were 0 "successful predictions."

This wide range of possible evaluations further confirms that VAN's statements are too vague to be considered "earthquake predictions." To obviate the recurrence of such arguments, Prof. Varotsos should promptly hold discussions with other researchers to agree on an unambiguous format for VAN's "predictions." VAN should also, in consultation with other researchers, introduce a scoring system based on a smooth function. It doesn't make physical sense to assign a score of 100% to, say, a case when $\Delta t = 10.99$ d, but to assign a score of 0% when $\Delta t = 11.01$ d.

8.1 Issues arising in a reexamination of VAN's claims

VAN's telegrams and faxes: Obtaining the "predicted" epicentral coordinates, magnitude, and origin time from the text of VAN's telegrams or faxes is frequently non-trivial and sometimes impossible. Also, VAN's telegrams or faxes sometimes fail to specify the time of an "SES," or the starting and ending times of "GVEF" or "SES electrical activity."

In some cases telegrams were sent by one member of the VAN group (e.g., Ms. Lazaridou) to another (e.g., Prof. Alexopoulos, who is the "A" in VAN). This makes it difficult to verify that the telegrams were sent as claimed, and impossible to verify that unsuccessful telegrams were not quietly discarded. Had all of VAN's telegrams been sent instead to government authorities or disinterested third parties, such questions would have been obviated.

VAN's explications regarding some telegrams are internally inconsistent. One notable case concerns telegram a,[b] which is discussed in detail in Section 8.2.3, below. Another questionable case involves telegram b (see Section 8.2.2, below), which Varotsos and Lazaridou [72] (p.340) and Dologlou [90] (p.194-195, p.198) claim was dispatched after a "GVEF" anomaly was observed, even though the "GVEF" anomaly is not mentioned in the telegram itself.

---

[b]Dologlou's (reference no. 90) telegram numbering is followed.

The ambiguity of VAN's "predictions" seems to be steadily worsening. Many of VAN's recent faxes state the time of the "predicted" earthquake as, for example, "according to Fig. 28A[c] of Varotsos and Lazaridou, Tectonophysics 188, 321-347, 1991." This diagram, which is essentially identical to the upper panel of Fig. 1 of this paper, shows "SES electrical activity" followed about 3 weeks later by some M5 earthquakes, followed 2-3 weeks still later by somewhat larger earthquakes. Some of VAN's recent faxes (e.g., that of April 27, 1995) say instead that "the time evolution of the current activity might follow the same time chart as that depicted in Fig. 22" of Varotsos *et al.*, [89] which is reproduced as Fig. 1 of this paper. As this figure has five different time charts from which to choose, VAN's "predictions" have become even more ambiguous. The interpretation of such diagrams is extremely subjective. As it is impossible to obtain a clear expiration date for one of VAN's "predictions" from such figures, VAN can claim "success" for almost every "prediction."

VAN's earthquake source parameters: Varotsos and Lazaridou [72] and Dologlou, [90] acknowledge that the earthquake source parameters (origin time, epicenter, and magnitude) cited by VAN are taken from "Ten Days Preliminary Seismogram Readings at Athens" rather than from the final bulletin.[d] As discussed above in Section 7.2, this is indefensible.

VAN cite data from other organizations when the Athens data are unfavorable. Varotsos *et al.* [81] (p.241) criticize Drakopoulos *et al.* [79] for publishing source parameters which "were in sharp and unusual disagreement" with the source parameters reported by Centre Sismologique Euro-Méditerranéen "in its official bulletin." It would have been far more constructive if VAN had instead relocated the events using the published arrival times (rechecking the arrival time picks if necessary), rather than arbitrarily choosing among source parameters published by various organizations.

Correlation of telegrams and earthquakes: If VAN made only a few "predictions" of large (say, $M \geq 7$) earthquakes there would be relatively little problem deciding whether or not VAN's "predictions" were "successful." However, as VAN issue large numbers of "predictions" of relatively small events, there are a variety of non-trivial problems. Let us consider as an example the period from Jan. 1987 through Oct. 1989, which was studied by Dologlou. [90]

---

[c]Strictly speaking, "Figure 28A" does not exist, but there is a Fig. 28 on p.346 of Varotsos and Lazaridou (reference no. 72). At the London meeting Prof. Varotsos explained that "Fig. 28A" refers to the upper panel of Fig. 28 of Varotsos and Lazaridou. Fig. 28 of Varotsos and Lazaridou is almost identical to the upper two panels of Fig. 22 of Varotsos *et al.* (reference no. 89) which is reproduced in Fig. 1 of this paper.

[d]Varotsos and Lazaridou (p.335 of reference no. 72) and Dologlou (p.193 of reference no. 90) respectively use the incorrect titles "Preliminary Seismological Bulletin (PSB) of SI-NOA" and "Preliminary Bulletin of Observatory of Athens."

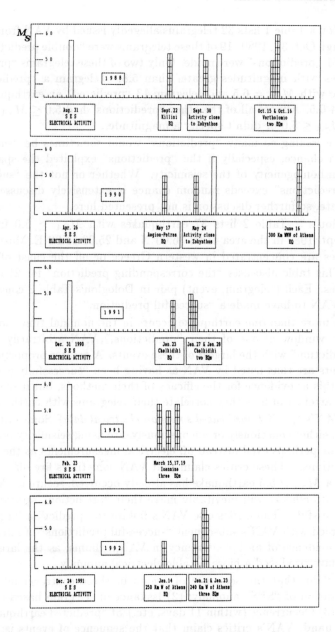

Figure 1: Time diagrams (after Fig. 22 of reference no. 89). The original caption reads "Schematic comparison of five significant SES electrical activities recorded during the last 4 yrs. Note that the first strong EQ occurs 3 weeks after the SES activity, while small magnitude events (not shown in the figure) start earlier (usually within approximately 11 days)."

Dologlou's Table 1 lists 32 telegrams allegedly issued by VAN from Jan. 1, 1987 through Oct. 31, 1989. 19 of these telegrams were "double predictions," so a total of 51 "predictions" were made. Only two of these telegrams "predicted" earthquakes with magnitudes greater than 5.8. Telegram a "predicted" an earthquake with $M_{SA} = 6.5$ and telegram 13 "predicted" an earthquake with $M_{SA} = 6.4\text{-}6.5$. Almost all of the other "predictions" had $5.0 \leq M_{SA} \leq 5.5$; a few had $M_{SA} \leq 5.0$ or didn't specify a magnitude.

Anyone issuing so many "predictions" would sometimes be "successful" by random chance, especially if the "predictions" exploited the spatial and temporal inhomogeneity of the seismicity. Whether or not the "success" of VAN's "predictions" exceeds random chance is extensively discussed in the *GRL* Debate, so further discussion is not presented here.

Dologlou's [90] Table 2 lists 37 earthquakes with $M_{SA} \geq 5.0$ from Jan. 1987 to Sept. 1989 in the area 41.00-36.00°N and 25.00-19.00°E. Most of these earthquakes are concentrated in western Greece or off the coast of western Greece. This table also lists "the corresponding prediction" for 21 of the 37 earthquakes. Each (telegram, event) pair in Dologlou's Table 2 constitutes a claim by VAN to have made a "successful prediction."

When more than one earthquake occurs in the nominal time, space, and magnitude window of one of their "predictions," VAN arbitrarily correlate their "prediction" with the largest of these events. A greater proportion of the larger events are thus listed by VAN as having been "successfully predicted." VAN cite this as evidence for the efficacy of their methods, but it seems to be simply an artefact of how they correlate their telegrams with earthquakes.

Are VAN's "predictions" based solely on electrical data? Some critics claim that VAN, either consciously or unconsciously, are using seismicity data in addition to, or perhaps even instead of, electrical measurements as the basis for its "predictions." These critics claim that VAN sometimes fire off a telegram after, say, a $M_{SA} \geq 5.5$ earthquake has already occurred, secure in the knowledge that aftershocks will inevitably make their "prediction" appear to have been "successful." The critics cite VAN's failure to "predict" a larger initial shock coupled with VAN's subsequent "successful predictions" of later smaller shocks as evidence of an inconsistency in VAN's claims, as the larger shock ought to have produced a more notable electrical anomaly.

VAN claim that the sequence of events in these controversial cases is: (1) Observation of "SES" anomaly; (2) Issuance of telegram based on "SES" anomaly; (3) Occurrence (within 11 days, etc.) of "predicted earthquake." On the other hand, VAN's critics claim that the sequence of events is: (1) Occurrence of mainshock or microearthquake swarm that was not "predicted" by VAN; (2) Issuance by VAN of a telegram purportedly based on an "SES" anomaly; (3) Occurrence (within 11 days, etc.) of aftershock or further microearthquake, claimed by VAN as a "successful prediction." Criticism regarding this point by Drakopoulos *et al.* [79] and VAN's reply [81] present both sides of the issue.

8.1.1 Case studies presented in this section

Five case studies of VAN's "predictions" are presented below. Dologlou's [90] study of VAN's telegrams in 1987-89 is reanalyzed in Section 8.2. Hamada's [92] study reported that certain of VAN's claims could be accepted with a 99.8% confidence level, but the analysis in Section 8.3 vitiates this conclusion. Three of VAN's "predictions" in 1994, as reported by a Japanese newspaper, are examined in Section 8.4 and shown to have been "failures." VAN's "predictions" for April-June 1995 are considered in Section 8.5 and shown to have been "unsuccessful." Finally VAN's claims to be particularly "successful" in "predicting" large ($M_{SA} \geq 5.8$) events are shown to be invalid in Section 8.6.

It is notable that the critical analyses reject all of VAN's claims of "successful predictions" of events with $M_{SA} \geq 5.0$. Those cases falling within the nominal time, space, and magnitude windows are shown either to have been insufficiently documented or to have been "ineligible" for "prediction," due either to their occurring during microearthquake swarms or to their having been aftershocks.

8.2 Dologlou's results

I begin by analyzing VAN's claims, as presented in Dologlou's [90] Tables 1 and 2, to have made "successful predictions" in 1987-89. Dr. Dologlou is a seismologist affiliated with the Solid State Section of the Department of Physics of the Univ. of Athens, which is also Prof. Varotsos's affiliation. Dologlou's paper therefore represents VAN's views.

Although it might seem tedious, a case by case analysis is essential to evaluate VAN's claims. The reader is advised to consult Varotsos and Lazaridou [72] and Dologlou [90] while reading this section.

Criteria for "success" or "failure": VAN call it a "success" if they "predict" an event with M5.0, and an event with M4.3 occurs within the nominal time and space windows. I don't think this should be recognized as a success from a physical point of view, as an M4.3 earthquake is incapable of causing significant damage. Also, an earthquake with $M_{SA} = 5.0$ will on average have $M_S = 4.13$ on the usual global scale (see eq. 3), so it is unlikely to cause significant damage.

The question I consider here is: Were earthquakes with $M_{SA} \geq 5.0$ "successfully predicted"? rather than: Were "predictions" with $M_{SA} \geq 5.0$ "successful"? The former is more meaningful when considering whether or not VAN's "predictions" are useful for society. In all other respects I follow Dologlou's statements, above, regarding the windows for time, space, and magnitude. I use $\Delta r \leq 100$ km and $|\Delta M| \leq 0.7$, using M_{SA} as the magnitude. I use $\Delta t = 11$ d for a single SES.

The time window for "electrical activity" is a trickier question. In both cases Δt is measured from the midpoint of the "activity." However, in the neutral analysis, I require $\Delta t \leq 40$ d, on the basis of Dologlou's [90] statement (p.196) that the time lag might be "on the order of one month." Having to interpret phrases like "on the order of one month" to decide on the "success" or "failure" of VAN's "predictions" is one of the most frustrating aspects of evaluating VAN's claims. It would be far preferable for VAN's "predictions" to specify a definite expiration date. Varotsos and Lazaridou [72] give the time window for "electrical activity" as "around 22 days." In the critical analysis I impose a strict requirement of $\Delta t \leq 22$ d for "electrical activity."

VAN's time window for a GVEF is "a few weeks." In the neutral analysis I interpret this as 60 d. VAN claim only one "GVEF" anomaly for the period of my reexamination. This is disqualified in the critical reexamination on the grounds of insufficient documentation, so it was not necessary to decide a time window for a GVEF.

8.2.1 Elimination of events falling outside windows

We eliminate events that Dologlou lists as falling outside the 11 d time window for a "single SES" (the events listed as corresponding to telegrams d, e, h; see Section 7.1) or the 100 km radius (the events listed as corresponding to telegrams c, d, 7, 13, 15, 18, and 21; see Section 7.3). The event for telegram d failed on both time and space. Each of the two events listed as corresponding to telegram 18 fell outside the 100 km radius (170 and 160 km respectively); the former could also have been eliminated due to $\Delta M = 0.9$. A total of 11 candidate events [e] and ten telegrams now remain. The remaining (telegram, event) pairs are (a, 4), (b, 5), (b, 6), (j, 16), (1, 17), (2, 18), (3, 19), (4, 20), (10, 22), (12, 24), (17, 34).

---

[e]The earthquakes in Dologlou's Table 2 are unnumbered, but for purposes of reference I have numbered them consecutively from 1 to 37.

8.2.2 Did telegram b report a "GVEF" anomaly?

Dologlou's (p.198) translation of telegram b, which was dispatched by Ms. Lazaridou to Prof. Alexopoulos on 26 Apr. 1987 reads:

> "Arrival at 04:00 on 26-4-87 from distance 50 from PIR-5.5
>
> "Mary Lazaridou"

The above text means that VAN observed an electrical anomaly at their PIR station at 04:00 and "predicted" that an earthquake with magnitude 5.5 would occur 50 km from PIR. Neither the above telegram nor the telegrams cited below specify the time, space, and magnitude windows of the "prediction."

The following "REMARK" was appended to the English language translation of telegram b on p.198 of Dologlou's paper, but does not appear in the original Greek text of the telegram:

> "A letter accompanied this telegram which stated that the anomalous electrical variations were of Sobolev type i.e. GVEF (gradual variations of electric field)."

P.195 of Dologlou's paper claims:

> "Telegram b was accompanied by a letter addressed to the Minister of Public Works explaining that this type of signal is usually followed by earthquake activity." [90]

But why wasn't the "GVEF" anomaly briefly mentioned in telegram b? If a letter was sent to the Minister of Public Works, why wasn't telegram b itself sent to the Minister? Why wasn't the letter mentioned in telegram b (e.g., "letter follows")? Why isn't this letter (or an excerpt) reproduced in Dologlou's Appendix I? And what, exactly, does "earthquake activity" mean?

In view of these unanswered questions, the critical analysis concludes that telegram b should be treated as announcing a "prediction" based on an ordinary "SES" rather than a "GVEF" anomaly. Events 5 and 6 are therefore disqualified as "successful predictions," as Δt is in excess of a month for each event. It contrast, the neutral analysis accepts without question VAN's claims for the GVEF anomaly, and scores the "prediction" of one of the event (arbitrarily chosen as event 5) as a "success." The neutral analysis disqualifies event 6, as only one "success" is allowed per "prediction."

8.2.3 Telegram a: Inconsistent documentation

The 27 Feb. 1987 ($M_{SA} = 5.9$) earthquake (Dologlou's event 4) is one of three significant main shocks during the period of Dologlou's study. The others are event 17 (18 May 1988, $M_{SA} = 5.9$) and event 24 (16 Oct. 1988, $M_{SA} = 6.0$). Furthermore, telegram a is one of only two "predictions" in Dologlou's Table 1 with $M_{SA} \geq 6.0$. [f]   Telegram a and event 4 thus warrant consideration in detail.

P.198 of Dologlou's paper states that telegram a was sent by Ms. Lazaridou to Prof. Alexopoulos. Dologlou's English translation reads:

"Arrival 20:33 on 26-2-87 from W 200 - 6.5. Probability 60%

"Mary Lazaridou"

The following REMARK is appended to Dologlou's English translation:

"The date and the time of issue appears as a stamp on the right top of the telegram (i.e 1987 II 27 10:35). The three black labels are stamped by Post Office."

An earthquake with $M_{SA} = 5.9$ (event 4 in Dologlou's Table 2) occurred on 27 Feb. 1987 at 23:34 at a distance (according to Dologlou) of 30 km from the "predicted epicenter." If Dologlou's claim that telegram a was dispatched by Ms. Lazaridou to Prof. Alexopoulos on 27 Feb. 1987 at 10:35 is accepted, telegram a should be regarded as a "successful prediction." As the neutral analysis accepts VAN's documentation without question, this is scored as a "success." However, on the basis of several inconsistencies which are noted below, the critical analysis concludes that VAN's documentation of their claims regarding this case is insufficient, and scores this case as a "failure."

Time and Date Stamp: Copies of telegrams a and b may be found on p.198 of Dologlou, [90] and on p.53 and p.59 respectively of Stavrakakis et al. [93] The latter apparently reproduced telegrams a and b as they were distributed to participants at the Athens meeting in Feb. 1990 (except for the addition of the English words "DAY" and "TIME" over "HMEPA" and "ΩPA" respectively in telegram a). The date and time stamp in the version of telegram a shown in Dologlou's paper is not visible in the copy of telegram a in the report by Stavrakakis et al. In contrast, the copy of telegram b in Dologlou's paper appears to agree exactly with that given in the report by Stavrakakis et al.

When the copy of telegram a on p.198 of Dologlou's paper is enlarged 1600% using 4 successive applications of 200% enlargement on a photocopier, what ought to be the year "1987" appears to read either "19:17" or "19 17." I have asked several people to check these characters, and no one has read

---

[f] The other was telegram 13, which was issued on 21 Oct. 1988. This was a "double prediction" with one of the two "predicted" events having $M_{SA} = 6.4$–6.5. Dologlou's (reference no. 90) Tables 1 and 2 list an event with $M_{SA} = 5.3$ as "corresponding" to the "prediction" with $M_{SA} = 6.4$–6.5, but based on Dologlou's own data this is a "failure" due to both $|\Delta M| > 0.7$ and $\Delta r > 100$ km. Hamada (reference no. 92) also scores telegram 13 as unsuccessful.

them as "1987." This appears to conflict with Dologlou's above REMARK. The reader is encouraged to replicate this experiment.

Internal inconsistency in Dologlou (1993): The date of telegram a is listed as "26/02/87" in Dologlou's Table 1. This is inconsistent with Dologlou's "REMARK" on p.198 stating that telegram a was sent at "1987 II 27 10:35." This is the only telegram for which the dispatch date in Dologlou's Table 1 does not agree with the dispatch date in Dologlou's Appendix I.

Addressee: VAN sent 13 "official telegrams" to the Greek government: 7 in 1986, none in 1987, and 6 in 1988.[79,93] Since telegram a "predicted" an event with $M_{SA} = 6.5$, why wasn't it addressed to the Greek government rather than to Prof. Alexopoulos? This would have made it much easier for VAN to have documented their claims.

Internal inconsistencies in Varotsos and Lazaridou (1991): A paper by Varotsos and Lazaridou[72] titled "Latest aspects of earthquake prediction in Greece based on seismic electric signals" (hereafter abbreviated VL91 in this subsection) was submitted for publication in *Tectonophysics* on 10 Oct. 1989. The revised version was accepted on 27 April 1990, and it was published in 1991.

VL91's only reference to the "successful prediction" of the 27 Feb. 1987 earthquake (Dologlou's event 4) is on p.340 in a footnote to Table 2, "Complete list of predictions issued from April 1, 1987 to May 15, 1987*." The footnote denoted by the asterisk reads:

> "*During the previous period of January 1, 1987 to April, 1987 *only one* prediction was issued on February 26; it announced a $M_S \approx 6.5$ earthquake with an epicenter at W300 and it was *actually* followed by a $M_S \approx 5.9$ earthquake on February 27 at W295." [72] (Italics follow original ms)

I note the following internal inconsistencies in VL91:

(1) The abstract says "A detailed list of the predictions officially issued in Greece during the past 3 years (January 1, 1987–November 30, 1989) is also given." Why is the Feb. 27, 1987 event included only as a footnote in the detailed list in Table 2 on p.340? Why isn't it instead listed in Table 2 itself? And why isn't it mentioned in the abstract or on p.340-341?

(2) P.335, middle of left column, says "Here a complete list of the predictions issued during the period Jan. 1, 1987 to Aug. 10, 1989 is presented." Why, then, is Table 2, on p.340, headed "Complete list of predictions issued from April 1, 1987 to May 15, 1988," rather than "...from January 1, 1987 to May 15, 1988" ?

(3) P.340, top left, has the heading "Predictions for the period April 1, 1987 to May 15, 1988." Why doesn't this say "...Jan. 1, 1987 to May 15, 1988" ? Ditto for the immediately following sentence, which reads "Table 2

lists in chronological order all the predictions issued during the period April 1, 1987 until May 15, 1988."

(4) The subheading on p.341 is "Comments on the results of the period April 1, 1987 to May 15, 1988." Why wasn't the period from Jan. 1, 1987 to May 15, 1988 discussed, since the largest event that was reported to have been "successfully predicted" was the M=5.9 event of Feb. 27, 1987?

(5) Fig. 23 (top of p.341) shows a time chart of "predictions" and earthquakes for the period April 1, 1987 to May 15, 1988. Why doesn't this figure start on Jan. 1, 1987? And why was the "successful prediction" of event 4 by telegram a omitted from this figure?

Discrepancy between VL91 and Dologlou (1993): As noted above, the footnote to Table 2 of VL91 (p.341) claims "*only one* prediction was issued on February 26; it announced a $M_S \approx 6.5$ earthquake ..." On the other hand, the "REMARK" on p.198 of Dologlou's paper claims: "The date and the time of issue appears as a stamp on the right top of the telegram (i.e 1987 II 27 10:35) ..." At least one of these statements must be wrong, as telegram a could not have been dispatched on both Feb. 26 and Feb. 27.

Conclusion: VAN have not presented sufficient documentation for their claim to have dispatched telegram a. The (telegram, event) pair (a, 4) is therefore scored as a "failure" by the critical analysis.

8.2.4   Telegram j: Was $\Delta t \leq 11$ d?

Dologlou's (p.199) English translation of telegram j, which was dispatched on 28 Apr. 1988 at 22:44, reads as follows:

> "Preseismic signal from W 300 - 5 or WNW 300 - 5 STOP. We clarify that it is not an aftershock.
>
> "Varotsos."

Dologlou lists event 16, which occurred on 9 May at 16:52 with $M_{SA} = 5.0$, as corresponding to telegram j. $\Delta M = 0.0$, and $\Delta r = 30$ km, but what about Δt? If the SES occurred after 16:52 on 28 Apr., then $\Delta t \leq 11$ d and the "prediction" was "successful," but if the SES occurred before 16:52 on 28 Apr., then $\Delta t > 11$ d, and the "prediction" would be a failure. As telegram j does not state the time of the SES, VAN have not clearly established that event 16 fell within the 11 d limit. The neutral analysis therefore scores this as "questionable," while the critical analysis scores this as a "failure."

8.2.5   Telegrams 2, 3, and 4: "predictions" of aftershocks

Event 17 occurred on 18 May 1988 at 38.22°N, 20.18°E at 05:17 with $M_{SA} = 5.9$. Following this event VAN issued three "predictions" for the same area.

Telegrams 2, 3, and 4 were issued on 21 May, 30 May, and 4 June respectively. (Telegrams 2 and 3 also listed a second possible epicenter, but this complication can be disregarded.) Dologlou lists the earthquakes that were "successfully predicted" by these telegrams as event 18 (22 May, 03:44, 38.36°N, 20.43°E, $M_{SA} = 5.5$), event 19 (2 June, 10:35, 38.27°N, 20.35°E, $M_{SA} = 5.0$), and event 20 (6 June, 05:57, 38.29°N, 20.36°E, $M_{SA} = 5.0$), respectively. The neutral analysis scores these as "successes." As all of these events are clearly aftershocks of the 18 May event, the critical analysis considers them to be ineligible for having been "successfully predicted."

8.2.6 Telegram 12: magnitude not specified

Telegram 12 was dispatched on 3 Oct. 1988 to B. Massinon *et al.* It was followed by event 24 on 16 Oct., which had $M_{SA} = 6.0$. This is one of the three main shocks during the period 1987-89. Dologlou's Table 2 lists event 4 as having been "predicted" by telegram 12. However, the text of telegram 12, given below, failed to mention the magnitude of the "predicted" earthquake:

> "OUR TELEGRAM OF SEPTEMBER 29-30 WAS ACTUALLY FOLLOWED BY EARTHQUAKES WITH MAGNITUDES 4.9 4.7 AND 3.8 WITH EPICENTER 235 KILOMETRES WEST OF ATHENS STOP HEWEYER [sic] THE NBR OF SIGNALS RECORDED UP [sic] IOA-STATION INDICATES THAT THE SEISMIC ACTIVITY FROM THE AFOREMENTIONED EPICENTRAL AREA HAS NOT FINISHED STOP FURTHERMORE A SES WITH COMPARABLE EFFECTIVE AMPLITUDE WAS RECORDED LATE IN THE NIGHT OF OCTOBER 1 1988 AT KOR-STATION LYING 70 KILOMETRES WEST OF ATHENS STOP THE LATTER EVENT MIGHT BE FELT AT ATHENS
>
> "PROF. P. VAROTSOS"

As no magnitude is stated, the above telegram is not a "prediction." Therefore event 24 is scored as a "failure" by both the neutral and critical analyses.

Ad hoc claim: VL91's [72] Table 1 (p.333) links telegrams 11 and 12 by braces, and claims that the magnitudes stated in telegram 11 ("W240-5.3 or NW330-5.0") apply to telegram 12 as well. This is the only instance in which VL91 link two telegrams in this way. However, if the magnitude wasn't specified in telegram 12 it didn't constitute a "prediction." The *ad hoc* explanation that the "predicted magnitude" was given in another telegram is unacceptable.

8.2.7 Telegram 17: $\Delta r > 100$ km

Telegram 17 was issued on 23 July 1989. The addressee was Prof. S. Uyeda. The text, as given by Dologlou (p.201), reads:

"SIGNAL WAS RECORDED AT KER - STATION STOP EXPECTED EVENT 40 KM FROM A WITH MAGNITUDE 5 STOP EPICENTER MIGHT BE AT NE 40 STOP PLEASE TO NOT ANNOUNCE

"PROF. P. VAROTSOS"

Table 1 of Dologlou's paper lists the "prediction" as "NE 40, 5.0," and lists event 34 with epicenter "N 130" and $M_{SA} = 5.0$ on 1 Aug. 1989 as having fulfilled this "prediction." Table 2 of Dologlou's paper lists the epicentral distance for this (event, telegram) pair as 90 km, but this seems incorrect. Using Athens as the origin, "N130" translates to Cartesian coordinates of (0, 130), while the "predicted epicenter" of NE40 translates to roughly (28.3, 28.3). This gives $\Delta r \approx 110$ km. A more exact calculation by Wyss and Allmann [94] using the actual epicenter gives $\Delta r = 116$ km. Thus the "prediction" of event 34 by telegram 17 is scored as a "failure" by both analyses, as $\Delta r > 100$ km.

"A" in the above telegram apparently means Athens. The point is moot here, but unless ambiguous "predictions" are interpreted unfavorably, there will be no incentive to make unambiguous predictions. The "predicted epicenter" ("40 km from A") should thus be set to the farthest point from the epicenter of event 34 (N130) that is also 40 km from Athens, namely S40.

8.2.8 Telegram 10: fuzzy time window

Dologlou's translation of telegram 10, which was dispatched on 1 Sept. 1988 at 11:30 reads:

"SIGNIFICANT ELECTRICAL ACTIVITY WAS RECORDED AT IOA - STATION ON AUGUST 31 1988 EPICENTER AT N. W. 300 OR W 240 WITH MAGNITUTES [sic] 5,3 AND 5,8

"PROFESSOR P. VAROTSOS"

An earthquake with $M_{SA} = 5.1$ at 12:05 on 22 Sept. 1988 at "W250" (event 22 in Dologlou's Table 2) is listed in Dologlou's Table 1 as having fulfilled this "prediction." In at least one case, as discussed in item (2) on p.195 of Dologlou,[90] VAN have claimed that a "series of signals" was observed over a three day period. It thus seems entirely possible that the midpoint of the electrical activity in the above telegram was at or before 12:00 on August 31. This would make the "prediction" of event 22 by telegram 10 a "failure" in the critical analysis, as $\Delta t > 22$ d. The neutral analysis, which requires $\Delta t \leq 40$ d, scores telegram 10 as a "success."

8.2.9 Telegram 1: swarm or success?

Dologlou's translation of telegram 1, which was dispatched at 21:56 on 15 May 1988 to the Greek government, reads:

Table 1: Seismicity preceding the dispatch of telegram 1 on 1988 May 15

YMD	HMS	°N	°E	Z	M_{SA}
88 05 06	03 06 01.7	38.17	20.66	1	4.0
88 05 07	17 13 22.6	39.00	21.62	4	3.8
88 05 07	19 34 27.3	38.33	20.41	1	4.1
88 05 07	20 41 33.0	38.40	20.47	1	4.2
88 05 11	17 46 39.1	38.38	20.33	1	4.1
88 05 11	23 43 39.8	38.36	20.36	1	4.0
88 05 13	04 06 54.3	38.21	20.29	1	4.2
88 05 14	05 33 26.8	40.57	19.63	1	4.6

"Recording of preseismic signal at Ioannina station on 15-5-88 STOP Expected event NW 330 - 5 or W 300 - 5.3 STOP. Preseismic session impossible due to absence abroad.

"Panayiotis Varotsos"

According to Dologlou's Table 1, event 17, which occurred at "W310" with a magnitude $M_{SA} = 5.9$ at 05:17 on 18 May 1988, fulfilled the second of the "predictions" in the above telegram ("W 300 - 5.3"). As the magnitude, time, and distance all fall within the nominal windows, this is scored as a "successful prediction" by the neutral analysis.

As pointed out by Stavrakakis *et al.*,[93] this general area is the most seismically active in Greece. An earthquake with $M_{SA} = 7.0$ occurred here on Jan. 17, 1983. As summarized by Stavrakakis *et al.*, the April-June 1988 sequence of seismicity in this area began with a foreshock with $M_{SA} = 4.1$ on 5 Apr. 1988. The sequence continued with a foreshock of $M_{SA} = 5.0$ on 24 April. The seismicity in the general area of the $M_{SA} = 5.8$ earthquake of 18 May in the ten days before VAN's telegram 1 was issued on May 15 is summarized in Table 1 (after Stavrakakis *et al.*). As pointed out by Stavrakakis *et al.*, there also were 11 similar small earthquakes in the other area ("NW330") flagged by VAN's telegram 1 in the ten days preceding the dispatch of this telegram; these are not listed here. In view of this seismicity, the critical analysis does not consider event 17 as eligible for having been "predicted."

8.2.10 Recap

The above analyses are summarized in Table 2. The first two columns are the telegram number (letter) and date in Dologlou's Table 1; the next two columns are the number assigned to the "corresponding events" in Dologlou's

Table 2 and the earthquake date; the fifth through seventh columns give the scoring assigned by Dologlou ("D"), the neutral analysis ("N"), and the critical analysis ("C") respectively; the eigth column is the section in this paper where each event is discussed; the last column summarizes the reason the "prediction" was not classified as "successful" by the critical analysis. "(D)" in the last column means that, as discussed in Section 8.2.1, Dologlou [90] acknowledged the event to have fallen outside the nominal window.

A plus sign in the fifth through seventh columns of Table 2 indicates a "success," a minus a "failure," and a question mark (for the neutral analysis only) a case where neither "success" nor "failure" could be clearly established. The letter "I" in the seventh column indicates an event which the critical analysis considers ineligible for "prediction," as it was a clear aftershock or occurred during a swarm. The final line of Table 2 gives the number of "successful predictions" determined by each of the three analyses. Whereas Dologlou claimed 21 "successful predictions," the neutral analysis obtains a total of only 7 "successes," and the critical analysis zero.

Dologlou's paper listed 21 "successful predictions," but acknowledged that 10 of these fell outside the nominal windows. The neutral analysis eliminates these 10 events. The neutral analysis also eliminated the (telegram, event) pair (12, 24), as no magnitude was specified in VAN's telegram, and (17, 34), as $\Delta r > 100$ km. The neutral analysis eliminated the pair (b, 6), as at most one "success" is allowed per "prediction." The neutral analysis also assigned a questionable rating to the pair (j, 16), as it could not be determined whether Δt fell just inside or just outside the 11 d window.

Of the 7 "successes" allowed by the neutral analysis, the "critical analysis" eliminated the pair (a, 4) due to VAN's insufficient documentation of telegram a, and eliminated the pair (b, 5) due to VAN's insufficient documentation of their claim that telegram b reported a GVEF anomaly. The pair (10, 22) was eliminated due to the fact that $\Delta t \le 22$ d had not been clearly established. The pair (1, 17) was eliminated due to its having followed a swarm of microearthquakes, and (2, 18), (3, 19) and (4, 20) were eliminated due to their having been aftershocks. The last four eliminations were not scored as "failures," but rather as events that were "ineligible" for "prediction." Many of the events that were eliminated by the neutral analysis due to parameters falling outside the windows could have been ruled ineligible as aftershocks or parts of a swarm by the critical analysis had they not already been eliminated for other reasons.

Table 2: Scoring of "predictions" of events with $M_{SA} \geq 5.0$

Tele.	Date DMY	Event	Date DMY	"Successes" D	N	C	Analysis	Comment
a	26?/02/87	4	27/02/87	+	+	-	8.2.3	Insuf. documnt.
b	27/04/87	5	29/05/87	+	+	-	8.2.2	Insuf. documnt.
"	"	6	10/06/87	+	-	-	8.2.2	$N_{success} \leq 1$
c	13/06/87	7	21/06/87	+	-	-	8.2.1	$\Delta r > 100$km (D)
d	01/02/88	13	18/02/88	+	-	-	8.2.1	$\Delta r > 100$km & $\Delta t > 11$ d (D)
e	10/03/88	14	26/03/88	+	-	-	8.2.1	$\Delta t > 11$ d (D)
h	07/04/88	15	24/04/88	+	-	-	8.2.1	$\Delta t > 11$ d (D)
j	28/04/88	16	09/05/88	+	?	-	8.2.4	$\Delta t > 11$ d (?)
1	15/05/88	17	18/05/88	+	+	I	8.2.9	Swarm
2	21/05/88	18	22/05/88	+	+	I	8.2.5	Aftershock
3	30/05/88	19	02/06/88	+	+	I	8.2.5	Aftershock
4	04/06/88	20	06/06/88	+	+	I	8.2.5	Aftershock
7	10/07/88	21	12/07/88	+	-	-	8.2.1	$\Delta r > 100$km (D)
10	01/09/88	22	22/09/88	+	+	-	8.2.8	$\Delta t > 22$ d (?)
12	03/10/88	24	16/10/88	+	-	-	8.2.6	No magnitude
13	21/10/88	25	08/11/88	+	-	-	8.2.1	$\Delta r > 100$km (D)
15	03/06/89	33	07/06/89	+	-	-	8.2.1	$\Delta r > 100$km (D)
17	23/07/89	34	01/08/89	+	-	-	8.2.7	$\Delta r > 100$km
18	16/08/89	35	20/08/89	+	-	-	8.2.1	$\Delta r > 100$ km & $\Delta M > 0.7$ (D)
"	"	36	24/08/89	+	-	-	8.2.1	$\Delta r > 100$km (D)
21	15/09/89	37	19/09/89	+	-	-	8.2.1	$\Delta r > 100$km (D)
Number of "successes"				21	7	0		

8.3 Hamada's results

Hamada[92] made an independent statistical evaluation of essentially the same "predictions" reported by Dologlou.[90] As VAN's "predictions" don't specify the windows, Hamada had to make *a posteriori* choices of the window parameters. Many of the participants in the *GRL* Debate argue that this invalidates a statistical test of a hypothesis. Hamada's particular choice of *a posteriori* window parameters was $\Delta t \leq 22$ days, $\Delta r \leq 100$ km, and $|\Delta M| \leq 0.7$. It is hard to justify Hamada's choice of a 22 day window when VAN only claims 11 days for "predictions" based on a "single SES."

VAN frequently cite Hamada's[92] study as strongly supporting their work. For example, in their reply to Mulargia and Gasperini,[75] Varotos *et al.*[76] cite

Hamada's work prominently:

> "By means of a detailed statistical treatment Hamada [then in press] has recently analysed *exactly* the same VAN data as Mulargia & Gasperini and has concluded that:
>
> " 'When the predictions were limited to expected $M_S \geq 5.3$, 10 were successful and 2 failed predictions were obtained. With a confidence level of 99.8% *it is rejected that this success rate (cf. VAN) can be explained by a random model of earthquake occurrence taking into account a regional factor which includes a high density of earthquakes in the prediction area.'* ...
>
> "It is therefore obvious that the detailed analysis of Hamada invalidates the conclusions of Mulargia & Gasperini." [76] (Italics follow original ms)

It therefore seems important to reexamine Hamada's work.

I cannot find the above quotation in Hamada's [92] paper. The discrepancy may be due to VAN's having quoted a preprint rather than the published ms. The following three passages in Hamada's [92] paper are the only places I could find any mention of the figure of 99.8%:

> "Twelve predictions were issued for earthquakes with expected $M_s \geq 5.3$. Following our criteria, ten were successful, two were not. The magnitudes of the predicted earthquakes varied from 4.1 to 5.5. For earthquakes with $M_b(USGS) \geq 5.0$, the ratio of the predicted to the total number of earthquakes is 6/12 (50%) and the success rate of the prediction is also 6/12 (50%), with a probability gain of a factor of 4. With a confidence level of 99.8%, the possibility of this success rate being explained by a random model of earthquake occurrence, taking into account a regional factor which includes high seismicity in the prediction area, can be rejected." (Abstract, p.203)

> "A statistical significance for the cases in Table 3 [Probabilities related to the prediction of earthquakes with an expected $M_s \geq 5.3$] was also obtained. The possibility that an SRP [success rate of the prediction] of 6/12 (50%) can be explained by the random model with EDF [earthquake distribution factor] = 3 for the case of $M_b \geq 5.0$, can be rejected with a confidence level of 99.8%, ..." (p.209)

> "When the predictions were limited to earthquakes with a predicted M_s of ≥ 5.3, ten successful and two failed predictions were obtained. The possibility that the 6/12 (50%) SRP for $M_b \geq 5.0$ can be explained by the random model with EDF = 3 is rejected with a confidence level of 99.8%." [92] (p.209)

The above quotations make it clear that the confidence level of 99.8% does not refer to the claim that 10 out of 12 "predictions" were successful. Rather, Hamada considered twelve cases in which he claimed that VAN had "predicted" events with $M_{SA} \geq 5.3$ in their telegrams. (He used the USGS catalog, and converted m_b to M_{SA} using the empirical relation $M_{SA} = m_b +$

Table 3: "Predictions" with $M_{SA} \geq 5.3$ considered "successful" by Hamada.

Tele.	Date DMY	Event	Date DMY	"Successes" H	N	C	Analysis	Comment
a	26?/02/87	4	27/02/87	+	+	-	8.2.3	Insuf. documnt.
1	15/05/88	17	18/05/88	+	+	I	8.2.9	Swarm
2	21/05/88	18	22/05/88	+	+	I	8.2.5	Aftershock
10	01/09/88	22	22/09/88	+	?	-	8.2.8	$\Delta t > 22$ d (?)
12	03/10/88	24	16/10/88	+	-	-	8.2.6	No magnitude
15	03/06/89	33	07/06/89	+	-	-	8.2.1	$\Delta r > 100$km (D)
Number of "successes"				6	3	0		

0.3.) Hamada then made an evaluation in which telegrams that "predicted" events with $M_{SA} \geq 5.3$ were scored as successful if an event with $m_b \geq 5.0$ occurred within the window. Hamada scored 6 of these 12 "predictions" as "successful." This is what Hamada is claiming to be significant at the 99.8% level.

A confidence level of 99.8% sounds impressive, but the sample is small ($N = 12$), so small changes in the number of "successes" will produce huge swings in the apparent statistical significance. Also, 99.8% is only about 3 standard deviations ($3\sigma = 99.7\%$). As noted by Anderson, "Any good experimentalist will doubt a surprising result with less than 5-6σ 'significance,' for instance." [69]

As the dataset is the same considered in the previous section, it is easy to reexamine Hamada's analysis. The reader may wish to consult both Hamada's [92] and Dologlou's [90] papers in conjunction with this subsection.

By comparing the twelve "Predict. of EQs(Ms≥5.3)" in Hamada's Fig. 3 to Dologlou's [90] Table 1, it can be inferred that the six "successes" cited by Hamada are the (telegram, event) pairs (a, 4), (1, 17), (2, 18), (10, 22), (12, 24), and (15, 33). We conduct both neutral and critical analyses of these six pairs. In this subsection both the critical and neutral analyses use a time lag of $\Delta t = 22$ d, following Hamada.

The analyses for the six (telegram, event) pairs considered "successful" by Hamada are shown in Table 3. The format is the same as Table 2, except that the "H" in the fifth column refers to Hamada's [92] analysis. The neutral analysis scores the pairs (a, 4), (1, 17), (2, 18) as "successful." The neutral analysis scores (10, 22) as questionable since $\Delta t \leq 22$ d is not clearly established, and scores (12, 24) and (15, 33). as failures. The former is rejected because telegram 12 doesn't give the magnitude, and the latter because

Table 4: Earthquake predictions and results (*Asahi Shinbun*)

	Actual earthquakes		Predicted earthquakes		
	Date (YMD)	Mag.	Date (YMD)	Mag.	Occurrence time
A	94/11/29	M5.3	94/10/16	M5.7	40~50 days later
	94/12/1	M5.2			
B	94/5/23	M6.1	94/5/14	M5.7~6.2	" "
C	94/4/17	M5.9	94/2/28	M5.5~6.0	" "

$\Delta r > 100$ km. Dologlou's Table 2 gives $\Delta r = 120$ km for (15, 33), while Wyss and Allmann [94] give $\Delta r = 114$ km and $\Delta r = 118$ for the PDE and Athens epicenters respectively.

The critical analysis scores the pair (10, 22) as a failure, following the principle of rejecting all questionable cases. The critical analysis also rejects all three cases scored as "successful" by the neutral analysis. (a, 4) is rejected due to VAN's insufficient documentation of telegram a; (1, 17) is ineligible, due to event 17's being part of a swarm; and (2, 18) is ineligible due to event 18's being an aftershock.

Hamada's claim of a 99.8% confidence level was based on the premise that 6 out of 12 telegrams "successfully predicted" events with $m_b \geq 5.0$ (i.e., events with $M_{SA} \geq 5.3$). But, as shown above, the neutral analysis reduced the number of "successes" to 3, and the critical analysis reduced this to 0.

In view of the ambiguity and uncertainty involving the "success" or "failure" of VAN's individual "predictions," the type of statistical analysis conducted by Hamada [92] does not seem meaningful. Hamada's claim of a 99.8% confidence level should therefore not be accepted.

8.4 VAN's 1994 "predictions"

VAN's "predictions" and "success *communiqués*" in 1994-95 were disseminated by fax. The verbatim text of these faxes is given in the Appendix. The salutations and signature blocks are omitted, and references are cited by superscripts. Boldface, italics, and underlining follow the original.

An article on Dec. 24, 1994 in the *Asahi Shinbun*,[95] Japan's leading quality newspaper, presented data on VAN's "predictions" in 1994. At least some of these data were presumably obtained from Prof. Varotsos.

Table 4 is a translation of a table from the *Asahi* article. Fig. 2 (upper left) is a map from the *Asahi* article.[95] The crosses show the four earthquakes listed in Table 4 (the cross marked A indicates two earthquakes with nearly the same epicenter; the other two crosses show one event each), the hatched areas are alleged to be the "predicted epicenters," and the solid circles are VAN's electrical stations.

Both the "predicted" and actual epicenters are labeled "A," "B," and "C" in both Table 4 and in Fig. 2 (upper left). This is likely to have led readers to believe that all of VAN's "predictions" were "successful," even though this was not stated explicitly in the *Asahi* article.

Each of VAN's three "predictions" in 1994 was a double "prediction." The "predicted" epicenters, which are indicated by hand-drawn curves, are shown in Fig. 2. The *Asahi* map (Fig. 2, upper left) misrepresents VAN's predictions by showing only one of the two "predicted epicenters" (II, I, and 1 for Faxes A, B, and C, respectively) for each of VAN's "predictions." Another quality newspaper, *Mainichi Shinbun*, presented a front page story on VAN on July 23, 1994. [96] The *Mainichi* also showed only one "predicted epicenter" – in both cases the same one showed by the *Asahi* – for Faxes B and C.

8.4.1 Greek seismicity in 1994

Dr. G. Stavrakakis (personal communication) sent me the SI-NOA final bulletin for 1994. Table 5 lists the 16 events in 1994 with $M_{SA} \geq 5.0$ $(M_L \geq 4.5)$ in the region $(19°\text{-}27°E, 35°\text{-}42°N)$. The last column of Table 5 was obtained using $M_{SA} = M_L + 0.5$. The four events cited in the *Asahi* article are labeled in the left column of Table 5. One additional event, labeled "D," is discussed below. The PDE (USGS Preliminary Determination of Epicenters) epicenters for the five labeled events are within about 20 km of the SI-NOA epicenters.

The 16 events with $M_{SA} \geq 5.0$ listed in Table 5 are plotted in Fig. 3 (left). Fig. 3 (right) shows all 139 events in the SI-NOA bulletin for 1994 with $M_{SA} \geq 4.3$. This figure confirms that VAN's "predicted epicenters" (Fig. 2) were all in areas with a high level of seismicity.

Only 4 of the 16 events listed in Table 5 are shown in the *Asahi*'s seismicity map (Fig. 2, upper left) and listed in the *Asahi*'s table (Table 4), even though the *Asahi*'s caption claimed all the events with $M_{SA} \geq 5.0$ were shown. This might have given readers the incorrect impression that VAN had "predicted" all of the major events in Greece in 1994. Note that the date (17 Apr.) for event C in Table 4 is Greek local time, while that in Table 5 (16 Apr.) is GMT.

The locations for events A and C in the *Asahi* map (Fig. 2, upper left) agree with those given in Table 5 and shown in Fig. 3 (left), but there is a discrepancy of about 200 km for event B. As the map in the *Mainichi* [96] also showed the same incorrect epicenter for event B, responsibility for the error might lie with the news source rather than the reporters.

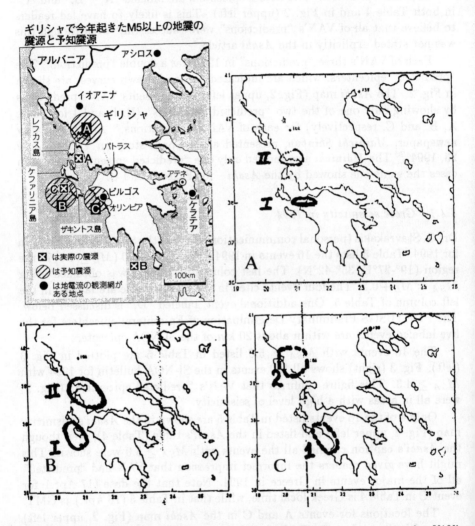

Figure 2: (Upper left) Figure from *Asahi* article. (Others) Figures accompanying VAN's 1994 faxes, labeled to correspond to the *Asahi* article.

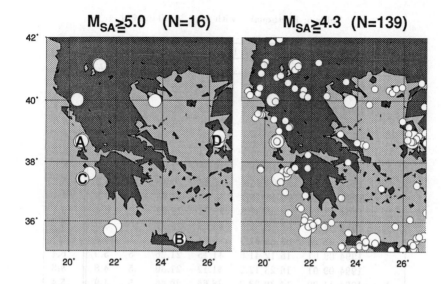

Figure 3: (Left) The 16 events with $M_{SA} \geq 5.0$ that occurred in 1994. Events B and C and the two events labeled A are referred to in the *Asahi* article. Event D is discussed in connection with the Yokohama meeting on 25 May 1994. (Right) All 139 events with $M_{SA} \geq 4.3$ in 1994. Smaller symbols denote events with $5.0 > M_{SA} \geq 4.3$.

8.4.2 Were VAN's 1994 predictions successful?

Event C occurred on April 16, 1994 at 23:09GMT. Prof. Varotsos disseminated a fax on April 17 claiming that Fax C was a "successful prediction" of Event C. In a postscript to a Fax dated Dec. 11, 1994, Prof. Varotsos claimed that Fax A was a "successful prediction" of Event A. To decide whether these these claims should be accepted we must ascertain the nominal "prediction" windows.

Time: Faxes A, B, and C all refer to Fig. 28A of Varotsos and Lazaridou,[72] which is the top chart in Fig. 1. I have no idea how one is supposed to obtain an unambiguous expiration date for one of VAN's "predictions" from this chart. If VAN want to claim they are making "predictions" that are useful for society, their faxes should unambiguously state the starting date and ending date of the window. It seems unacceptable for the time chart to be the sole means by which VAN announce the "predicted time" of earthquake occurrence.

In my analyses I rely on VAN's statements rather than their time chart to obtain the window. Faxes A, B, and C all report "SES activity." As quoted

Table 5: Earthquakes with $M_{SA} \geq 5.0$ in 1994.

		SI-NOA ($M_{SA} \geq 5.0$)					
Fax	YMD	HMS	°N	°E	Z	M_L	M_{SA}
	1994 01 11	07 22 53.3	35.84	21.95	40	5.4	5.9
	1994 01 11	08 59 52.4	35.85	21.97	26	4.8	5.3
	1994 01 12	19 29 33.3	35.69	21.76	5	4.6	5.1
	1994 01 14	06 07 48.7	37.61	20.88	5	4.5	5.0
	1994 02 03	18 52 52.1	40.02	20.33	69	4.5	5.0
	1994 02 25	02 30 49.7	38.73	20.58	5	5.3	5.8
	1994 04 10	19 46 21.0	39.98	23.67	13	4.5	5.0
C	1994 04 16	23 09 36.4	37.43	20.58	30	5.3	5.8
B	1994 05 23	06 46 16.3	35.40	24.73	81	5.6	6.1
D	1994 05 24	02 05 37.6	38.71	26.32	12	5.6	6.1
	1994 05 24	02 18 35.5	38.69	26.37	10	5.2	5.7
	1994 05 24	03 35 31.9	38.84	26.41	5	4.6	5.1
	1994 09 01	16 12 41.6	41.15	21.26	5	5.9	6.4
	1994 09 01	16 23 12.2	41.12	21.30	5	4.8	5.3
A	1994 11 29	14 30 28.2	38.66	20.46	5	4.9	5.4
A	1994 12 01	07 17 35.8	38.69	20.55	5	4.8	5.3

		PDE (events A, B, C, D only)					
Fax	YMD	HMS	°N	°E	Z	m_b	M_S
C	1994 04 16	23 09 33.9	37.430	20.617	26.0	5.3	5.0
B	1994 05 23	06 46 16.1	35.559	24.727	76.0	6.0	
D	1994 05 24	02 05 36.2	38.664	26.542	17.5	5.0	5.3
A	1994 11 29	14 30 28.4	38.707	20.484	21.0	4.9	4.8
A	1994 12 01	07 17 36.0	38.694	20.519	30.0	4.6	4.4

in Section 7.1, Varotsos and Lazaridou [72] say the window is "around 22 days" for the case of electrical activity, while Dologlou [90] says the window is "of the order of one month." Faxes C and B mention "3-4 weeks for the first big event, etc.," while Fax A just cites the two time charts.

These statements are both ambiguous and mutually contradictory. "Around 22 days," "of the order of one month," or "3 to 4 weeks for the first big event, etc.," all fail to yield an unambiguous expiration date. As the largest of these values is "around one month," I use a window of $\Delta t = 31$ d for the critical analysis, as opposed to the value $\Delta t = 22$ d that was used above in the critical analysis of the "predictions" reported by Dologlou. [90] As above, I use $\Delta t = 40$ d for the neutral analysis. VAN and their supporters might claim these choices are arbitrary. But their failure to state the value of Δt forces anyone who wants to review their work to make some arbitrary choice.

Distance: As VAN do not explicitly discuss the spatial window, I assume it is still $\Delta r = 100$ km. However, the "predicted epicenters" in the figures from Faxes A, B, and C shown in Fig. 2 are crudely drawn handwritten curves. It is not clear how these curves can be used to distinguish between "success" and "failure." To obviate this ambiguity, VAN's faxes should give numerical values for the "predicted" epicentral region.

Conclusion: The results of the neutral and critical analyses are the same for Faxes A and C. As $\Delta t > 40$ d in both cases, both analyses score each of these "predictions" as a "failure."

Fax B was discussed at the 25 May 1994 Yokohama meeting of SC-IDNDR (see Section 4). Prof. Varotsos was scheduled to give a talk, but canceled his attendance because of the issuance of a "prediction" which was announced by Fax B on May 14. The "predicted epicenters" (bottom left panel of Fig. 2) were along the western coast of Greece.

The Yokohama meeting of SC-IDNDR was informed that two magnitude-6 earthquakes that took place in Greece on 23 May and 24 May were sensational yet not isolated proof of the predictive power of VAN's methods. These two events are respectively events B and D. Their epicenters are given in Table 5 and shown in Fig. 3 (left). Event B is located at a depth of 80 km beneath Crete, and event D is located along the coast of Turkey on the eastern coast of the Agean Sea. The epicenters of both events B and D are about 500 km from the "predicted epicenters" for Fax B shown in Fig. 2. As $\Delta r \gg 100$ km, both the neutral and critical analyses score these two events as not having been predicted.

It is difficult to understand the criteria that VAN and their supporters used to justify claiming events B and D as having been "successfully predicted" in their presentations to the Yokohama meeting of SC-IDNDR. VAN's list of 13 large earthquakes from Feb. 1987 to May 1995, nine of which they claim as "successful predictions," is analyzed in Section 8.6. Neither event B nor event D is now claimed by VAN as a "successful prediction," despite their having been so described to the Yokohama meeting of SC-IDNDR on 25 May 1994.

8.4.3 *Inverse correlation*

As shown in Table 5, an earthquake with $M_{SA} = 5.8$ occurred on 25 Feb. 1994 at (38.73°N, 20.58°E). This earthquake is listed by VAN's own reckoning (see Section 8.6) as one of four large earthquakes they failed to "predict" in Jan. 1987-May 1995. On Feb. 28, three days after this event, VAN disseminated

Fax C. As shown in Fig. 2 (lower right) the epicentral area of the Feb. 25 event lies between the two "predicted epicenters" of Fax C.

VAN claim that Fax C was a "successful prediction" of Event C, which occurred over 40 days after Fax C was disseminated. The opposite hypothesis, namely that Fax C was issued on Feb. 28 in response to the earthquake of Feb. 25 cannot be ruled out. Possible mechanisms are (i) the "SES activity" could be a post-seismic effect of the Feb. 25 event; or (ii) conscious or unconscious fudging.

8.5 April-June 1995: "Predictions" and earthquakes

VAN issued three "predictions" in April-May, 1995, and later claimed that three earthquakes with $M_{SA} \geq 6$ in May-June 1995 were "successfully predicted." The following analyses show that these claims should not be accepted.

8.5.1 4 May 1995 Thessaloniki earthquake

An earthquake close to Thessaloniki with $M_{SA} = 6.0$ and epicenter (USGS PDE) (40.673°N, 23.466°E) occurred on 4 May 1995 at 00:34 GMT. A list of large earthquakes prepared by Prof. Varotsos (see Section 8.6) lists this event as having been "successfully predicted" with an error $\Delta M = +0.2$. As the magnitude was $M_{SA} = 6.0$, VAN is apparently claiming that the "predicted magnitude" was 5.8. However, this is inconsistent with VAN's fax of April 7, 1995, which says: "If the epicenter lies in area 'c' (i.e. around *Thessaloniki*) then Ms(ATH) ≈5.2. For larger epicentral distances, then M will be larger (e.g. south of, but very close to, Chalkidiki peninsula Ms(ATH)≈5.8)."

Fig. 1 (p.270) of Varotsos *et al.*, [97] Fig. 2 (p.404) of Varotsos *et al.*, [98] and Fig. 4 of this paper all show the "selectivity area" of VAN's ASS station, marked "c" in all figures. I enlarged each of these figures with a photocopier and plotted the USGS epicenter. I found that the epicenter lay 2 km outside area "c" for the first of the above figures, but just inside area "c" for the second and the third. (The reader is encouraged to replicate this experiment.) I therefore conclude that the event of 4 May 1995 lies "in area 'c' (i.e. around *Thessaloniki*)" and that the "predicted magnitude" of "Ms(ATH) ≈5.2" should apply. Hence VAN's claim of "success" is inadmissible, as $\Delta M = 0.8$.

8.5.2 13 May 1995 Kozani-Grevena earthquake

VAN issued a "prediction" on 27 April 1995. VAN claim that a slightly different version of this "prediction" was sent to the Greek government on 30 April. Both versions are given in the Appendix.

The Kozani-Gravena earthquake occurred on 13 May with $M_{SA} = 6.6$. The USGS PDE source parameters were epicenter: (40.144°N, 21.674°E), origin time: 08:47 GMT. VAN's faxes of April 27 and 30 give three possible "predicted epicenters": (i) 200 km west of Athens, (ii) the Vartholomio area (37.90°N, 21.11°E), or (iii) "~50 km from IOA station" (April 27 fax) or "a few tens of km NW from IOA station" (April 30 fax). (IOA station is at approximately 39.8°N, 20.7°E.) Only (iii) applies to the earthquake of 13 May.

Although VAN claims the Kozani-Gravena event as a "successful prediction," their claim is unacceptable due to $|\Delta M| > 0.7$. The fax of April 27 fax says: "(An alternative solution is $M_s(\text{ATH}) \approx 5.5$ with an epicenter ~50 km from IOA station)." As the May 13 event had $M_{SA} = 6.6$, we have $|\Delta M| = 1.1$, so the "prediction" was a "failure."

VAN's "predictions" usually state a single "predicted magnitude," with an implied window of ±0.7. However, VAN's fax of April 30 says: "An alternative solution is $M_s(\text{ATH}) \approx 5.5 - 6.0$ with an epicenter a few tens of km NW from IOA station." There are three possible interpretations. (a) The magnitude should lie between 5.5 and 6.0. (b) The magnitude should lie within ±0.7 of the midpoint of this range. (c) The magnitude should be bracketed by the maximum value plus 0.7 and the minimum value minus 0.7. When expressed as inequalities these three interpretations are respectively: (a) $5.5 \leq M_{SA} \leq 6.0$; (b) $5.05 \leq M_{SA} \leq 6.45$; (c) $4.8 \leq M_{SA} \leq 6.7$.

If either of the first two inequalities is used the "prediction" was a "failure," but if the last inequality is used the "prediction" was a "success." A list prepared by Prof. Varotsos, which is discussed below in Section 8.6, claims "success," listing an error of $\Delta M = 0.6$, apparently on the basis of interpretation (c). This claim should not be accepted, as it exploits ambiguity.

8.5.3 15 June 1995 Egion earthquake

A damaging earthquake near Egion occurred on 15 June 1995. This earthquake, which had $M_{SA} = 6.1$, caused 26 deaths. [99] VAN's *communiqué* of June 15 claims this event was "successfully predicted." Fig. 4, which was included in this *communiqué*, shows the epicenter at about (38.5°N, 22.6°E), which is just outside the "selectivity area" for VAN's IOA station. (This determination was made by measuring an enlarged copy of Fig. 4. The reader is encouraged to repeat this test.)

The epicenter claimed by VAN appears to be incorrect; both the USGS PDE epicenter (38.401°N, 22.269°E) and that of Carydis *et al.*, [99] (38.36°N, 22.15°E) fall within the "selectivity area" for IOA. VAN's fax of May 19 specifically excluded the "areas belonging to IOA selectivity map,"

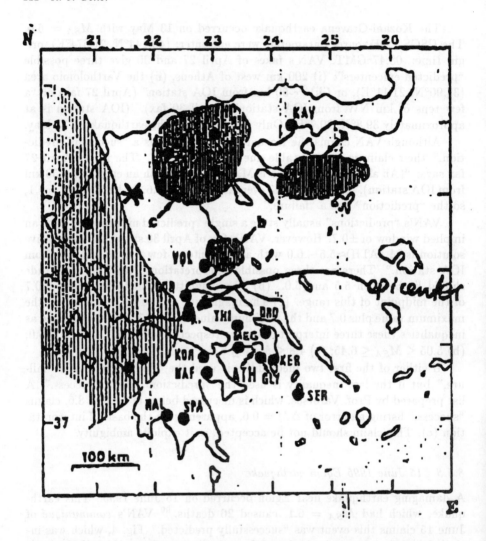

Figure 4: Figure accompanying VAN's fax of June 15, 1995 which claimed the Egion earth-quake of June 15, 1995 had been "successfully predicted." The epicenter of the May 13, 1995 event is indicated by the hand drawn asterisk at about (40°N, 21.5°E). The above figure shows the epicenter of the June 15, 1995 event (labeled "epicenter") as being at about (38.5°N, 22.6°E), which is outside the "selectivity area" for VAN's IOA station. But the epicenter given by Carydis et al. (reference no. 99) is (38.36°N, 22.15°E), and the epicenter given by the USGS PDE is (38.401°N, 22.269°E). Both of the latter fall within the selectiv-ity area for IOA (the large hatched area at left, center), which was specifically excluded by VAN's "prediction" of 19 May 1995. The "prediction" is thus scored as a "failure."

because the "SES activity was not recorded at IOA." This "prediction" is thus scored as a "failure."

A Greek colleague kindly sent me translations of some news stories. According to a Cosmos news (an electronic bulletin board) story dated June 20, 1995, Prof. Varotsos filed a complaint with the public prosecutor which accused government officials of having been negligent in failing to take action based on his "prediction" dated May 19. On the other hand, Mr. D. Papanikolaou, the Director of OASP (the Organization of Anti-Seismic Planning) is quoted as saying VAN's prediction was not of any use, as the prediction included as a probable epicenter 2/3 of the area of Greece.

Mr. Papanikolaou told *The News* (June 19, 1995) that Egion (the epicentral region of the June 15 event) was in the area excluded by VAN's prediction. This agrees with the conclusion above. In summary, the Greek officials responsible for disaster mitigation apparently did not regard VAN's "prediction" as having been of any practical use and regard it as having been unsuccessful.

8.6   Large earthquakes, 1987-1995

Prof. Varotsos has claimed, [10] as quoted in Section 1, that VAN "successfully predicted" 9 out of 13 large ($M_{SA} \geq 5.8$) events from Jan. 1987 through May 1995. At the request of Prof. H. Kanamori, Prof. Varotsos prepared a list of these 13 events. Prof. Kanamori kindly sent me a copy of this list, which is reproduced in Table 6. Let us now analyze VAN's claims of success.

Of the events claimed by VAN as "successes," events 1-3 and 11-13 have already been considered in this paper. Unfortunately I do not have access to copies of VAN's faxes for event 9. Let us now consider events 6 and 8.

Event 6 is discussed in Table 1 (p.271) of Varotsos *et al.*[97] The last column, "deviation of the prediction," lists "$\Delta r \leq 120$ km." As the listing for the event immediately above is $\Delta r \leq 100$ km, I infer that the deviation for event 6 was 100 km$< \Delta r \leq 120$ km. Thus the entry "~ 100 km" in Table 6 really means $\Delta r > 100$ km. Event 6 is therefore scored as a "failure" by both the critical analysis and the neutral analysis.

We now consider event 8. "SES activity" was observed on Jan. 27-29, 1993, and VAN's fax dated Jan. 30 was sent on Feb. 1. Measuring from the midpoint of the "electrical activity" we have $\Delta t = 36$ d for event 8. The critical analysis thus scores this as a "failure"; the neutral analysis scores this as a "success."

VAN's analysis, the neutral analysis, and the critical analysis for 12 of the 13 events in Table 6 (excepting event 9) are given in Table 7. All events scored as "failures" by VAN are also scored as "failures" by the critical and neutral analyses. The neutral analysis recognizes three "successes." Of these

Table 6: "Successful predictions" of large events claimed by VAN

No	DMY	M_{SA}	Δr (km)	ΔM
1	27 02 87	5.9	< 50 km	-0.6
2	18 05 88	5.8	< 50 km	+0.5
3	16 10 88	6.0	< 30	+0.7
4	19 03 89	5.8	missed (Conference Aussois)	
5	20 08 89	5.9	unsuccessful because $\Delta r \approx 140$ km	
6	16 06 90	6.0	~100 km	+0.2
7	21 12 90	5.9	missed	
8	05 03 93	5.8	~80 km	0
9	13 06 93	5.9	≪40 km	+0.1
10	25 02 94	5.8	missed	
11	16 04 94	5.8	≪ 30	0
12	04 05 95	6.0	~40	+0.2
13	13 05 95	6.6	~ 70	+0.6
Number of "successes"				9
Number "missed" or "unsuccessful"				4

the critical analysis disqualifies event 1 due to insufficient documentation and event 8 due to the stricter time window (31 d). Event 2 is ineligible due to its having been preceded by a microearthquake swarm.

In summary, VAN claims 8 of the twelve events shown in Table 7 were "successfully predicted," but the neutral analysis reduces this to 3, and the critical analysis to 0. Thus VAN's claims to be especially successful in predicting larger events should not be recognized.

9   VAN and the Media

VAN has been extensively discussed by newspapers, magazines, and television in Japan. Prof. Varotsos, Prof. Uyeda, and Dr. Nagao have been the main sources. A few news stories were cited above. Headlines and some additional information (my translation) for a number of other stories are cited below:

"Earthquake prediction ... 60% successful ... Joint research by Dr. Nagao in Greece," *Asahi Shinbun*, evening ed., Aug. 25, 1992.

"Earthquakes can definitely be predicted! ... The world's only successful prediction method, VAN," *Bart*, Feb. 27, 1995, p.24-25. A figure on p.25 of this article shows the same incorrect "predicted epicenters" and same incorrect epicenter for event B as Fig. 2, upper left. The date of event C (see Section 8.4) is incorrectly given as April 14.

Table 7: Scoring of "predictions" of events with $M_{SA} \geq 5.8$

Event	Date DMY	"Successes" V	N	C	Analysis	Comment
1	27 02 87	+	+	-	8.2.3	Insuf. documnt.
2	18 05 88	+	+	I	8.2.9	Swarm
3	16 10 88	+	-	-	8.2.6	No magnitude
4	19 03 89	-	-	-	8.6	missed (conference)
5	20 08 89	-	-	-	8.6	$\Delta r \approx 140$ km
6	16 06 90	+	-	-	8.6	$100 < \Delta r \leq 120$ km
7	21 12 90	-	-	-	8.6	missed
8	05 03 93	+	+	-	8.6	$\Delta t > 31$ d
10	25 02 94	-	-	-	8.6	missed
11	16 04 94	+	-	-	8.4.2	$\Delta t > 40$ d
12	04 05 95	+	-	-	8.5.1	$|\Delta M| > 0.7$
13	13 05 95	+	-	-	8.5.2	$|\Delta M| > 0.7$
No. of "successes"		8	3	0		

"The VAN method ... Greek earthquake prediction technique ... 60% successful," *Sankei Shinbun*, evening ed., May 8, 1995. The date of event C (see Section 8.4) is incorrectly given as April 14.

"70% successful in Greece ... the VAN method ... Earthquakes can definitely be predicted! ... Earthquake prediction in Japan cannot be permitted to repeat the tragedy of Kobe." *Bart*, May 8, 1995, p.84-89.

"VAN ... 60% successful in Greece ... 'meaningless', [say] critics" *Yomiuri Shinbun*, June 1, 1995, evening ed. This story is notable as both supporters and critics are featured.

"The VAN method ... 15 years of success in Greece ... 60% success rate" *Tokyo Shinbun*, Sept. 5, 1995, evening ed.

"The VAN method—70% successful," *Quark*, Sept. 1995. Features 4 page dialog between Prof. Varotsos and Prof. Uyeda.

Many of the above stories contain incorrect information. The epicenter of the May 23, 1994 event (event B) is consistently misrepresented. VAN's predictions for 1994 are consistently misrepresented; several of the stories incorrectly give the date of event C as April 14 rather than the correct April 16 (GMT) or April 17 (Greek local time). VAN should publish a paper on their 1994-95 "predictions" in a scientific journal, or, at the very least, as a technical report issued by their institute. It is not desirable to have to rely on newspaper reports to learn the details of VAN's recent claims.

A brief interview with Prof. Varotsos was featured on a two hour special, "The Kobe earthquake could have been predicted!?", on the TBS/MBS television network on July 16, 1995. Prof. Varotsos, speaking in English (my transcription from tape), said "13 earthquakes – nine were predicted .... The success rate is about 70% when you speak from [sic] earthquakes around 5.5 or 5.8."

VAN and their supporters are depicting VAN to the media and the public as an operational earthquake prediction method which has a proven track record of 70% (or 60%) success in Greece. However, 12 of the 13 events on which this claim is based were reanalyzed above in Section 8.6. VAN claim 8/12 "successes," but the neutral analysis recognized only 3/12, and the critical analysis recognized 0 "successes." The claim that VAN has achieved "70% success" rates is thus highly misleading. VAN and their supporters should desist from making such claims.

9.1   VAN – The comic book

VAN's work has been featured recently in yet another mass medium, comic books (*"manga"*). An 81 page two-part series on VAN, "Earthquakes can be predicted!", was published in issues no. 27 (June 21, 1995) and no. 28 (June 28, 1995) of the weekly comic *Shonen Sunday*. The title page of part 1 of this series is shown in Fig. 5. Dr. Nagao is listed as "editorial supervisor."

The comic is billed as non-fiction, but the fine print adds that certain episodes have been embellished for dramatic reasons. Prof. Varotsos and Dr. Nagao cooperated fully with the artist. Following the (happy) ending, in which VAN is depicted as having been internationally recognized, is a two page essay "Earthquakes can be predicted in Japan!!"

In this essay Dr. Nagao is quoted as calling for the installation of a network of 200 VAN-stations throughout Japan, with an equipment budget of 10^9 Yen and an annual operating budget of 2×10^8 Yen. The essay says such a network would allow earthquake forecasts to be announced as routinely as weather forecasts are presently announced. The essay concludes by calling on scientists and politicians to "abandon their prejudices and make the correct choices." Two photos of Prof. Varotsos, one together with the artist, are featured in this two page spread.

10   Summary

The purpose of this paper is to evaluate VAN's claims, some of which are quoted in Section 1, to be making "successful predictions" of earthquakes.

Figure 5: Title page of "Earthquakes can be predicted!" Dr. Nagao is listed (lower left) as "editorial supervisor."

Earthquake prediction research over the last 110 years is reviewed in Section 2. There was a brief flurry of optimism in the mid-1970's, but apparently promising lines of research ultimately failed. Section 3 discusses results in non-linear dynamics and earthquake source theory which suggest earthquakes might be inherently unpredictable. These results, and a long history of previous failures, suggest that stringent standards of proof should be required for claims, including VAN's, of success in earthquake prediction.

The formidable accuracy and reliability requirements for an operational earthquake prediction system are discussed in Section 5. Unreliable prediction systems could easily exacerbate, rather than ameliorate, casualties and damage. Generally accepted definitions of earthquake prediction are summarized in Section 6. VAN's telegrams and faxes fall significantly short of meeting these standards. (i) The windows for time, epicenter, and magnitude are not specified in VAN's "predictions," and an expiration date is never specified. (ii) The retrospectively stated windows for VAN's "predictions" (e.g., $|\Delta M| = \pm 0.7$, $\Delta r = 100$ km, $\Delta t = 11$ d to 40-50 d) are too wide for VAN's "predictions" to be of any practical use. (iii) Many of VAN's "predictions" involve small shocks which are unlikely to cause serious damage. (iv) VAN's "predictions" are often ambiguous, and VAN frequently claim "success" even when one or more of the source parameters lies outside the (retrospectively stated) window. (v) VAN will claim "success" if they "predict" an M5.0 earthquake and an M4.3 earthquake occurs, but will not acknowledge "failure" if an M4.3 earthquake occurs but no "prediction" was issued. This inconsistency leads to an inflated "success rate" for VAN's "predictions." (vi) It is questionable whether VAN's "predictions" outperform random chance.

VAN's statements about their time, space, and magnitude windows are reviewed in Section 7. In 1981 VAN claimed to be detecting precursors "a few minutes before each earthquake" and another type of precursor that occurred "about seven hours before the impending earthquake." In 1984 they claimed to be detecting SES (seismic electrical signals) that occur "6-115 h" before earthquakes. By 1991 this widened to "between 7 hours and 11 days." Most of VAN's faxes now seem to report "electrical activity" (or "SES activity") rather than "SES" anomalies. The maximum time lag for "electrical activity" is unclear, but VAN sometimes claim "success" for earthquakes with a time lag of over 40 days.

Section 7.2.1 showed that VAN's magnitude window of ± 0.7 implies a dynamic range of 1000 (between a factor of 0.03 and 30) for the energy of the "predicted earthquake." This seems physically implausible.

Some of VAN's telegrams were sent by one member of their group to another, and VAN's faxes are not consecutively numbered. Such problems make it difficult to verify VAN's claims. Particular earthquakes VAN claims to have "successfully predicted" are discussed in Section 8. A neutral analysis disqualified most of these claims for having violated the windows, or due to incomplete specification of the "predicted" parameters. A critical analysis disqualified all of VAN's claims of "success," without exception, for various reasons. For example, VAN's claim (Section 8.2.3) that an earthquake on 27 Feb. 1987 had been "successfully predicted" was disqualified because VAN's claim to have dispatched telegram a before the earthquake was insufficiently documented. The critical analysis also disqualified several events that were aftershocks or part of an ongoing swarm, and hence not "eligible" for "prediction."

In summary, the various detailed analyses in Section 8 show that VAN's claims to be making "successful predictions" of earthquakes should not be recognized. Section 8.3 showed that a statistical analysis by Hamada[92] claiming certain of VAN's claims could be accepted with a 99.8% confidence level was vitiated. Finally, it was suggested in Section 9 that VAN and their supporters are overzealous in using mass media, including comic books, to disseminate their claims to be making "successful predictions."

10.1 Lessons for the future

It appears that anyone who claims to be making "successful predictions" of earthquakes and who has at least one credible supporter can attract the attention of the media and government authorities. The scientific community is then compelled to evaluate these claims. Such evaluations require much time and effort, especially if the "predictions" are ambiguous and poorly documented. Predictors should therefore be required to document their claims so they can be readily and objectively evaluated.

The predicted source parameters should be stated numerically; "time charts" or hand drawn maps should be considered unacceptable. The windows should also be stated numerically as an integral part of the predictions. The earthquake catalog to be used to test the predictions should be clearly specified, and should be a final, not preliminary, catalog. The physical data on which each prediction is based (e.g., electrical observations) should be summarized in the announcement of the prediction, and a complete data set should be made available, perhaps by anonymous ftp. Seismicity data for the region of the predictions should also be made available. All predictions should be issued in a standard format. Predictions should be consecutively numbered, and deposited with a disinterested third party who will supply verified copies on request. "Predictions" that fail to meet these minimal requirements should be ignored.

Acknowledgments

I am grateful to the readers who critically reviewed earlier drafts and suggested many valuable improvements. I thank Dr. H. Igel for helping me meet the final deadline by printing this ms. in the U.K. I thank Mr. N. Takeuchi for preparing Fig. 3 and reformatting the data for Table 5.

Appendix: VAN's Faxes in 1994–95

Fax C – February 28, 1994

We have recorded a **SES activity** with the following characteristics:
Date: February 25, 1994 (and earlier dates).
Station: IOA
Components: NS - short dipole arrays and on the long dipoles.
Polarity: Negative

Impending earthquake: Magnitude: Ms(ATH) 6.0 or 5.5 for cases 1 and 2 respectively.
                    MB(USGS) 5.5 or 5.0   "   "   "
Epicenter: See the attached map [Fig. 2 –RJG] (two alternative solutions*).
Time: According to Fig.28A of Varotsos and Lazaridou[72] (i.e. 3-4 weeks for the first big event, etc.). [Top graph in Fig. 1 –RJG]

*P.S. No definite discrimination can be made (based on SES characteristics) between the two alternative epicentral areas. However, as the current SESs seem to be similar to those recorded on Jan. 27, 1993, the solution 1 might be more probable.

FAX – April 17, 1994 (Success communiqué)

It is the scope of this FAX to inform you on the results of our prediction that was sent to you on March 1, 1994.

For your convenience a copy of the prediction is attached in which a solid square indicates the epicenter of a Ms=6.0 EQ that occurred today at 02:10 local time (i.e. at 23:10 GMT on April 16, 1994). A time chart is also attached [Omitted due to space limitations –RJG] for the sake of comparison of the present activity with those observed in 1988 and 1993. We draw your attention to the fact that the top time-chart corresponds to the Fig. 28A of Varotsos and Lazaridou[72] which, as mentioned in our prediction, was expected to be followed in the present case.

The authorities publicly confirmed our prediction and also explained that the strongest earthquake was actually expected during the last week (in accordance to [sic] the aforementioned Fig. 28A of Varotsos and Lazaridou[72]).

P.S. Except of [sic] the SES mentioned in our prediction, some additional SESs (not included in the attached time-chart) have been observed but they had smaller amplitudes.

Fax B – May 14, 1994

We have recorded a **SES activity** with the following characteristics:
Date: May 12, 1994 (cf. the activity started since May 3, 1974 [sic]).
Station: IOA
Components: NS - short dipole arrays and on the long dipoles.
Polarity: Negative

Impending earthquake: Magnitude: Ms(ATH) 6.2 or 5.7 for cases I and II
respectively.
MB(USGS) 5.7 or 5.2   "   "   "
Epicenter: See the attached map [Fig. 2 –RJG] (two alternative solutions*).
Time: According to Fig.28A of Varotsos and Lazaridou[72] [Top graph of Fig. 1 –RJG]
(i.e. 3-4 weeks for the first big event, etc.).

*P.S. 1) No definite discrimination can be made (based on SES characteristics) be-
tween the two alternative epicentral areas shown in the map. However, it seems
that **the solution II might be more probable.** 2) SESs have been also recorded
at KER station (not simultaneous with those at IOA and with appreciably smaller
amplitudes. (They might correspond to a M_s=4.5 - 5.0 EQ at a distance 80 km from
ATH, probable direction NE to W).

Fax A – October 16, 1994

We have recorded a **SES activity** with the following characteristics:
Date: October 16, 1994 (cf. A smaller activity started on September 28, 1994).
Station: IOA
Components: Mainly on NS - short dipole arrays (and on the long dipoles).
Polarity: Positive

Impending earthquake: Magnitude: Ms(ATH) 5.7 or 5.0 for cases I and II
respectively.
        MB(USGS) 5.0
Epicenter: See the attached map [Fig. 2 –RJG] (two alternative solutions*).
Time: According to Fig.28A of Varotsos and Lazaridou [72] or Fig.22 of Varotsos et
al[89] [Fig. 1 –RJG]

*P.S. No definite discrimination can be made (based on SES characteristics) between
the two alternative epicentral areas shown in the map.. However, it seems that **the
solution I might be more probable.**

FAX – December 11, 1994 (Success communiqué)

[Only postscript is given; time chart and map omitted. –RJG]

P.S. For your convenience I attach the time-chart (and the map of the epicenters) of
the EQs that followed our previous prediction on October 16, 1994[.]

FAX – April 7, 1995

We have recorded a **SES activity** with the following characteristics:*
Date: The activity started on April 6, 1995
Station: ASS (i.e. close to *Thessaloniki*)
Components: Both short dipole arrays (and on the long dipoles).
Polarity: Positive

Impending earthquake:
Magnitude: Ms(ATH) $\approx$ 5.2, if the epicenter lies a few tens of km around ASS
(but see below).
$$\text{MB(USGS)} \approx 4.7 \quad " \quad " \quad "$$
Epicenter: For the selectivity map of ASS, see Fig.1 of Varotsos et al.[97] If the
epicenter lies in area "c" (i.e. around *Thessaloniki*) then Ms(ATH) $\approx$5.2. For larger
epicentral distances, then M will be larger (e.g. south of, but very close to, Chalkidiki
peninsula Ms(ATH)$\approx$5.8). A more precise determination of epicenter is not possible,
because the other stations around Thessaloniki are still out of operation.
Time: According to Fig.28A of Varotsos and Lazaridou[72] or Fig.22 of Varosos et
al.[89] [Fig. 1 –RJG]

*P.S. Another SES activity was recorded at IOA the same day, i.e. on April 6, 1995,
which is not simultaneous with the aforementioned activity at ASS. It might correspond to Ms(ATH)$\approx$5.0 at 200 km west of Athens.

FAX – [April 27, April 30], 1995

Note: Where only a small segment differs between the April 27 and April 30 versions,
both are shown in brackets, with the April 27 version coming first. The differences
are so extensive near the end that the two versions are given separately.

RECENT SEISMIC ELECTRIC SIGNAL ACTIVITIES IN GREECE
by
P. Varotsos and M. Lazaridou
Dept. of Physics, University of Athens, Greece

Abstract: *Three seismic electric signal activities were recently recorded at IOA
station. They might indicate that a pronounced series of earthquakes will occur [in
western Greece, in Greece with $M_s \approx$6.0 units].*

The most catastrophic earthquake (EQ) in Greece during the last eight years
seriously damaged the village of Vartholomio on October 16, 1988. The epicenter
was at 37.90°N, 20.96°E with a surface wave magnitude (according to the National
Observatory of Athens) M_s(ATH)=6.0. It was preceded by a M_s(ATH)=5.5 EQ on
Sept. 22, 1988 at 37.99°N, 21.11°E, which destroyed Killini harbour. The prediction
of these EQs was publicly announced by Haroun Tazieff (through AFP and Antenne 2
of the French television) after the receipt of related telegrams sent by the first author.
Three seismic electrical [signal, signals] (SES) activities were recorded on Aug. 31,

Sept. 29 and Oct. 3, 1988 (at IOA station, located in northwestern Greece) which were extensively discussed by Varotsos and Lazaridou[72] and by Varotsos et al.[89]

On April 6, 1995, a SES activity was recorded at IOA station with character-istics (polarity, form, etc.) [similar, almost similar] to those of the aforementioned activities on Sept. 29 and [Oct. 16, Oct. 3], 1988, but with a smaller amplitude. By following the procedure described by Varotsos and Lazaridou[72] we concluded that a seismic activity is expected with $M_s(ATH) \approx 5.0$ units at an epicenter 200 km west of Athens. This information was sent on April 7, 1995 to the Greek authorities and to 29 Institutes abroad. In the meantime, however, two new SES electrical activities were recorded (on 18 and 19 April, 1995) at IOA station [,(see Fig.1)] that have comparable amplitudes and almost the same characteristics with those of Sept. 29 and Oct. 3, 1988. [**April 27:**] Therefore it seems probable that a $M_s(ATH) \approx 6.0$ EQ might occur close to the aforementioned Vartholomio area. (An alternative solution is $M_s(ATH) \approx 5.5$ with an epicenter $\sim$50 km from IOA station). The time evolution of the current activity might follow the same time chart as that depicted in Fig. 22 of Varotsos et al.[89]

[**April 30:**] Therefore a $M_s(ATH) \approx 6.0$ EQ might occur close to the aforementioned Vartholomio area. An alternative solution is $M_s(ATH) \approx 5.5 - 6.0$ with an epicenter a few tens of km NW from IOA station. Note that the latter solution seems to be more compatible with the experimental fact that $\Delta V/L$ differs between the two long dipoles at IOA having unequal lengths (case of nearby source). The two alternative candidates are shown in Fig.2. The time evolution of the current activity might follow the same time chart as in Fig.22 of Varotsos et al.[89]

FAX - May 19, 1995

CONTINUATION OF THE SEISMIC ELECTRIC SIGNAL ACTIVITIES IN GREECE

by **P. Varotsos and M. Lazaridou**

Abstract: *The 6.6 earthquake of May 13, 1995 was preceded by seismic electri-cal activities recorded at IOA. A similar seismic electrical activity was subsequently recorded at VOL. It seems probable that a new strong earthquake (EQ) might hit Greece. This EQ should occur at a different epicentral area but with comparable mag-nitude. The present prediction is not equally reliable with the previous one, as VOL station is not yet calibrated (because it is operating during the last 6 months).*

On April 27 and 30, 1995 a prediction was issued based on seismic electrical signal (SES) activities recorded at IOA on April 18 and 19, 1995. This prediction was actually followed by a M_s=6.6 earthquake with a USGS epicenter at 40.0°N, 21.6°E (labelled with asterisk in Fig.1). This area was previously considered to be *aseismic* because such EQ had not occurred there for a period more than 1000 years.

Since last September an experimental station is operating at VOL. The inspec-tion of its records indicated that an SES activity was recorded on April 30 (i.e. not simultaneous with those earlier recorded at IOA) on both short and long dipole arrays. Although this station is not yet calibrated we might guess that it is precursor of an EQ similar to that of May 13, because its amplitude (10 mV/kM) is comparable to those at IOA. The epicenter should be different due to th following facts.

(i) The two long dipoles installed at VOL (with lengths 22 km and 5 km) show comparable ΔV/L-values thus indicating a non nearby source (and hence the neighbouring area of VOL should be excluded).

(ii) As the SES activity was not recorded at IOA, the areas belonging to IOA selectivity map (large hatched area in Fig.1 [See Fig. 4 of this paper, which is identical to the figure cited in this fax except that it also includes the epicenter of the 15 June 1995 earthquake as marked by Prof. Varotsos. -RJG]) should be excluded. As the epicenter of May 13 seems to belong to IOA selectivity map, the regions lying in its immediate vicinity (and especially those in its western side) should also be excluded.

(iii) As the SES activity was not recorded at ASS, the area belonging to ASS selectivity map (Fig.1), i.e. that surrounded by the regions a, b, and c, (and that of Chalkidiki peninsula and the neighbouring sea) should be excluded. (Note however that the VER area still remains as a candidate area as it is not well verified that it belongs to the ASS selectivity map. Unfortunately VER is out of operation).

(iv) As the SES activity was not recorded at KER, the area lying in the vicinity of Athens (and those in Peloponese) should be excluded. GOR is not operating and its immediate vicinity cannot be excluded.

(v) The central Aegean sea (recorded at ASS or at KER) should be excluded. However the area around Skiros and Alonisos islands cannot be excluded (they are of smaller probability).

By summarising: The new EQ might occur in the remaining part (i.e. after deleting the areas due to the points (i) to (v)) of continental Greece. The spectrum of the SES activity at VOL is **strikingly similar** to those recorded at IOA (on April 18, 19) thus indicating that the EQ of May 13 and the expected EQ might belong to the **same** tectonic process which, according to our opinion, is still going on. The time evolution might follow Fig. 22 of Varotsos et al.[89]

FAX – June 15, 1995 (Success communiqué)

Subject: To the recipients of our prediction issued on May 19, 1995.

After the 6.6 earthquake (EQ) that occurred in northern Greece on May 13, 1995, we sent you a **new** prediction indicating that: "...*a new strong EQ might hit Greece. This EQ should occur at a different epicentral area but with comparable magnitude* ... ". The prediction which was issued on May 19, 1995, it [sic] was based on a strong SES activity recorded on April 30, 1995 at VOL-station (however we clarified that: "... *indicating a non nearby source and hence the neighbouring area of VOL should be excluded* ... ").

Today, i.e. June 15, 1995 at 00:15 GMT, a new strong EQ actually occurred. Preliminary announcement stated that the magnitude was 6.2 (in comparison to the predicted value of 6.6) and the epicenter 40-50 km South (or SSW) of GOR-station (see the triangle in the map attached [Fig. 4 of this paper -RJG]). Please recall that our prediction explicitly stated: "... *GOR is not operating and its immediate vicinity cannot be excluded* ... " (thus GOR area remained as candidate area for future epicenter). As for the time, we attach (for your convenience) the **expected** time chart (upper) in comparison to what actually happened in the current case (lower). Please note the striking similarity of the two time-charts.

By summarising: The 6.2 EQ today is the **third** one that hitted [sic] Greece within a short time, i.e. (a) the 6.0 EQ on May 4 close to Thessaloniki (which was predicted by our prediction issued on April 7), (b) the 6.6 EQ on May 13, 70 km ENE of Ioannina (preceded by our prediction issued on April 30) and (c) the 6.2 EQ on June 15 (preceded by our prediction issued on May 19, 1995). This unusual fact, i.e. the occurrence of three EQs with $M_g \geq 6.0$ within 1.5 month at three quite different epicentral areas, was recognised well in advance by our group (please recall that each time we forecasted [sic] a **different** epicentral area).

References

1. P. Varotsos *et al.*, *Prakt. Akad. Athenon* **56**, 418 (1981).
2. P. Varotsos and K. Alexopoulos, *Tectonophysics* **110**, 73 (1984).
3. P. Varotsos and K. Alexopoulos, *Tectonophysics* **110**, 99 (1984).
4. K. Meyer *et al.*, *Tectonophysics* **120**, 153 (1985).
5. P. Varotsos and O. Kulhanek (eds), *Tectonophysics* **224**, vii (1993).
6. "Quake-forecast method said 75% accurate," *Japan Times* (Feb. 22, 1990).
7. H. Tazieff, *Earthquake Prediction* (McGraw-Hill, New York, 1992).
8. S. Uyeda and T. Nagao, *Proc. Eq. Pred. Res. Symp.* (Tokyo), (in Japanese, with English abstract), 7 (1994).
9. E. Masood, *Nature* **375**, 617 (1995).
10. "Interview of Prof. P. Varotsos," (in Japanese) *Asahi Shinbun*, morning edition (June 22, 1995).
11. Y. Kagan and D. D. Jackson, *J. Geophys. Res.* **100**, 1943 (1995).
12. J. C. Savage, *Bull. Seism. Soc. Am.* **81**, 862 (1991); *Geophys. Res. Lett.* **19**, 709 (1992).
13. J. C. Savage, *Bull. Seism. Soc. Am.* **83**, 1 (1993).
14. J. Milne, *Trans. Seis. Soc. Japan* **1**, 10 (1880),
15. W. H. Hobbs, *Earthquake*, 308 (D. Appleton, New York, 1907).
16. A. Imamura, *Theoretical and Applied Seismology*, 346 (Maruzen, Tokyo, 1937).
17. J. Milne, *Nature* **85**, 515 (1911).
18. H. F. Reid, *The California earthquake of April 18, 1906, The mechanics of the earthquake*, vol. 2, 31 (Carnegie Institution, Washington, 1910).

19. *Nature* **87**, 16 (1911); **128**, 155 (1931); **129**, 26 (1932); **130**, 969 (1932); **149**, 640 (1942).
20. *Nature* **90**, 340 (1912); **109**, 361 (1922); **112**, 538 (1923); **119**, 684 (1927); **120**, 619 (1927); **127**, 796 (1931); **135**, 1078 (1935). H. F. Reid, *Proc. Nat. Acad. Sci. U.S.* **6**, 559 (1920). K. Sassa and E. Nishimura, *Trans. Am. Geophys. Un.* **32**, 1 (1951). E. Nishimura and K. Hosoyama, *Trans. Am. Geophys. Un.* **34**, 597 (1953).
21. *Nature* **132**, 817 (1933); **132**, 964 (1933).
22. *Science* **81**(2105, suppl.), 11 (1935).
23. H. Landsberg, *Science* **82**, 37 (1935).
24. J. B. Macelwane, *Bull. Seism. Soc. Am.* **36**, 1 (1946).
25. C. F. Richter, *Elementary Seismology* (W. H. Freeman, San Francisco, 1958).
26. C. F. Richter, *Calif. Inst. Tech. Quarterly*, 2 (Jan. 1964).
27. C. F. Richter, *Bull. Seism. Soc. Am.* **67**, 1244 (1977).
28. R. J. Geller, *Nature* **352**, 275 (1991).
29. C. Tsuboi *et al.*, *Prediction of Earthquakes: Progress to date and Plans for Further Developments* (Earthquake Prediction Research Group, Tokyo University, 1962).
30. F. Press and W. F. Brace, *Science* **152**, 1575 (1966).
31. "Earthquake prediction—skeptical mood in America," (in Japanese) *Asahi Shinbun*, evening edition, (Feb. 25, 1994).
32. C. H. Scholz *et al.*, *Science* **181**, 803 (1973).
33. F. Press, *Sci. Am.* **232**(5), 14 (1975).
34. "Forecast: Earthquake," *Time* **106**(9), 38 (Sept. 1, 1975).
35. H. Kanamori and G. Fuis, *Bull. Seism. Soc. Am.* **66**, 2017 (1976).
36. P. Leary and P. Malin, *J. Geophys. Res.* **87**, 6919 (1982).
37. J. S. Haase *et al.*, *Bull. Seism. Soc. Am.* **85**, 194 (1995).
38. R. A. Kerr, *Science* **200**, 419 (1978).
39. C. R. Allen, *Bull. Seism. Soc. Am.* **72**, S331 (1982).
40. D. Swinbanks, *Nature* **364**, 370 (1993); *Nature* **367**, 670 (1994).
41. D. L. Turcotte, *Ann. Rev. Earth Planet. Sci.* **19**, 263 (1991).
42. T. Rikitake, *Earthquake Prediction* (Elsevier, Amsterdam, 1976).
43. B. E. Smith and M. J. S. Johnston, *J. Geophys. Res.* **81**, 3556 (1976).
44. Z. Suzuki, *Ann. Rev. Earth Planet. Sci.* **10**, 235 (1982).
45. Mituko Ozima *et al.*, *J. Geomag. Geoelectr.* **41**, 945 (1989).
46. S. K. Park *et al.*, *Rev. Geophys.* **31**, 117 (1993).
47. Report of the ICSU SC-IDNDR, Contribution to ICSU's Annual Report for 1993.
48. Y. Kagan, *Physica D* **77**, 160 (1994).

49. M. Wyss(ed.), *Evaluation of Proposed Earthquake Precursors* (Am. Geophys. Un., Washington, 1991).
50. L. M. Jones and P. Molnar, *J. Geophys. Res.* **84**, 3596 (1979).
51. K. Mogi, *Pure Appl. Geoph.* **122**, 765 (1984).
52. R. J. Geller, *Proc. Eq. Pred. Res. Symp.* (Tokyo), (in Japanese, with English abstract), 131 (1994).
53. W. Thatcher, in *Earthquake Prediction— An International Review*, ed. D. W. Simpson and P. G. Richards, *Ewing Series* 4, 394 (Am. Geophys. Un., Washington, 1981).
54. R. O. Castle *et al.*, *Science* **192**, 251 (1976).
55. D. D. Jackson *et al.*, *Science* **210**, 534 (1980).
56. Y. Bock *et al.*, *Nature* **361**, 337 (1993).
57. S. Gruszow *et al.*, *Géophysique Interne* **320**, 547 (1995).
58. H. Kanamori, *Ann. Rev. Earth Planet. Sci.* **22**, 207 (1994).
59. M. Otsuka, *Phys. Earth Planet. Inter.* **6**, 311 (1972).
60. J. N. Brune, *J. Geophys. Res.* **84**, 2195 (1979).
61. P. Bak and C. Tang, *J. Geophys. Res.* **94**, 15,635 (1989).
62. D. P. Hill *et al.*, *Science* **260**, 1617 (1993).
63. M. Wyss and J. N. Brune, *Bull. Seism. Soc. Am.* **57**, 1017 (1967).
64. D. Harvey and M. Wyss, *Pure Appl. Geophys.* **124**, 957 (1986).
65. C. Mendoza *et al.*, *Bull. Seism. Soc. Am.* **84**, 269 (1994).
66. D. J. Wald and T. H. Heaton, *Bull. Seism. Soc. Am.* **84**, 668 (1994); B. P. Cohee and G. C. Beroza, *Bull. Seism. Soc. Am.* **84**, 692 (1994).
67. A. Franklin, *The Rise and Fall of the Fifth Force* (Am. Inst. Phys., New York, 1993).
68. G. Taubes, *Bad Science—The Short Life and Weird Times of Cold Fusion* (Random House, New York, 1993).
69. P. W. Anderson, *Physics Today* **45**(1), 9 (1992).
70. I. Langmuir, *Physics Today* **42**(10), 36 (1989).
71. B. V. Derjaguin, *Sci. Am.* **223**(6), 52 (1990); *Sci. Am.* **229**(3), 52 (1993).
72. P. Varotsos and M. Lazaridou, *Tectonophysics* **188**, 321 (1991).
73. P. Varotsos and K. Alexopoulos, *Tectonophysics* **161**, 58 (1989).
74. Report of the ICSU SC-IDNDR, Contribution to ICSU's Annual Report for 1994.
75. F. Mulargia and P. Gasperini, *Geophys. J. Int.* **111**, 32 (1992).
76. P. Varotsos *et al.*, Comment on "Evaluating the statistical validity beyond chance of 'VAN' earthquake precursors,"(by F. Mulargia and P. Gasperini), *Geophys. J. Int.* **111**(1), separate page (1992).
77. P. W. Burton, *Nature* **315**, 370 (1985).
78. J. Drakopoulos *et al.*, *Tectonophysics* **161**, 55 (1989).

79. J. Drakopoulos *et al.*, *Tectonophysics* **224**, 223 (1993).
80. P. Varotsos *et al.*, *Nature* **322**, 120 (1986).
81. P. Varotsos *et al.*, *Tectonophysics* **224**, 237 (1993).
82. R. J. Geller, *Nature* **373**, 554 (1995).
83. W. Leith, *EOS* **76**, 257 (1995).
84. H. Wood and B. Gutenberg, *Science* **82**, 219 (1935).
85. C. R. Allen, *Bull. Seism. Soc. Am.* **66**, 2069 (1976).
86. U.S. National Research Council, Panel on earthquake prediction of the Committee on Seismology, "A scientific and Technical Evaluation—with implications for society," U.S. National Academy of Sciences, Washington, D.C. (1976).
87. "Guidelines for earthquake predictors," *Bull. Seism. Soc. Am.* **73**, 1955 (1983).
88. "Code of Practice for Earthquake Prediction," *IUGG Chronicle*, No. 165, 26 (Feb. 1984).
89. P. Varotsos *et al.*, *Tectonophysics* **224**, 1 (1993).
90. E. Dologlou, *Tectonophysics* **224**, 189 (1993).
91. M. Wyss, Inaccuracies in seismicity and magnitude data used by Varotsos and coworkers, *Geophys. Res. Lett.*, in press (1995).
92. K. Hamada, *Tectonophysics* **224**, 203 (1993).
93. G. Stavrakakis *et al.*, "Evaluation of the correlation of SES with earthquakes in Greece time period: 1987-1989," Natl. Obs. Athens, Geodyn. Inst., Publ. No. 1 (1990).
94. M. Wyss and A. Allmann, Probability of chance correlations of earthquakes with predictions in areas of heterogeneous seismicity rate: the VAN case, *Geophys. Res. Lett.*, in press (1995).
95. H. Kabata, "Earthquake prediction: look at Greece" (in Japanese), *Asahi Shinbun*, morning edition (Dec. 24, 1994).
96. "Earthquake prediction: the decisive breakthrough" (in Japanese), *Mainichi Shinbun*, evening edition (July 23, 1994).
97. P. Varotsos *et al.*, *Tectonophysics* **224**, 269 (1993).
98. P. Varotsos *et al.*, *Tectonophysics* **188**, 403 (1991).
99. P. G. Carydis *et al.*, "The Egion, Greece, Earthquake of June 15, 1995," *Eq. Engr. Res. Inst. Newsletter (Spec. Eq. Rept.)*, 1 (July, 1995).

FORESHOCKS PRECEDING VAN SIGNALS (SES)

KEN SUDO

International Institute of Seismology and Earthquake Engineering, 1 Tatehara, Tsukuba, 305, Japan

32 earthquakes in the period of February 1987 to October 1989 were reportedly predicted by the VAN method. Investigation of the seismicity in this period reveals that some of the earthquakes were preceded by small earthquakes which had occurred within 11 days before the emission of the Seismic Electrical Signals (SES). This fact would be helpful to elucidate the VAN prediction, because so far very few seismological analyses have been made on earthquakes predicted by VAN.

1 Introduction

Until recently, the present author was a member of the United Nations Secretariat for IDNDR (International Decade for Natural Disaster Reduction) and, moreover, took part in those meetings of the Special Committee for IDNDR of ICSU (International Council of Scientific Unions) at which some discussions about VAN led to the decision[1] to hold the present review meeting. In the mean time the Secretariat for IDNDR at Geneva was receiving inquiries about VAN from many people, such as school teachers, journalists, bankers and others, as well as people involved in earthquake disaster reduction in several countries.

In order to provide the inquirers with relevant information, currently the Secretariat is collecting information or views on VAN. Stavrakakis of the National Observatory at Athens Seismological Institute, Greece, provided the Secretariat with a 357-page publication[2]. He investigates 32 earthquakes in the period from February 1987 to October 1989 in Greece which were reportedly predicted by the VAN method, referring to activities of earthquakes with the magnitude of 3.5 or larger (a local magnitude, ML, determined by the Observatory) which took place from 11 days before the SES emission to 11 days after the SES emission in and around Greece (Appendix). This short note is to discuss the data for the 32 earthquakes.

2 Foreshocks

It is seen in the Appendix that most earthquakes which were predicted by VAN were preceded by one or more earthquakes close to their hypocenters. For convenience of reference, the preceding earthquakes are called hereafter foreshocks and the earthquake predicted by VAN are called the mainshocks. It should be noted that the foreshocks can be found before the emission of the SES. The following criteria for foreshock identification are used: (1) it occurs between SES and 11 days before SES; (2) its magnitude is less than 5.0; and (3) its location is within 0.3 degree in both the latitude and longitude around the forthcoming mainshock. As per the criteria, 19 SES out of the 32 SES were preceded by foreshocks. 8 SES were emitted during

239

active earthquake swarms or aftershock activities so that foreshocks could not be identified uniquely.

5 earthquakes which reportedly were predicted by VAN were not accompanied by any foreshocks within 11 days before the SES. Instead, however, 3 couples of earthquakes which occurred at almost the same locations sandwiched the SES emission in terms of its occurrence time.

3 Preliminary Analyses of the Seismicities

The following analyses are made to find clues for elucidation of the VAN signals, although the number of cases is not large:

(i) Two time intervals are measured in hours: the interval, T1, between the foreshock and the corresponding SES; and the interval, T2, between the SES and the mainshock. No clear correlation between T1 and T2 is seen (Fig. 1);

(ii) Magnitude differences between fore- and mainshocks against T1 are investigated (Fig. 1). It seems that the magnitude of the mainshocks would increase with the time interval T1 .; and

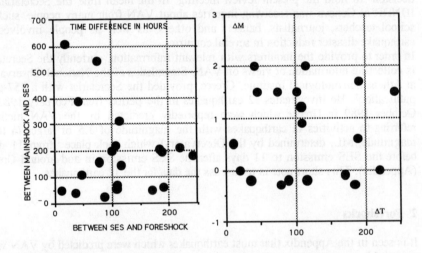

Fig.1 Time differences between SES and mainshocks (left) and magnitude differences between mainshocks and foreshocks (right) are plotted against time differences between SES and foreshocks.

(iii) Supposing that (a) SES be emitted not from the earthquake source but from a site close to observation sites and (b) the emission be triggered by the arrival of the disturbance which was caused by the foreshock(s), the apparent speed of the propagating disturbance is estimated (Fig. 2). The speeds are between 10 cm/sec and 100 cm/sec which are almost same as the speeds with which earthquake activities migrate or trigger geophysical activities at remote sites[3,4].

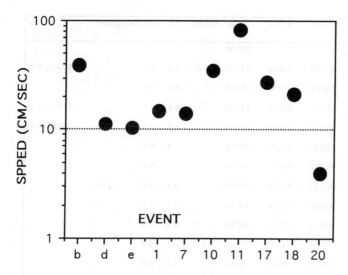

Fig. 2 Apparent speeds of disturbance propagation which was generated at the source of foreshock and reached sites close to observatories are estimated for 10 SES events(horizontal axis denotes event code used in the publication by Stavrakakis)

4 Conclusion

The objective of this note is to emphasize the following: (1) Apparently the above primitive study provides us with an additional clue to understand what SES reflects in terms of the physical process of earthquake occurrence; and (2) at the IDNDR Secretariat, the present author considered that world wide collaboration between seismologist groups and VAN groups should be urged to elucidate VAN so that the availability of VAN for earthquake disaster reduction would be clarified[5,6].

Appendix

The information about 32 earthquakes which were reportedly predicted by VAN is compiled into 11 columns of the following table: (1) event code used in the publication by Stavrakakis; (2) the time difference in hours between the foreshock and the SES; (3) the time difference between the SES and the mainshock; (4) the occurrence time of the foreshock, month, day(2 digits), and hour(2 digits); (5) the time of SES incidence; (6) the occurrence time of the mainshock; Station(s) recording SES; (7) the local magnitude of the foreshock; (8) the local magnitude of the mainshock; (9) the background activity around the SES incidence. Any high background activity made it difficult to identify foreshocks; and (10) earthquakes which were not predicted.

Event	T1	T2	Fore	Signal	Main	Station	Mf	Me	Background	No-Predicted
a					22620					
b	20	606	42508	42604	52110	PIR	3.6	3.6		52918(5.5)
c	89	23	60907	61300	61323		3.8	3.6		
d	220	206	12306	20120	21010		4.1	4.1		21811(5.1)
e	189	194	30222	31019	31821		3.8	4.4		32620(5.4)
f	78	154	33010	40216	40902		4	3.7		
g		35	40320	40319	40506			3.9		
h	170	50	33114	40716	40918		4.3	4.2	many	
i	36	40	42006	42118	42310		3.7	3.5		
j	185	201	42102	42819	50704		3.8	3.5	?	
1	190	59	50720	51518	51805	IOA	4.1	5.8		
2		27	52114	52100	52203		4.1	5.5		
3	110	67	52601	53015	60210		4.1	5		
4	11	48	60319	60405	60605		4.5	5		
5					61005				swarm	
6	117	141	61603	62106	62703		4.3	4.1		
7	109	51	70412	70923	71202	PIR	3.8	5		
7	109	51	70412	70923	71202	PIR	3.8	5		
8				71302		KOR		5	aftershock	71601(5.0)
9			71423	71800	72309	NAF			aftershock	
10	75	540	82721	83100	92212	IOA	4	5.5		
11	42	386	92816	93010	101612	IOA	4	6		
12	115	314	92816	100311	101612	IOA	4	6		
13				102021		IOA			Aftershocks	110808(5.3)
14				30209		IOA				No earthq.
15	46	110	60207	60305	60719	IOA	3.7	5.2		
16				61308		IOA				No earthq.
17	102	199	71913	72319	80102	KER	3.5	5		
18	108	218	81012	81500	82402	IOA	4.5	5.7		
19				82400					Foreshock	
20	243	179	90117	91120	91907	IOA	3.5	5		
21				91509		ASS			Swarm	
22				101600		IOA			Swarm	

References:
1. J. Lighthill in 1994 ICSU Annual Report (International Council of Scientific Unions, Paris, 1995).
2. Stavrakakis, G.N. et. al. Evaluation of the Correlation of SES with Earthquakes in Greece, 1987-1989, National Observatory of Athens Seismological Institute (1990).
3. Mogi, K. Bull. Earthq. Res. Inst, 46, 53-67 (1968)
4. Rydelek, P.A. & Sacks, I.S. Nature, 336, 234-237 (1988)
5. Geller, R. Nature, 373, 554 (1995).
6. Swinbanks, D. Nature, 374, 754 (1995).

BEHIND VAN: TECTONIC STRESS CHANGES OR EARTHQUAKE INDUCED ALERTNESS?

F. MULARGIA

Dipartimento di Fisica, Settore di Geofisica, Università di Bologna,
Viale Berti Pichat 8, I-40127 Bologna, Italy

P. GASPERINI

Dipartimento di Scienze della Terra, Università di Firenze, Via La Pira, 4,
I-50121 Firenze, Italy

The largest earthquakes which occurred in Greece in the period 1987–1992 with epicenter in the operational region of the VAN network have been identified on the basis of the average recorded values for m_b, M_S and M_L. About half of these earthquakes associate with predictions following them while less than one third associate with the ones preceding them. We propose two possible explanations, neither appearing entirely satisfactory. The first one assumes that VAN signals have a physical basis similar to that of the electric signals measured in the laboratory by Hadjicontis & Mavromatou [1], i.e. they are originated by changes in stress, and occur therefore more after than before earthquakes. Alternatively, the prevalence of backward associations is originated by an unconscious increase in "sensitivity" of the observers after an earthquake with large felt effects.

1 Introduction

We study here the association of VAN signals with the largest earthquakes in Greece in the period 1987 – 1992, for which a complete list of predictions has been published. Restricting attention to the largest events hopefully helps to better understand and evaluate VAN predictions: if a proposed precursor fails to associate with the largest earthquakes it is both a) difficult to justify in terms of physics, and b) of little practical use.

An obvious prerequisite to this attempt is the identification of the largest earthquakes which occurred in Greece, the only region in which VAN predictions have been claimed successful. More specifically, we consider events with epicenters only in the areas of Greece which the VAN network covers. Wyss[2] called attention to the fact that magnitudes were not correctly treated by VAN and by papers which tried to evaluate the efficiency of the VAN method. Essentially, a bias had been introduced in translating magnitudes from the measured value into the scale used by VAN, which, in spite of being declared as M_S, is not the surface wave magnitude.

2 The largest earthquakes in Greece in the period 1987–1992

Magnitude is a fairly inefficient way to parametrize the size of an earthquake and different magnitude definitions yield somehow different numbers. Yet, it is still the easiest, the most widely used, and in many cases the only available standard to gauge earthquakes. We therefore restrict our analysis to the set of events with the largest magnitude. As a source we use the NEIC PDE catalog, which includes both local and global records, and consider i) the body wave magnitude m_b; ii) the surface wave magnitude M_S; iii) the local magnitude M_L.

A comparison of the largest events according to the three magnitude scales shows that both the M_L and the M_S files are affected by major incompleteness: for example, the M_L scale lacks the 1990/12/21 earthquake, which according to the other two scales is the largest of the whole period ($m_b = 5.8$ and $M_L = 5.9$), while the M_S scale lacks the 1988/05/15 earthquake, which the other two scales file both with magnitude 5.4. Nor is there apparent any obvious origin for this incompleteness, like for example a deeper hypocenter. We might have been inclined to base our event list on the m_b scale. However, in order to get a more reliable estimate of size, we take the arithmetic average M_{ave} of the magnitude values in the different scales.

3 Scoring VAN prediction: the alarm rate for large earthquakes in 1987–1992

We take into account the VAN claimed selective antenna effect by assuming as operational the region of the Greek territory sited within the epicentral indetermination Δr (which was recently confirmed to be 120 km by Varotsos *et al.*[7]) from at least one prediction; the rationale for this is that since earthquakes falling within this region would be counted if correctly predicted, they must be counted also when they are not (cf. the *non–homogeneous region* of Mulargia & Gasperini[3 4 5 6]).

We aimed at studying the 25 and the 10 largest earthquakes, but ties in the average magnitude values prevent these numbers from being exactly reached. The closest numbers are the 27 events which occurred in the operational region with $M_{ave} \geq 4.9$ and the 11 with $M_{ave} \geq 5.1$. The list of the 27 largest events in the period 1987-1992 within the operational region is reported in table 1, which shows also the top 11 in bold face. The right part of the table shows the 42 testable predictions issued in this period as identified by Mulargia & Gasperini[5 6].

We associate VAN predictions with earthquakes as follows:

a) since VAN calibrated their empirical scheme for estimating magnitude on a different scale, we score as correct a prediction independently of the difference between predicted and real magnitude (i.e. we disregard ΔM);

b) for the same reason we score as correctly predicted events within 22 days from all predictions;

c) in order to better understand the physical nature of VAN signals we not only score the association of predictions with earthquakes following them, but also with the ones preceding them within the same time–distance window.

4   Results and discussion

The most important result emerging from Table 1 seems the stability of performance with respect to magnitude increase: the forward alarm rate for the 27 largest events is 7/27, while the same for the top 11 is 3/11; a symmetrical coherence is exhibited by the backward association in time, which is 9/27 for the basic set and 5/11 for the top 11. How should this evidence, which was identically apparent for smaller events (cf. Mulargia & Gasperini[3 4 5 6]), be interpreted? The marked predominance of the backward on the forward association stands for a prevalent postseismic nature of VAN signals. In turn, this prompts another question: where does it come from?

We propose two different interpretations, neither appearing entirely satisfactory.

Let us assume that VAN predictions do have a physical basis, an issue which has recently been offered some laboratory support by Hadjicontis & Mavromatou[1], who measured electric signals reminiscent of VAN in a controlled rock straining experiment. However, in agreement with all the (qualitative) VAN models so far proposed, the measured signals appear merely related to changes in stress and not to impending failure.

Combining this with the fact that perturbations to the stress field in and around the focal zone are much larger after an earthquake than before, (an indirect evidence of this being that aftershocks largely outnumber foreshocks) one should observe more associations in reverse than in forward time. However, to fully accept this view one should explain the origin of the apparent detection threshold which, absent in the laboratory, prevents *in situ* to measure electric signals before 8 of the 11 top earthquakes and after 6 of them. Note that following this reasoning one should expect VAN signals in concomitance with all stress variations above this given threshold, be them preceding, following or independent of a sizable earthquake.

The alternative explanation is different. Let us assume that VAN predictions are due to an unconscious reaction that makes VAN researchers more "sensitive" when looking at their records after a sizable earthquake. A similar effect is sometimes observed in the discrepancies of the written daily reports on erratic animal behaviour written before an earthquake with the reports describing pre-earthquake behaviour compiled by the same person after the event. We have then to account for the fact that the VAN researchers are neither expert seismologists, nor do they run any seismic network. Their real–time information on earthquake occurrences is likely to rely on the news, which, in turn, are essentially based on damage and felt effects. VAN is therefore expected to issue more predictions after the earthquakes which appeared large in terms of felt effects. While this issue would require a cross–check with the Greek press of the time, the generic (though by no means strict) correspondence between the felt effects of an earthquake with its magnitude should explain the observed prevalence of the backward time association. The correct predictions would then be essentially lucky strikes favoured by the high frequency of both predictions and earthquakes, the large indetermination allowed for the predicted epicenter, and the fact that both real and predicted magnitudes are approximately constant (cf. table 1). However, this reasoning can hardly explain the successful prediction of the 1987/2/26 event, which occurred in a seismically quiet period and with very infrequent predictions.

References

1. V. Hadjicontis and C. Mavromatou, *Geophys. Res. Lett.* **21**, 1687 (1994).
2. M. Wyss, *Geophys. Res. Lett.*, in press (1995).
3. F. Mulargia and P. Gasperini, *Geophys. J. Int.* **111**, 32 (1992).
4. F. Mulargia and P. Gasperini, *Geophys. J. Int.* **115**, 1199 (1993).
5. F. Mulargia, and P. Gasperini, *Geophys. Res. Lett.*, in press (1995a).
6. F. Mulargia, and P. Gasperini, *Geophys. Res. Lett.*, in press (1995b).
7. P. Varotsos, *et al.* , *Geophys. Res. Lett.*, in press (1995).

Table 1. Arrows connect associated predictions and earthquakes in forward and reverse time. Below them is shown the distance in km between predicted and real epicenter.

	Earthquake				Prediction		
Date	Coordinates		M_{ave}	Date	Coordinates		M
				1987/ 2/26	37.94	20.32	6.5
1987/ 2/27	38.47	20.27	5.5 _(59)_				
				1987/ 4/27	37.67	21.46	5.5
1987/ 5/29	37.52	21.57	4.9				
1987/ 6/10	37.22	21.45	5.2 _(83)_	1987/ 6/13	37.97	21.46	5.2
1988/ 1/ 9	41.23	19.62	5.5				
				1988/ 2/ 1	39.25	25.39	5.0
				1988/ 3/10	40.18	20.83	5.0
					38.85	20.97	5.0
1988/ 3/26	40.17	19.87	4.9 _(81)_				
				1988/ 4/ 2	37.95	20.89	5.0
					36.06	21.39	5.5
			_(74)_	1988/ 4/ 3	38.89	23.75	5.0
				1988/ 4/ 7	38.82	21.08	5.0
					40.24	20.75	5.0
				1988/ 4/28	37.94	20.32	5.0
					38.98	20.54	5.0
				1988/ 5/15	40.05	21.00	5.0
					37.94	20.32	5.3
1988/ 5/18	38.42	20.47	5.4 _(55)_				
			_(55)_	1988/ 5/21	37.94	20.32	5.3
					40.18	2..o3	5.0
1988/ 5/22	38.40	20.45	5.0 _(53)_ _(53)_	1988/ 5/30	37.94	20.32	5.4
					40.18	20.83	5.0
				1988/ 6/ 4	37.94	20.32	5.0
				1988/ 6/10	36.70	22.16	5.1
				1988/ 6/21	37.94	20.32	5.0
				1988/ 7/10	37.13	21.24	5.2
				1988/ 7/13	37.98	22.95	5.0
				1988/ 9/ 1	37.96	21.01	5.8
					39.87	21.26	5.3
				1988/ 9/30	37.96	21.01	5.3
					40.05	21.00	5.0
				1988/10/ 3	37.96	21.07	5.0
					37.98	22.95	5.0
1988/10/16	37.93	20.90	5.6 _(13)_				
			_(8)_	1988/10/21	37.96	21.01	6.4
					40.48	20.40	5.5
				1989/ 3/ 2	37.94	20.32	5.4
					40.05	21.00	5.0
1989/ 3/17	41.23	19.87	5.0				
1989/ 3/19	39.25	23.50	5.3				
				1989/ 6/ 3	37.94	20.32	5.5
					40.18	20.83	5.0
1989/ 6/ 7	38.03	21.62	5.0 _(114)_ _(16)_	1989/ 6/13	37.97	21.46	5.2

Table 1. (continued)

Date	Earthquake Coordinates		M_{ave}		Date	Prediction Coordinates		M
					1989/ 7/23	38.24	24.07	5.0
					1989/ 8/16	38.66	21.62	5.0
1989/ 8/20	37.25	21.20	5.5					
				15	1989/ 8/24	38.62	21.73	5.2
						37.13	21.24	5.8
1989/ 8/24	37.98	20.15	5.1					
1989/ 9/ 5	40.20	25.08	5.0					
					1989/ 9/11	38.62	21.73	5.2
						37.13	21.24	5.8
					1989/10/18	37.96	21.01	5.5
					1990/ 4/26	37.96	21.35	5.8
1990/ 5/14	40.67	19.78	5.0					
					1990/ 5/28	37.96	21.35	5.8
1990/ 6/16	39.25	20.50	5.4					
1990/ 7/31	37.23	21.45	4.9					
1990/ 8/ 4	39.28	20.42	4.9					
					1990/10/20	37.95	20.78	5.6
1990/12/21	41.00	22.27	5.9					
				61	1991/ 1/ 2	40.90	23.00	5.5
1991/ 2/19	37.53	20.88	4.9					
					1991/ 2/23	40.00	21.50	6.0
1991/ 3/15	39.32	20.50	4.9	113				
1991/ 3/19	39.23	20.45	4.9					
1991/ 6/26	38.43	21.08	5.0					
					1991/12/27	37.96	21.01	5.7
						40.50	20.00	5.0
1992/ 1/23	38.33	20.32	5.1					
1992/ 1/23	38.33	20.28	4.9	75				
					1992/ 1/27	37.96	21.01	5.2
					1992/ 2/22	38.00	20.50	5.0
					1992/ 2/22	38.62	21.73	5.0
					1992/ 2/22	37.17	21.35	5.4
						36.51	21.93	6.0
					1992/ 3/13	37.98	25.23	5.4
						40.05	26.49	5.4
					1992/ 3/31	37.96	21.01	5.3
					1992/ 4/22	37.40	21.25	5.5
						38.20	21.40	5.0
1992/ 5/30	38.05	21.42	4.9					
1992/ 6/21	39.13	19.78	4.9					

BRIEF SUMMARY OF SOME REASONS WHY THE VAN HYPOTHESIS FOR PREDICTING EARTHQUAKES HAS TO BE REJECTED

M. WYSS

Geophysical Institute, University of Alaska Fairbanks, AK 99712, USA,
max@giseis.alaska.edu.

The hypothesis that seismic electric signals (SES) are precursors to earthquakes cannot yet be tested rigorously because the parameter of precursor time, Δt, is inadequately defined as "of the order of 1 month" and "a few weeks (e.g. ≤ 22 days)" for two signal types, respectively. Assuming that $\Delta t = 22$ days, magnitude uncertainty $\Delta M = 0.7$ and location uncertainty $\Delta R = 100$ km, and using the declustered earthquake catalog of the Seismological Institute National Observatory Athens (SINOA) to estimate the background rate for each specific area of prediction (except for predictions during times of aftershock activity when we used an appropriately elevated rate for the background estimate), we find that the predictions listed by Varotsos and Lazaridou [1] lead to a poor prediction performance record. During the period covered in which VAN made 23 predictions, analyzed by Wyss and Allmann [2], 466 earthquakes eligible for prediction occurred. 456 (98%) of these were missed and 11 correlated with predictions. The remaining 12 predictions were false alarms. Given the high probability that earthquakes occur at random in the prediction window, $p = 0.62$, this performance is likely to be achieved by chance at the 96% confidence level. Since the average probability of an event as specified to occur at random is larger than 50%, earthquakes occur too frequently to effectively test the hypothesis that VAN perform better than random guessing. This problem could be corrected by raising the magnitude level.

VAN's current hypothesis (Δt measured in weeks) was preceded by at least three different versions ($\Delta t = $ a few minutes in early 1981, $\Delta t = 6$ to 11 hours in 1981/83 and $\Delta t \leq 1$ week in 1985/87). At the time of these earlier versions of the hypothesis hundreds of successful correlations and predictions were claimed, and the hypothesis was vigorously promoted by VAN on television and in the newspapers, but these versions have been abandoned as apparently incorrect. It follows that these hundreds of predictions were incorrect.

The changes in magnitude scale used by VAN have contributed to the confusion on how to evaluate the performance of the VAN predictions. In the early work (1981/83) ML was used. In 1984 a new scale, $Ms(VAN) = ML(ATH) + 0.4$, was introduced. In 1985 $ML(ATH)$ was used again. At an unknown time $Ms(VAN) = ML(ATH) + 0.5$ was adopted. For evaluations of the method it is important to use the same scale for defining background rates, missed events and successes. It is of course essential to use the same minimum magnitude for all of these estimates: If an earthquake of $Ms(VAN) = 4.3$ is accepted as a success, then the background rate for estimating the probability of occurrence at random must be based on $Ms(backgr) = 4.3$ also, and the same is true for estimating the number of missed events. Varotsos et al [3] appear to violate this simple requirement of logic.

The amount of precursory slip that might occur on a fault plane before the typical earthquake "predicted" by VAN is estimated as less than 1 millimeter. This result is derived from estimating the slip in a $Ms(VAN) = 5.0$ earthquake as approximately 5 cm (dividing the seismic moment by an estimate of the rupture area and an assumed shear modulus). Based on observations elsewhere we know that precursory slip must be smaller than a few percent of the coseismic slip, if it occurs. It seems unlikely that this small slip could generate an electric signal that can be measured at hundreds of kilometers distance. Also Bernard [4] showed that long distance electrotelluric precursors are implausible. In addition the absence of coseismic signals are an argument in favor of rejecting the existence of a connection between SESs reported by VAN and earthquakes.

Comparing the current stage of the VAN hypothesis with 40 other proposals [5] for successful precursors I conclude that its development lags far behind some precursor hypotheses and that it is at a preliminary stage of development similar to many other as yet poorly supported hypotheses. Currently I see no evidence that suggests that the VAN hypothesis is correct.

1   Lack of Precise Definition of Prediction Parameters

The lack of adequate definition of the parameters of the hypothesis that SESs are earthquake precursors is a major problem. Several authors have argued that the VAN hypothesis cannot and should not be evaluated because it is too poorly defined (e.g. [2,6,7,8,9,10,11]). The problem that some parameters in prediction hypotheses are not precisely defined has also surfaced in work other than that of VAN. For example, the debate about the usefulness of seismic gaps to identify locations of increased probability for main shocks showed that a tightening of definitions improved the understanding of the problem [12,13,14,15]. In the case of VAN, no progress has been made because the authors did not respond to the criticism; their latest definitions are still unclear [3]. In this section three aspects of VAN's imprecision are discussed.

(1) The magnitude scale used is not specified in the telegrams sent by VAN, which are said to be the basis of VAN's prediction record. This leads to confusion. Supporters and critics of the VAN work are not clear to this day what exactly the word "magnitude" means in these telegrams. Varotsos pointed out to me that Varotsos and Alexopoulos [16] defined $Ms(VAN)$ in a footnote according to equation (1) and later [1] defined $Ms(VAN)$ according to equation (2). So what should the reader assume is the "magnitude" referred to in a telegram sent in 1987/1988 (between these two definitions) and reported by Dologlou [17]? First, the reader has to assume that "magnitude" or "M" refers to surface wave magnitude, and second the reader has to make a choice of whether to use the definition in (1), which had appeared in print at the time of the telegram, or whether the definition of (2) is to be applied retroactively. This lack of clarity in defining magnitude leads to problems of interpretation and is unsatisfactory. Additional problems introduced by the choice of magnitude

scale by VAN are discussed in a later section.

(2) The refusal of VAN to clearly define the range of precursor time Δt poses the most significant problem for testing their hypothesis. In spite of criticism by nearly everyone who has tried to evaluate their work, they insist on vague statements as "Δt is of the order of 1 month", "$\Delta t \leq$ a few weeks (e.g. ≤ 22 days)" [3]. These unclear statements are compounded by the VAN proposal that there are three different types of precursory signals, each of which has its own Δt, different from the others. Examples of the confusion created by the lack of precise definitions of Δt were shown at the London conference in May 1995, when Uyeda, along with other speakers, tested the VAN hypothesis assuming $\Delta t \leq 22$ days. Varotsos objected, saying one should use $\Delta t \leq 11$ days, except for some cases so identified by him. This type of fluidity of definition is unacceptable, and renders the hypothesis untestable.

(3) The maximum separation distance, R, between predicted location and epicenter of the "correlated" earthquake also has varied in different publications of VAN. Currently, most people assume that $R = 100$ km is the value to use.

Varying the precursor parameters is acceptable during the time in which a hypothesis is formulated. During this initial research stage the authors are entitled to optimize the parameters to achieve a maximum correlation of precursors and main shocks. Changes made in the definition of parameters during this phase of the work should be identified as being part of the development stage of the hypothesis and a result of the optimization process. During testing of the performance of a hypothesis, and especially during its application phase to real time predictions, the parameters cannot be changed any more. The continued changing of the prediction parameters by VAN up to the present includes several retroactive adjustments (e.g. [8]). This is, of course, not part of testing a hypothesis; this is part of developing it, and cannot be accepted at this stage when VAN wish to have their method implemented.

2   Developing a Hypothesis

After an initial formulation of a hypothesis, it may be that additional information becomes available that forces a revision. If this happens, the revision, the reasons for it and the consequences for previous claims must be stated. VAN have changed their hypotheses more than once, and significantly, without acknowledging that they did so. In 1981 they wrote "In a former paper (Praktika of Academy of Athens 56, 277, 1981) precursor changes of the telluric currents were described; they occur[r]ed a few minutes before each earthquake. Subsequently, electric signals of another type have also been found; they occur about

Table 1: Changes in the VAN Hypothesis.

Hypothesis No.	Date	Precursor Time	No. of Cases	Ref.
I	early 1981	a few minutes		20
II	late 1981	7 hours and 20 min	hundreds	18
IIa	1983	6 to 11 hours	143	21
III	1985/87	$\leq 1 week$		22, 23
IVa	1991/95	order of a month		1, 3
IVb	1991/95	$\leq 11 days$		1, 3
IVc	1991/95	$\leq 22 days$		1, 3

seven hours before the impending earthquake." ([18], page 417). They continue in that paper to claim that a one-to-one correlation between SESs (then called ESs) and earthquakes existed with a time lag of seven hours, and they state "This has been ascertained in hundreds of cases" (page 418). However, Wyss [19] showed that the earthquake record on which this claim was based was incorrect. It was not the record as reported by SINOA. It omitted major events and introduced events for which no evidence existed. For example, a seismic signal registered at Athens on 4 October, 1981 at 00:21, which was identified as a teleseism in the "Seismogram Readings" of SINOA, and which was located in Papua-New Guinea, was mistakenly assumed to be a local event and claimed as a successful correlation with an SES by Varotsos et al [18].

Varotsos et al [18] introduced the first major change in their hypothesis by modifying the precursor time to seven hours, whereas in their earlier paper [20] they proposed that precursory electric signals preceded earthquakes by "a few minutes." After 1981 there seem to be no further references by VAN to precursors occurring "a few minutes" prior to earthquakes. The reader is thus forced to assume that the initial hypothesis, as formulated by Varotsos et al [20], was wrong and was abandoned.

In 1983, the second version of the hypothesis was modified slightly as the precursor time was said to be "6 to 11 hours" [21]. This version of the hypothesis was claimed to have been successful in 143 out of 150 attempts. In 1985, Δt was expanded to one week [22,23]. In 1991, the definition of Δt was further widened to include three categories (Table 1) and the precursor time henceforth extends over weeks and months, no longer over hours. The current definitions (IV) technically still include the definition of stages I and II, but

Varotsos and Lazaridou [1] had no cases in their tables in which the precursor time was as short as 7 hours, except for one possible case on 21/10/88. So I conclude that another major revision of the hypothesis took place between 1987 and 1991. Additional revisions of which I am not aware may also have taken place.

The reader is to conclude, then, that hundreds of predictions by VAN and their verifications of precursors have been wrong (those of stages I, II and III, Table 1), because these hypotheses are abandoned now, although they had, at the time, been vigorously claimed to be correct in the scientific literature [18,20,21], in the newspapers, on radio and on television.

3   Magnitude Scales Used by VAN Add Confusion to the VAN Prediction Record

The magnitude, M, was introduced by Gutenberg and Richter [24] with the purpose of classifying earthquakes by a single parameter proportional to the energy radiated by them. Because earthquakes are complex processes, the measurement of an amplitude at a single frequency (the usual basis for M) is not entirely adequate to define the phenomenon. Today, several magnitude scales are in use. These scales are based on different types of waves, and are valid for different distance ranges. The various scales were designed to coincide, but these attempts were not entirely successful. Consequently, one has to empirically find the conversion equations for the various scales for the various regions in which these scales are used. Unfortunately, magnitudes may not be equivalent in all regional seismograph networks, and they also change as a function of time, chiefly due to modernization of observation techniques.

In the case of VAN, a problem first arose because they did not specify the scale in their telegrams [17]. This is a serious omission because it introduces an uncertainty that ranges for most cases from 0.5 to 1.0 magnitude units. Second, most readers did not realize until recently [19] that VAN used the expression "surface wave magnitude" and the symbol "Ms" to mean

$$Ms(VAN) = ML(ATH) + 0.4 \qquad (1)$$

$$Ms(VAN) = ML(ATH) + 0.5 \qquad (2)$$

([16,1], respectively). In the professional literature the surface wave magnitude scale is usually used for large earthquakes ($Ms > 6$). When the scale is used for moderate events ($Ms = 5$) it is understood to be the scale used in the Preliminary Determination of Epicenters world-wide catalog (PDE). In 1983, Varotsos et al [21] had used an obscure identification of magnitude "5.5 R," which I interpret to mean Richter magnitude, which would mean ML. It appears that the first mention of the unusual magnitude scale used by VAN (equation 1) occurs in 1984 in a footnote by Varotsos and Alexopoulos [16]. In 1985 $ML(ATH)$ is used again for VAN predictions [22]. Thus it seems that VAN may have switched the scale they used sometime between 1984 to 1987 from ML to their initial definition of Ms (equation 1), and then at an undisclosed subsequent time, on or before 1991, they switched once more, this time to the definition according to equation (2).

Basically, it does not matter which scale one uses, provided it does not saturate for the range in question. For moderate earthquakes $(4 < M < 6)$ the local magnitude scale is most appropriate. The locations for earthquakes in this range are most accurate, and the list is most complete, in bulletins based on local and regional seismograph networks. In the case of Greece, the PDE listing is incomplete in the magnitude range in question, whereas the Bulletin of SINOA is complete. So, in the case of VAN, the authoritative source of the earthquake record is the Bulletin of SINOA. These locations and the $ML(ATH)$ should be used without modification to determine whether a correlation exists between SESs and earthquakes.

Therefore, VAN were correct in using $ML(ATH)$ as the basic information on the size of the earthquakes they claimed as predicted. However, by adding 0.5 and calling the result Ms, when they should have subtracted 0.5 on average (the scales are about equivalent at $M = 5.5$ but below that $Ms < ML$) since

$$Ms(PDE) = 1.5ML(ATH) - 2.62, \qquad (3)$$

they caused their $Ms(VAN)$ estimate to be larger by about an order of magnitude than the $Ms(PDE)$ values [19]. They also opened the door for two types of errors:

(1) Scientists who did not realize that $Ms(VAN) = Ms(PDE) + 1.0$ on average excluded most earthquakes contained in the PDE from the estimates of background rate and missed events rate, when evaluating the Van prediction performance. When VAN announce that they predict an "Ms=5" earthquake, it means that they announce an $ML(ATH) = 4.5$ event, which corresponds to a $Ms(PDE) = 4.1$ quake (using equation 2). If one estimates the background rate, and hence the probability of such an event occurring at random, by counting $Ms(PDE) \geq 5$, instead of $Ms(PDE) \geq 4.1$ earthquakes, one underestimates the background rate by about an order of magnitude and the estimate of the probability of chance correlation will be entirely wrong.

(2) By speaking of an $Ms = 5$ earthquake, one creates the false impression that the source radius may be about 3 km, when in reality it should be estimated as approximately 1 km. The size of the earthquake that is thought to

be associated with a precursory SES has important implications regarding the plausibility of a mechanism that could generate precursory signals (see below).

4   Equal Yardsticks for Measuring Success, Failure and Background Rate

It is a basic requirement of logic that the minimum magnitude above which we count successful predictions ($Mpred$) be identical to that above which we count failures ($Mfail$), as well as to that which is used for estimating the normal background rate ($Mbackgr$).

$$M(pred) \equiv M(fail) \equiv M(backgr) \tag{4}$$

In turn, the background rate forms the basis for calculating the probability that the "predicted" event may have occurred at random. Surprisingly, Varotsos et al [3] seem not to appreciate the elementary point expressed in (4).

VAN stated that in their recent work they predict $M \geq 5$ earthquakes. However, they list as successes small earthquakes down to $M = 4.3$, and even $M = 4.0$, fulfilling their predictions of $M \geq 5.0$ earthquakes ([1,25]). VAN also published $\Delta M = \pm 0.7$, and in one instance $\Delta M = \pm 1.0$ ([25], page 271) as the allowable error in their magnitude estimates. At times, VAN predicted earthquakes smaller than $Ms(VAN) = 5$, accepting correspondingly smaller shocks as successes [2]. This means that VAN is claiming that precursory SESs are emanated by small earthquakes, down to $Ms(VAN) \approx 4.0$ at least, and VAN are attempting to establish a success history on the basis of correlating their predictions with these small earthquakes.

There is nothing wrong with defining the events eligible as successes as $Melig = Mpred \pm \Delta M$ and accepting success on this basis. What is wrong is VAN's attempt to force others to estimate the background and the failure rate on the basis of events with $M \geq Mpred$, without counting the vast majority of events eligible for successes in the range $Mpred > M \geq (Mpred - \Delta M)$.

Clearly, if an earthquake of $M = 4.3$ is claimed as a fulfillment of a VAN prediction, the probability that this event could have happened by chance in the window of a VAN prediction must be evaluated based on the rate of $M \geq 4.3$ events, not based on that of $M \geq 5.0$ earthquakes, as demanded by Varotsos et al [3]. Also, any earthquake that occurs within the area and time in which VAN attempts predictions and has a magnitude larger than the minimum magnitude eligible for successes, $M \geq Melig$, has to be counted as a failure if it was not predicted.

If this simple rule of logic is observed, one finds that on average the probability is $p=0.62$ for an earthquake to occur as specified by VAN within 100 km from their predicted epicenters ([2], using the SINOA catalog). With these odds it is more likely than not that a random earthquake fulfills the predictions by VAN in most cases. Therefore, the data set contained in the tables of Varotsos and Lazaridou [1] is not suitable for testing the hypothesis that SESs may be correlated with earthquakes. The magnitude threshold of predictions, background and failures should be raised such that $p < 0.5$ for testing the hypothesis.

Recognizing that the typical earthquake predicted by VAN has $Ms(VAN) = 4.7$, and accepting VAN's usual error in magnitude, $\Delta M = \pm 0.7$, one finds that 456 earthquakes eligible for prediction occurred during the period in which VAN issued 23 predictions [1], 11 of which correlated with one of the eligible events ([2], using the SINOA catalog). This means that about 98% of those earthquakes which VAN is attempting to predict were missed.

5 The Probability of Chance Correlation of Predictions With Earthquakes

The burden of furnishing evidence that a particular hypothesis works better in anticipating earthquakes than random guessing rests with those proposing the hypothesis [8,26,27]. In most of their papers VAN have not put forward arguments of statistical or other quantitative nature supporting their hypothesis of correlation of SESs with earthquakes. The only attempt I am aware of in which they argued their case of correlation quantitatively [22] was incorrect. They used a single interevent time between two $ML(ATH) = 4.0$ earthquakes in a specific location to estimate the seismicity rate of relevant events. However, in their 12 prediction attempts they listed two predictions as small as $ML(ATH) = 3.6$, and they allowed $\Delta M = 0.4$. Therefore, they should have used the entire Greek catalog for approximately the last 15 years to estimate the local seismicity rates for events with $ML(ATH) \geq 3.2(3.6 - 0.4)$. If they had performed this calculation correctly they would have seen that the probability of such small events in one-week intervals in Greece is large, and that their correlations occurred by chance. Unless they furnish quantitative evidence linking SESs and earthquakes, their claims are mere assertions.

Most authors who have studied this problem have concluded that the correlation of VAN predictions with earthquakes is most likely by chance (e.g. [2,6,8,28,29]). In some earlier papers the evaluations of the probability were wrong because the wrong magnitude for eligible and background earthquakes was used ($Ms(PDE)$ instead of $Ms(VAN) = ML(ATH) + 0.5 \approx Ms(PDE) + 1.0$). The reasons why Aceves et al [30] come to a partially different conclusion are not entirely clear to me. One of the reasons is that they use the PDE catalog which is demonstrably incomplete for the magnitude range in question, instead of the SINOA catalog, which is complete. Also, the location errors in the PDE catalog lead these authors to overestimate the successes by VAN. Another factor that contributes to their overestimate of the significance of the VAN prediction record is that they neglect to correctly model the background rate during prediction windows that include an ongoing aftershock sequence. During seismic crises in Greece the seismicity rate is elevated by a factor of 2.3 to 5. This elevated rate should be used for the estimate of the chance probability of correlation during these time intervals [2,8,28]. This is important because many of the VAN predictions are issued for areas that experience a seismic crisis and just after the increased activity started [28,31,32,33]. This fact was confirmed by Mulargia and Gasperini [29], who found a significant correlation of VAN predictions with earthquakes backward in time.

We estimated the probability that the VAN prediction record [1] resulted by chance [2]. Using the SINOA catalog, we find that 12 of 23 predictions are false alarms. We estimated the background seismicity rate separately for each area containing a prediction, based on the declustered SINOA catalog. For predictions issued during seismic crises, an appropriately elevated seismicity rate was used as background rate [2]. We found that the overall performance, as reported by Varotsos and Lazaridou [1], is most likely due to chance at the 96% confidence level.

Because VAN ignored the rule of logic in equation (4) they did not realize that their hypothesis will likely be rejected if they include small earthquakes as claimed successes, and that their record of missed events is dismal.

VAN's belief that the electric signals they measure are connected to earthquakes is apparently due to the fact that earthquakes happen so frequently that correlations occur by chance. I believe that they fell into this trap in 1981 as well as 1991 and 1995, because they did not evaluate the chance probability of a correlation. In 1981, they believed that $ML \geq 2.6$ earthquakes correlated with their electric signals [34]. At that time they used a precursor time of seven hours and 20 minutes [18]. One of the reasons VAN had the impression that these small earthquakes correlated with their signals is that, extrapolating the frequency-magnitude relationship for all of Greece (their area of study at that time [18]), one finds that Greece produces approximately one $ML \geq 2.6$ earthquake per seven-hour interval. This means one has approximately a 60% chance to catch at least one $ML \geq 2.6$ earthquake in any randomly chosen seven hour interval. I conclude that in the now abandoned hypotheses I, II and III, as well as in the current hypothesis IV, VAN's observed correlation between electric signals and earthquakes is due to chance.

6 Plausibility of a Physical Process that Could Generate Electrical Precursors Observable at Hundreds of Kilometers Distance

There are three objections I want to mention briefly: (1) No co-seismic signals are observed, (2) the possible precursory slip on the future faulting plane is minuscule (typically less than 1 millimeter), and (3) the distances are too large for the signal to be transmitted.

6.1 Co-seismic signals

For most models explaining the claimed SES precursors, effects due to piezo-electric, piezomagnetic or fracturing phenomena are invoked. For these models co-seismic signals should be observed because the slip on the fault plane and the stress (strain) changes are much larger during the main shock than at any other time during the seismic cycle (e.g. [7]). The absence of co-seismic electric changes renders the hypothesis suspect because it is difficult to construct a sensible special model that does not depend linearly on stress changes.

6.2 Slip in precursory events

The amount of slip in precursory events claimed to emanate SESs has not been estimated, to my knowledge. I estimate this slip as follows: It is assumed that precursory slip, or the corresponding stress-drop, on the fault plane that will rupture in the following main shock generates the SES in some unspecified way. I calculated the amount of slip that could be reasonably expected as a fraction of the slip in the main shock. I assumed that the ratio of slip during the precursor to that of the main shock is approximately the same as documented elsewhere.

When VAN claim to have measured electrical signals from precursors to $Ms \geq 5$ earthquakes an incorrect image of the approximate size of the rupture is generated in the minds of most seismologist readers. From the expression "surface wave magnitude," or the symbol Ms, seismologists will mistakenly be led to believe that this refers to the surface wave magnitude as issued by the PDE. In that case an equation relating magnitude to rupture dimensions would yield an approximate radius of 3 km for an $Ms(PDE) = 5.0$ event (e.g. [35]). As discussed earlier, however, $Ms(VAN) = 5$ actually corresponds to $Ms(PDE) = 4.1$. This is also born out by the facts that the $Ms(VAN)$ values given by Varotsos and Lazaridou [1] for the earthquakes they claim to have predicted are on average an order of magnitude larger than the corresponding $Ms(PDE)$ values (Figure 2 of [19]), and that about half of these "predicted" earthquakes measured $Ms(PDE) < 4$. Therefore, an

$Ms(VAN) = 5.0$ earthquake has a magnitude $Ms(PDE) = 4.1$, for which the rupture area is on average about A $= 1$ km^2 [35]. The scalar moment Mo for an $ML = 4.5(M(VAN) = 5, Ms(PDE) = 4.1)$ earthquake is calculated as $Mo(ML4.5) = 1.5 \cdot 10^{16}$ Nm [36]. Using a shear modulus of $\mu = 3 \cdot 10^{10}$ N/m^2, and the equation $Mo = mDA$, one estimates the amount of slip

$$D(ML4.5) = 5cm \qquad (5)$$

for earthquakes with $Ms(VAN) = 5.0$. Although relationships between source parameters are not constant over all magnitude bands, we can accept as approximately correct the estimate of the displacement in relationship (5). For smaller earthquakes, as typically claimed by VAN to fulfill their predictions, the displacement is even smaller. Upper limits on the amount of possible precursory slip have been proposed based on the total absence of strain precursors in the case of earthquakes located near strain meters [37,38]. These authors concluded that precursory slip could not exceed "a few percent" of the slip in the main shock because no evidence of precursory strain could be seen on the records, whereas the coseismic strain change was measured. Taking "a few percent" to mean 0.04, we estimate that the precursory slip, Dpr, to an $Ms(VAN) = 5.0$ earthquake claimed to be related to SESs could not exceed a couple of millimeters

$$Dpr(ML4.5) \le 0.2cm. \qquad (6)$$

Since most earthquakes claimed as successes by VAN have magnitudes smaller than $Ms(VAN) = 5.0$, the typical precursory slip would be even less than that estimated in relationship (6). That means the typical precursory slip for the VAN-predicted earthquakes is less than 1 millimeter.

The upper limit for the stress-drop in a precursor may alternatively be estimated as a few per cent of the stress-drop in the main shock. Using 30 bars for the average earthquake stress-drop we arrive at approximately 1 bar as the maximum stress-drop for possible precursors. It seems implausible that such a small amount of slip or stress-drop could generate an SES that can be measured at a distance of several hundred km from the source, as claimed by VAN.

6.3   Transmission of an electric signal

The problem of transmission of an electric signal also seems to be insurmountable. I believe the best account of various ways in which an electric signal related to an earthquake could arrive at a station is given by Bernard [4], who concludes that the typical VAN signal is more than an order of magnitude too large to be connected to a distant earthquake.

In summary, there are three physical reasons to reject the VAN hypothesis: 1) Main shocks do not generate SES signals; 2) the pre-cursory slip, hence stress change, is minute (typically less than 1 millimeter); 3) and the observed signals are more than an order of magnitude too large to have been transmitted from the source or generated locally by a precursor.

7 Responsibility in Earthquake Predictions and Ethics of Predictions

All suggested rules of ethics to be observed by those believing themselves to be in possession of information concerning upcoming earthquakes call for a review by experts of any prediction before it is made public (e.g. [39]). Since there is no law governing the behavior of earthquake predictors, adherence to any rules can only be achieved by peer pressure. In the United States this pressure is high. Reputable scientists do not announce predictions without thorough reviews by colleagues in their home institution as well as by scientists outside. Allen [39] also specified that the information making up a prediction should contain an estimate of the probability that the predicted event may happen by chance. Had VAN made such probability calculations, they would have seen that their claims were invalid.

VAN's false alarm, announced in Greek newspapers, on Greek television and in the French newspaper Le Mond, for a destructive earthquake near Thessaloniki between January 8 and 22, 1991, illustrates the difficulties that can be generated when the ethics guidelines are not followed. Stavrakakis and Drakopoulos [32] and Papazachos et al [33] give a detailed account of the panic that resulted and prompted some people to sleep outside of their homes, but no earthquake happened. Because false alarms will occur with any prediction method it is essential that no alarms be announced by individual researchers, in order to avoid high costs to the population and erosion of credibility of scientists.

The complete and accurate reporting of telegrams issued and their contents is another issue. Many cases have been documented in which VAN claimed successful predictions when in fact the supposedly correlating earthquake fell outside the prediction window (e.g. [2,28,31,32,33]). This type of error can be verified and corrected. However, the incorrect translation of a telegram text [40] and the appearance of a telegram without a date after the event [31,32] cannot be corrected and suggest that the record of predictions claimed to have been issued by VAN is neither complete nor accurate.

It is essential that reviews of predictions be conducted in a timely manner and with the best expertise available. Various nations have established earthquake prediction councils, which can evaluate statements on seismic hazard and even intermediate to short term predictions. The International Association of Seismology and the Earth's Interior (IASPEI) conducts evaluations of precursor hypotheses through its Sub-committee on Earthquake Prediction (SCEP). Of the 40 cases reviewed by the SCEP, five were accepted [5] for the Preliminary List of Significant Precursors, which is intended to accumulate the best cases of precursors during the International Decade of Natural Disaster Reduction. Six proposals for precursors were placed in a list of undecided cases and the rest were not accepted. The work presented to the SCEP covered a wide spectrum of maturity. Some of the proposals that were not accepted contained highly sophisticated work that affords a great deal of insight into earthquake and faulting processes, but they were judged not to include sufficient evidence to demonstrate their usefulness in predicting earthquakes. Other proposals contained more enthusiasm for the goal than solid advances in prediction research. In my estimation the VAN work compares with the latter class of proposals. As many have pointed out (e.g. [2,6,7,8,9,10,11]) the VAN hypothesis is not well enough defined to be tested. In similar cases of insufficiently defined hypotheses submitted for evaluation to SCEP, the chairman was instructed by the SCEP membership to return the proposal without evaluation, with the request to resubmit it with a tight and unique definition of the hypothesis.

Although VAN have made progress in their measuring technique since the beginning of their experiments in 1981, and in their method of identifying signals, they still have no clear definition of the precursor parameters such as the time interval. In addition they believe that three different classes of precursors with three different precursor times (all measured in units of days to weeks though) exist. At the London conference VAN disputed the impression of the majority of the scientists present that the usual precursor was ≤ 22 days, saying that 11 days was their standard. The currently practiced retroactive fitting of parameters by VAN [8] suggests that the hypothesis is still in the development stage. The hypothesis must be clearly and uniquely defined, before it can be implemented. Once real time testing has begun the prediction parameters can no longer be adjusted. VAN have not yet reached this stage. Rejecting the VAN hypothesis does not mean that electro-magnetic effects related to earthquakes may not exist.

Acknowledgments

I thank J. Lahr, D. Booth, S. McNutt and P. Bernard for helpful criticism on the manuscript. This work was supported in part by the Wadati Foundation at the Geophysical Institute of the University of Alaska.

References

1. P. Varotsos and M. Lazaridou, *Tectonophys.* **188**, 321 (1991).
2. M. Wyss and A. Allmann, *Geophys. Res. Lett* **22**, in press (1995).
3. P. Varotsos et al, *Geophys. Res. Lett* **22**, in press (1995).
4. P. Bernard, *J. Geophys. Res.* **97**, 17531 (1992).
5. M. Wyss, *Amer. Geophys. Union* **pp. 94**, Washington (1991), and M. Wyss, *Pageoph* **143**, in press (1995).
6. Y. Honkura and N. Tanaka, *Geophys. Res. Lett* **22**, in press (1995).
7. D.D. Jackson, *Geophys. Res. Lett* **22**, in press (1995).
8. Y.Y. Kagan and D.D. Jackson, *Geophys. Res. Lett* **22**, in press (1995).
9. F. Mulargia and P. Gasperini, *Geophys. Res. Lett* **22**, in press (1995).
10. D.A. Rhoades and F.F. Evison, *Geophys. Res. Lett* **22**, in press (1995).
11. H. Utada, *Geophys. Res. Lett* **22**, in press (1995).
12. S.P. Nishenko, *Pageoph* **135**, 169 (1991).
13. Y.Y. Kagan and D.D. Jackson, *J. Geophys. Res.* **96**, 21419 (1991).
14. S.P. Nishenko and L.R. Sykes, *J. Geophys. Res.* **98**, 9909 (1993).
15. Y.Y. Kagan and D.D. Jackson, *J. Geophys. Res.* **100**, 3943 (1995).
16. P. Varotsos and K. Alexopoulos, *Tectonophys.* **110**, 73 (1984).
17. E. Dologlou, *Tectonophys.* **224**, 189 (1993).
18. P. Varotsos et al, *Praktika of the Academy of Athens* **56**, 417 (1981).
19. M. Wyss, *Geophys. Res. Lett.* **22**, in press (1995).
20. P. Varotsos et al, *Praktika of the Academy of Athens* **56**, 277 (1981).
21. P. Varotsos et al in *Proceedings:*VIII IUGG General Assembly 144 Hamburg, Germany, 15-27 August, 1983.
22. K. Meyer et al, *Tectonophysics* **120**, 153 (1985).
23. P. Varotsos and K. Alexopoulos, *Tectonophysics* **136**, 335 (1987).
24. B. Gutenberg and C.F. Richter, *Ann. Geof.* **9**, 1 (1954).
25. P. Varotsos et al, *Tectonophys.* **224**, 1 (1993).
26. K.S. Riedel, *Geophys. Res. Lett* **22**, in press (1995).
27. P.B. Stark, *Geophys. Res. Lett* **22**, in press (1995).
28. Y.Y. Kagan, *Geophys. Res. Lett* **22**, in press (1995).
29. F. Mulargia and P. Gasperini, *Geophys. Res. Lett* **22**, in press (1995).
30. R.L. Aceves et al, *Geophys. Res. Lett* **22**, in press (1995).
31. G.N. Stavrakakis et al *Publication No.1, National Observatory of Athens, Athens, Greece, 1990.*
32. G.N. Stavrakakis and J. Drakopoulos, *Geophys. Res. Lett* **22**, in press (1995).
33. B.C. Papazachos et al, *personal communication* , (1995).

34. P. Varotsos et al in *Internat. Conference on Basement Tectonics, Oslo, Norway, 10-14 August, 1981.*

35. M. Wyss, *Geology* **7**, 336 (1979).

36. L.B. Kvamme et al, *Geophys. J. Int.* **120**, 525 (1995).

37. M.J.S. Johnston et al, *Tectonophysics* **144**, 189 (1987).

38. M.J.S. Johnston et al, *Geophys. Res. Lett* **17**, 1777 (1990).

39. C.R. Allen, *Bull. Seis. Soc. Amer.* **66**, 2069 (1976).

40. G. Papadopoulos, *Internat. Workshop on Electromagnetic Phenomena Related to Earthquake Prediction, oral presentation, Chofu, Japan, 1993.*

Part IV
Arguments in Favour of the VAN Approach

SOME OBSERVATIONS ABOUT THE STATISTICAL SIGNIFICANCE AND PHYSICAL MECHANISMS OF THE VAN METHOD OF EARTHQUAKE PREDICTION, GREECE

S.K. PARK

Institute of Geophysics and Planetary Physics

D.J. STRAUSS

Department of Statistics

R.L. ACEVES

Institute of Geophysics and Planetary Physics
University of California, Riverside, California, 92521, USA

Statistical tests reviewed here of the VAN results between 1987-1989 show that they are formally significant at a time lag of 22 days, although the level of significance for the prediction rate (ratio of correct predictions to number of predictions issued) varies from 0.07% to 7.96% depending on whether and how aftershocks are removed from the catalog. Comparison of these results to previous statistical tests reveal that differences result from choices of parameters by the evaluators. These parameters include uncertainties in magnitude, location, and time lag, as well as region of sensitivity. Future predictions should include these parameters in order to eliminate the dependence of the statistical tests on evaluators' choices. Viable physical mechanisms for generation of seismoelectric signals must explain both the signal amplitude and the site selectivity. Quantitative modeling of responses from three-dimensional models shows that seemingly high current levels are needed if the source is in the hypocentral region. Alternative mechanisms which involve generation of signals near sensitive sites therefore seem more probable.

1 Introduction

Varotsos, Alexopoulos, and Nomicos (VAN) observed that variations in the Earth's electric field appeared to correlate with the occurrence of earthquakes in Greece[1,2]. These signals, termed seismoelectric signals (SES), have amplitudes of up to 250 mV/km, durations of several minutes, and precede earthquakes by a few days to a month or more[3]. Through a process of correlating SES amplitudes and durations with earthquakes, the VAN team has developed a technique which allows them to issue earthquake predictions which specify location, time, and magnitude. This method has been used since 1984 to issue predictions for earthquakes in Greece. One particularly well-documented set of predictions spanned the period from January 1, 1987 to November 30, 1989[3,4].

As a result of an international workshop in 1992 on low frequency electrical precursors to earthquakes, Park et al.[5] identified several criteria which were essential for a valid precursor. First, the measurement system should be thoroughly characterized in terms of its response and its noise level. Second, techniques to differentiate between natural background variations and tectonically significant signals should be used. Third, objective criteria for the identification of precursors should be specified. Fourth, a statistically significant correlation between the precursors and earthquakes must be demonstrated. Finally, a physically plausible mechanism for generation of the precursory signals is needed. These last two steps are perhaps the most critical because they establish the cause-and-effect relationship between a precursor and an earthquake, which is essential for earthquake prediction. After this cause-and-effect relationship is established, statistical measures of success can be designed which evaluate the utility of a precursor for practical earthquake prediction.

The VAN method has already met several of these criteria. The system response has been adequately characterized in numerous papers by the VAN group[6] and improvements such as digital data acquisition, multiple sensors, and nonpolarizing electrodes have led to the discrimination of SES from telluric signals, electrode noise, telephone line noise, and known local noise sources[4] (see Park et al.[5] and Park[7] for discussion). Objective criteria for identification of SES signals[3] have been used to identify 8 SES in a set of 338 anomalous electric field variations during a three month period from July 19-October 18, 1988[8]. These eight SES coincided with the SES which prompted the VAN group to issue predictions. Thus, the ability to distinguish SES from other electric field variations using objective criteria appears to be established.

The question of statistical significance is difficult to resolve because the predictions issued by the VAN group are incomplete. A complete prediction would contain not only time, location, and magnitude, but also uncertainties for these three quantities[9]. The VAN predictions do not specify any of these uncertainties, but they can be deduced from attempts by the VAN group to assess their own predictions[3,4]. Jackson[9] clearly documents that the spatial and temporal uncertainties have varied over time, so any attempt to deduce stable implicit estimates fails. He argues that it is premature to perform any statistical tests until the predictions are completely specified. That argument aside, previous tests of statistical significance have resulted in contradictory conclusions. Several researchers[10,11,12,13] have concluded that the VAN results are formally statistically significant, while others[14,15] have concluded that they are not significant. Kagan[13] argues that this formal significance is probably due either to retroactive selection of the uncertainties or to the prediction of aftershocks, however. The fundamental problem in any statistical evaluation of the VAN method is that the evaluator must select which uncertainties to use because they are not provided by the VAN group.

In this paper, we review the statistical tests presented in Aceves et al.[12] and discuss the differences between this test and the others. As we will show, most of the differences result from choices of uncertainties in evaluating the VAN predictions. In addition, we will discuss some constraints on the mechanism creating the SES.

2 Statistical Considerations

2.1 Formal Significance

In a formal test of significance of a method's ability to predict earthquakes, the observed outcome (such as a prediction rate) is compared to its probability distribution under some null hypothesis of randomness. If the probability of achieving or exceeding the observed prediction rate is sufficiently small, then we reject the null hypothesis. There are many interpretations of randomness however, and thus many potential null hypotheses exist. The more stringent (and hence more interesting) tests incorporate information on spatial and other heterogeneities into randomness. We presented a model in which heterogeneous spatial and magnitude distributions derived from the historic earthquake catalog (NEIC catalog) are sampled randomly[12]. Use of the historic earthquake data to formulate the null hypothesis is justified because the VAN method was developed by empirically correlating SES with earthquakes. The VAN method would have an advantage over any uniformly random model because of this learning period, and the more powerful statistical tests would have this same advantage. Indeed, we showed that use of these distributions resulted in an approximate two-fold increase in the prediction rate (ratio of successfully predicted earthquakes to number of predictions issued) over a uniformly random model[12]. Additionally, earthquakes cluster in time (foreshock-mainshock-aftershock sequences) and we attempt to incorporate this clustering by declustering the catalog with aftershock models[16].

Aceves et al.[12] tested the predictions from Varotsos and Lazaridou[3] and Varotsos et al.[4] which spanned the period from Janauary 1, 1987 to November 30, 1989 (Table 1). A total of 32 predictions were issued during this time period, but 20 of these were double predictions. A double prediction is one for which an alternative location and magnitude were given. Of these, predictions 9, 15b, 16b,

17a, 19a, 19b, 22, 26b, 31, and 32a were excluded because they were missing a location or magnitude or because the predicted magnitude was below 5.0 (Table 1). This floor on predicted magnitude is consistent with other statistical tests[11,13] and is the currently accepted minimum predicted magnitude[17]. With these eliminations, we are left with a total of 14 double and 14 single predictions to be tested. Our random model issues the same number of double and single predictions in order to mimic the VAN process as closely as possible.

The magnitude reported in Table 1 is a predicted surface wave magnitude computed by adding a constant factor to the local magnitude reported in Athens[1]: $M_s(VAN) = M_L(Athens) + 0.5$. The use of more standard magnitudes might be preferable[15], but we had to use $M_s(VAN)$ because the empirical relationship between magnitude and SES voltage was based on this definition.

One of the most unusual features of the VAN method is that some sites record SES prior to earthquakes while others, perhaps much closer to the future epicenter, record nothing. Varotsos et al.[4] describe these as sensitive sites and have presented maps of sensitivity regions for two of their sites. Ideally, a statistical test would exclude all earthquakes outside these regions from the analysis in order to be as powerful as possible. Maps of the sensitive regions are available for only two sites[4], while the predictions we are testing are based on five sites[18]. We have therefore left the incorporation of the sensitive regions for future work. We note that other researchers have been frustrated by this same complication and have attempted to approximate these sensitive regions by using circles with radii equal to the uncertainty in predicted location[13,19]. This approach limits the ability to test the effect of varying the spatial uncertainty, a limitation we did not want to apply to our random model.

The VAN predictions are incomplete because they do not specify spatial or magnitude uncertainties or a time window over which the prediction is valid. These are adjustable parameters in a statistical test of significance[13]. We therefore chose to develop a random model for prediction from the historical data which allowed the use of variable uncertainties. The basic question was whether there are optimal values of these uncertainties for which the significance of the VAN method is maximized. As an example of this analysis, Aceves et al.[12] showed the effect of varying the time lag between SES and earthquake. This test showed that, with the full catalog, all values of time lags resulted in high levels of significance. This test was performed with a spatial uncertainty of 100 km and a magnitude uncertainty of 0.7 units, following most other statistical tests[11,14]. Future work will evaluate the effects of different spatial and magnitude uncertainties.

Table 1 - VAN Predictions

Pred #	Date	Lat.[14]	Long.[14]	M_s(VAN)	Station
1	26/02/87	37.937	20.324	6.5	PIR
2	27/04/87	37.667	21.458	5.5	PIR
3	13/06/87	37.965	21.464	5.2	PIR
4	01/02/88	39.247	25.388	5.0	KER
5a	10/03/88	40.177	20.833	5.0	IOA
5b	10/03/88	38.849	20.972	5.0	
6a	02/04/88	37.953	20.814	5.0	IOA
6b	02/04/88	36.055	21.386	5.5	
7	03/04/88	38.886	23.746	5.0	KER
8a	07/04/88	38.817	21.080	5.0	IOA
8b	07/04/88	40.239	20.747	5.0	
9*	21/04/88			4.3	IOA
10a	28/04/88	37.937	20.324	5.0	
10b	28/04/88	38.976	20.539	5.0	IOA
11a	15/05/88	40.054	21.005	5.0	IOA
11b	15/05/88	37.937	20.324	5.3	
12a	21/05/88	37.937	20.324	5.3	IOA
12b	21/05/88	40.177	20.833	5.0	
13a	30/05/88	37.937	20.324	5.4	IOA
13b	30/05/88	40.177	20.833	5.0	
14	04/06/88	37.937	20.324	5.0	IOA
15a	10/06/88	36.704	20.833	5.1	KER
15b*	10/06/88	37.987	23.746	4.7	
16a	21/06/88	37.937	20.324	5.0	IOA
16b*	21/06/88	40.177	20.833	4.8	
17a*	10/07/88	37.971	21.806	4.7	PIR
17b	10/07/88	37.134	21.247	5.2	
18	13/07/88	37.984	22.947	5.0	KOR
19a*	18/07/88				NAF
19b*	18/07/88				
20a	01/09/88	37.955	21.008	5.8	IOA
20b	01/09/88	39.869	21.260	5.3	
21a	30/09/88	37.955	21.008	5.3	IOA
21b	30/09/88	40.054	21.005	5.0	

Table 1 - VAN Predictions (continued)

Pred #	Date	Lat.[14]	Long.[14]	M_s(VAN)	Station
22*	03/10/88				IOA
23a	21/10/88	37.955	21.008	6.4	IOA
23b	21/10/88	40.484	20.402	5.5	
24a	02/03/89	37.937	20.324	5.4	IOA
24b	02/03/89	40.054	21.005	5.0	
25a	03/06/89	37.937	20.324	5.5	IOA
25b	03/06/89	40.177	20.833	5.0	
26a	13/06/89	37.965	21.464	5.2	IOA
26b*	13/06/89	40.177	20.833	4.8	
27	23/07/89	38.241	24.070	5.0	KER
28	16/08/89	38.656	21.618	5.0	IOA
29a	24/08/89	38.624	21.725	5.2	IOA
29b	24/08/89	37.134	21.244	5.8	
30a	11/09/89	38.624	21.725	5.2	IOA
30b	11/09/89	37.134	21.144	5.8	
31*	15/09/89				ASS
32a*	18/10/89	39.869	21.260	4.8	IOA
32b	18/10/89	37.955	21.008	5.5	

*Excluded- missing locations or magnitudes

The magnitude uncertainty of 0.7 units was derived from linear regressions between the logarithm of the SES voltage and the magnitude[1]. It is clearly a two-sided error (see Figure 15 in Varotsos and Alexopoulos[1], and we treat it as such. Several researchers[13,19] have truncated the earthquake catalog with magnitude floor at M_s(VAN)=5.0 because the predictions were issued only for earthquakes with expected magnitudes greater than 5.0[17]. A prediction is issued only if the magnitude determined from the SES voltage using the linear regression is greater than 5.0. However, the scatter in the relationship between magnitude and SES voltage results in the magnitude uncertainty. That minimum SES voltage could correspond to an earthquake with $4.3 \leq M_s(VAN) \leq 5.7$, based on a standard definition of experimental error. Thus, we apply a lower floor of 4.3 to the catalog for scoring both the VAN predictions and the random model. Note that, following recommendations of the VAN group[17], we exclude all predictions with magnitudes below 5.0 but permit successes for observed magnitudes below 5.0 and within 0.7 units of the predicted magnitude.

We assumed no apriori statistical distributions for evaluation of the prediction rate, but instead used a Monte Carlo model with 10,000 simulations for the random model[12]. Kagan[13] showed the importance of aftershock clustering and recommended that we adopt similar models. Therefore, we scored the random models and the VAN predictions (Table 1) against three different catalogs derived from the PDE (Preliminary Determination of Epicenters) catalog. The first was the full catalog and had 316 earthquakes with $M_s(VAN) \geq 4.3$ within the region 19-25°N and 35-42°E. The second was declustered with the following equations[13,16]:

$$r(M_s(VAN)) = r_c \; 10^{0.4M_s(VAN)-1.33} \; , \qquad (1)$$

and

$$t(M_s(VAN)) = 3.33 \bullet 10^{0.667(M_s(VAN)-4-T_c)} \; , \qquad (2)$$

where R_c and T_c are empirically determined parameters. While R_c has units of distance, T_c is a dimensionless parameter[13,16]. Each earthquake in the complete catalog was assumed to be a mainshock, and all earthquakes with lesser magnitudes within the distance given by (1) and follwing the mainshock within the time given by (2) were removed from the catalog[13]. Through this process, all aftershocks should be removed from the complete catalog. R_c was set to 2.5 km in both cases, and T_c was equal to 1.0 for the first and 0.0 for the second. The aftershock zones are 6.1 km for an $M_s(VAN)=4.3$ and 29.1 km for $M_s(VAN)=6.0$ for this value of R_c. The time limit for $M_s(VAN)=5.0$ increased from 3.33 days for $T_c=1.0$ to 7.18 days for $T_c=0.0$. For the latter two catalogs, we also declustered the list of predictions with the same aftershock parameters so that consistency between the catalog and predictions was maintained.

Significance levels for the VAN prediction rate are given at four different time lags between 5.5 and 22 days (Table 2), and show that the results are highly significant at all lags when the full catalog was considered (Table 3). P-values (the percentage of random simulations with prediction rates greater than or equal to the VAN prediction rate) of less than 0.07% are found. Results for the catalog declustered with $T_c=1.0$ are still formally significant, although P-values increase from 0.1% at 22 days to 3.59% at 5.5 days. The increased P-value results in a slight loss of significance compared to the full catalog. This aftershock model removed 60 earthquakes from the catalog and one single prediction (Table 4). The final test with a catalog declustered with $T_c=1.0$ resulted in the loss of an additional 20 earthquakes, two single predictions, and a double prediction. In addition, three double predictions were converted to single ones (Table 4). For this final test, the P-value dropped from 7.96% at 22 days to 24.88% at 5.5 days. The former result is marginally significant (P < 10%); the latter is not, although 5.5 days is perhaps an unreasonably short time lag.

While Kagan[13] reaches the same conclusions about the formal significance of the VAN method with different catalogs, his P-values are substantially larger than ours. For example, he derives a P-value of 3.6% using a Monte Carlo model with a time lag of 22 days and the full PDE catalog. Our corresponding value is 0.07%. One reason for the difference might be that Kagan[13] does not permit earthquakes with $Ms(VAN) < 5.0$ to be counted as successful predictions, while we do (Table 3). Fourteen of the VAN successes result from predicting an earthquake with magnitude below 5.0, but those earthquakes were chosen so as to minimize the time lag.   Upon rescoring the VAN predictions, we found earthquakes with $M_s(VAN) \geq 5.0$ for five of these 14 predictions. Thus, only five of the successes remain if we limit observed magnitudes to greater than 5.0. The VAN prediction rate drops from 67.9% (Table 2) to 39.3%. The average prediction rate for our

Table 2 - VAN Prediction Rates and P-values for Catalog Comparisons

	Full		$T_c = 1.0$		$T_c = 0.0$	
Δt, days	VAN	P-value	VAN	P-value	VAN	P-value
5.5	35.7%	0.06%	25.9%	3.59%	16.0%	24.88%
11.0	53.6	0.03	44.4	1.07	28.0	22.70
16.5	64.3	0.01	55.6	0.43	40.0	9.83
22.0	67.9	0.07	66.7	0.10	48.0	7.96

Monte Carlo simulation drops from 37.2% to 12.8% however, so the VAN prediction rate still has a P-value of 0.01% with all magnitudes below 5.0 excluded. This test was performed with the full catalog only; declustering should preferentially remove smaller earthquakes (aftershocks) and reduce the influence of the magnitude floor. The difference in magnitude floor apparently cannot explain the differences in significance levels.

Another possibility is that the use of the VAN locations and magnitudes in the simulation of Kagan[13] or just the locations in Mulargia and Gasperini[19] results in differences in the simulations that yield differing levels of significance.   The average prediction rate for a time lag of 22 days increased from 37.2% to 42.1% in our simulation when VAN locations and magnitudes were used in lieu of random predictions for these parameters[12].   An increase in average prediction rate will decrease the significance of the VAN method because the average from the random

Table 3 - Full Catalog Comparison

Pred. #	Δt,days	ΔM	Δr, km
1	1	-0.6	59.5
3	8	-0.6	82.7
4	9	-0.7	95.4
5a	6	-0.4	99.3
6a	7	-0.7	66.8
8a	15	-0.5	7.2
10a	9	-0.6	58.2
11b	3	-0.5	56.9
12a	1	+0.2	53.7
13a	3	-0.4	47.6
14	2	0.0	51.8
16a	6	-0.7	37.4
18	1	-0.5	99.1
20a	21	-0.7	10.2
21a	14	-0.7	99.3
25b	1	-0.1	89.6
26a	5	-0.7	97.3
28	15	-0.1	64.0
30a	8	-0.3	97.3

14 double, 14 single predictions.

model is closer to the method's prediction rate. We do not have the probability distribution for the prediction rate from Kagan[13], but estimate that a 4.9% increase in the probability of our random model could increase our P-value by 0.73% to 5.4%. The amount of increase is dependent on the starting prediction rate, but it is consistent with the difference in P-values between our 0.07% and Kagan's 3.5%[13] for the full catalog. We speculate that substitution of the VAN locations and magnitudes for two of the three random variables (space, time, and magnitude) may decrease the significance of the VAN results and argue that the more powerful statistical tests will treat location and magnitude as random variables.

The results reviewed here[12] show that success or failure of the null hypothesis is dependent on the choice of catalog and on uncertainties in the prediction. Specifically, a highly significant result can be turned into an insignificant one at a time lag of 5.5 days by simply eliminating aftershocks with $T_c = 1.0$. The significance for predictions with a time lag of 22 days is also dependent on the choice of aftershock model, but much more weakly so. In any case, these tests illustrate the need for careful selection of spatial, temporal, and magnitude uncertainties in order for proper statistical tests. The random model used by Aceves et al.[12] offers the greatest flexibility for varying all of these uncertainties during simulations because it simultaneously permits selection of time, location, and magnitude. Once the sensitivity regions are built into our simulation, we will have a superior way to account for the selectivity as compared to other simulations[13,19] which attempt to approximate the sensitive regions by using 100 km circles centered at each prediction.

Another improvement in the analysis would be to use a cost function to evaluate predictions from the VAN method and the random model. One of the most contentious aspects of the VAN method is the impreciseness with which spatial, magnitude, and temporal uncertainties are given[9,13,15,19]. All of the statistical tests performed so far use a binary system for evaluating success or failure, thus leading to debates such as whether an earthquake at a distance of 102 km counts as a success when the spatial uncertainty is given as 100 km. A cost function would permit a spectrum of successes, with events close to the predicted location resulting in greater weight. Such functions would also facilitate the next statistical step of measuring the utility of the VAN method for earthquake prediction.

2.2 Decision Theory

All of the tests described so far, including our own, have used as criterion the number of correct predictions (or, equivalently, the proportion of correct predictions). Some authors[13] have proposed various ad hoc measures of the performance of a prediction scheme. In fact the choice of suitable criterion can be a complex issue, and one that should ideally be made on the basis of explicit principles. Under very special conditions, an "optimal" test can be identified on the basis of classical statistical principles[20]. This approach does not generally apply however, and the user must decide what to optimize. The choice will depend on issues such as:

(1) Which is more serious, a "miss" (an earthquake that was not predicted) or a "false alarm" (where a prediction was issued and no earthquake occurred).

Table 4 - Declustered Catalog Comparisons

Pred. #	$T_c=1.0$			$T_c=0.0$		
	Δt,days	ΔM	Δr,km	Δt,days	ΔM	Δr,km
1	1	-0.6	59.5	1	-0.6	59.5
3	8	-0.6	82.7	DECLUSTERED		
4	9	-0.7	95.4	9	-0.7	95.4
5a	6	-0.4	99.3	6	-0.4	99.3
6a	7	-0.7	66.8	7	-0.7	66.8
8a	15	-0.5	7.2	15	-0.5	7.2
10a	9	-0.6	58.2	9	-0.6	58.2
11b	18	+0.3	47.6	DECLUSTERED		
12b	21	-0.6	60.5	21	-0.6	60.5
13a	7	-0.4	51.8	DECLUSTERED		
14	DECLUSTERED			DECLUSTERED		
16a	6	-0.7	37.4	DECLUSTERED		
18	3	-0.1	58.7	3	-0.1	58.7
20a	21	-0.7	10.2	21	-0.7	10.2
21a	14	-0.7	99.3	DECLUSTERED		
25b	1	-0.1	89.6	1	-0.1	89.6
26a	5	-0.7	97.3	5	-0.7	97.3
28	15	-0.1	64.0	15	-0.1	64.0
30a	8	-0.3	97.3	DECLUSTERED		

14 double, 13 single   10 double, 15 single
predictions   predictions

What is the precise ratio of the losses associated with these two types of error? Similarly, how does the "reward" associated with a correct prediction compare with these costs?

(2) Should a prediction be judged as success or failure according to a sharp cut-off? For example, if a true earthquake occurs within 99 km versus 101 km of a

predicted site, is the former to be judged an unqualified success and the latter an unqualified failure? If so, what is the "right" choice of cut-off? Does it have to be symmetrical, or could it be one-sided? It is usually agreed, for example, that only the size of the error matters for location, but there is dispute over whether the sign of magnitude errors should be taken into account.

If a "smooth" criterion is used, so that, for example, the smaller the distance between predicted and actual location the larger the reward, what function of distance should be used?   [Some workers have suggested that the function should be related to an assumed probability distribution of the errors, but in fact there is no compelling reason for this.]

(3) The rewards for success or costs for failure need not be the same for all earthquakes.   In particular, large earthquakes should perhaps weighted more heavily than small ones.   Possibly, too, the weight should depend to some extent on the importance of the spatial location.

A useful general framework for such issues is <u>decision theory</u>, according to which an explicit criterion is selected and optimized.   In the present context, the criterion is a <u>loss function</u>:

that defines an overall loss or cost associated with every possible set of predictions $\{p_i\}$ and actual earthquakes $\{q_j\}$. Each of these predicted and actual earthquakes is viewed as a point in a four dimensional space-time-magnitude domain.

To illustrate the use of this approach, we show how the common reliance on the number of correct predictions may be justified under certain explicit assumptions. Thus let C_m be the cost associated with a miss (an actual earthquake that was not predicted), let C_f be the corresponding cost for a false alarm, and let C_h be the "cost" (actually a gain, or negative cost) for a hit (successful prediction). For a given set of predictions and pattern of actual earthquakes, let n_m, n_f, and n_h be the numbers of events of each of these three types. Clearly, this framework will only apply when errors are taken to have a "sharp cut-off" of success or failure. We have also implicitly assumed that the costs do not depend on location, magnitude, etc. In this case, the loss function (3) reduces to

$$L = n_m C_m + n_f C_f + n_h C_h \ . \tag{4}$$

Next, suppose we wish to compare the observed outcome with those from a large number of random simulations, the latter generated with the <u>same total number of</u>

predictions as the actual (observed) scheme. This is the testing procedure that has previously been adopted. In this case, both the total actual number of earthquakes

$$N_q = n_m + n_h ,$$ (5)

and the total predicted number

$$N_p = n_f + n_h ,$$ (6)

are fixed throughout. Thus we may substitute for n_m and n_f from (5) and (6) into (4) and conclude that L may be viewed as a function of n_h alone. This justifies the use of the number of hits as test criterion under certain well-specified conditions.

The discussion so far has focused on the choice of a criterion for a significance test. A logically quite different issue is the assessment of the degree of success achieved by a particular prediction method. This issue is, however, a natural one to consider once it is agreed that the null hypothesis--that the prediction scheme is in some sense no better than "chance"--has been rejected.

Suppose, for example, that method A makes three predictions, all of which prove correct, and that method B makes 10 predictions of which eight are correct. Which is better? Evidently one would not argue that B is better because eight is larger than three; neither is it clear that A is better on the grounds that 100% (3/3) is larger than 80% (8/10). The point is that neither the number nor the percentage of correct predictions can claim to be the "right" measure.

Once again, the issue is only properly resolved by an explicit choice of loss function. If, for example, we are willing to work with the formulation given by (4), it is clear that with different choices of the costs C_m, C_f, and C_h, method A can seen as performing either better or worse than method B. We would emphasize that prediction methods can neither be compared or evaluated unless a loss function is specified. It would be preferable that this function be made explicit, so that the researcher and others may decide whether they accept the particular choice, and also explore the consequences of difference choices. We believe that further exploration of these issues, in the context of real prediction data, would be valuable.

3 Constraints on Mechanisms

Mechanisms for the generation of SES include electrokinetic phenomena associated with fluid flow[21,22], displacement of charged dislocations[23], and piezostimulated currents[24,25]. Any mechanism must explain the magnitude of the observed signal, which is typically an order of magnitude larger than the natural telluric signal, and the selectivity of sensitive sites. Much of the debate about the possible mechanism has focused on the question of whether a source mechanism at the hypocentral region can create enough current to generate an observable electric field at a distance of 100 km[4,21]. Assuming a homogeneous earth, the electric field decays as $1/r^3$ away from the source. This decay requires unrealistically large source currents in order to account for the size of the SES[21], thus leading to a conceptual model in which a narrow conductive channel focuses current at the sensitive site and provides a much slower field decay with distance[26].

Alternatively, local sources close to the sensitive sites could be triggered by regional stress changes[21,22]. However, Bernard[21] shows that nonlinear effects are necessary to explain the magnitude of the SES signal unless precursory strain is larger than the coseismic strain, a physically unrealistic condition. Such nonlinear effects may result from fluid reservoirs stressed almost to rupture[21], but still seem to be highly unusual. Either hypocentral generation or local generation of SES, then, poses problems which must be solved by invoking special conditions which is unsatisfactory. Here, we attempt to add quantitative constraints based on the conceptual models proposed[21,26] through numerical simulation of responses to current sources in 3-D models using an algorithm[27] modified by one of us (Park).

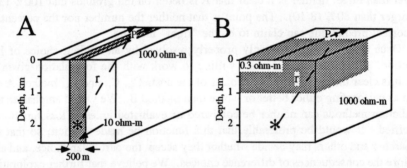

Figure 1 - 3-D models of narrow, conductive fault zone (A) and ocean-continent contact (B). Both models extend to 20 km depth, and distance (r) from current source (asterisk) to potential measurement at P ranges from 4-130 km.

Specifically, we examine the case of a narrow conductive fault channel (Figure 1A) and channeling by the conductive ocean (Figure 1B). We examine electric fields at distances of 80-120 km to determine whether current levels are low enough at the hypocentral region to be physically realistic. The current source in both models is buried at a depth of 2 km.

The fault in Figure 1A is 500 m wide and extends to 20 km depth. The resistivity contrast of 100:1 (Figure 1A) is large, but not unreasonably so. The simulation clearly shows that the potential within and outside the fault zone decays more slowly than in a homogeneous model (Figure 2). However, the electric fields within and outside the fault zone are small and enhanced over that of the homogeneous model by factors of only 3-34% (Table 5). While the potentials outside the fault zone are slightly larger than within the fault zone, the electric fields are not appreciably different.

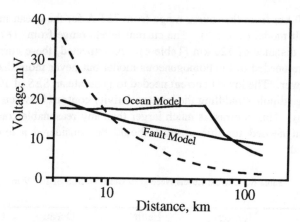

Figure 2 - Profiles of potential versus distance from source for models in Figures 1A and 1B within the fault zone (Fault Model) and across the ocean-continent contact (Ocean Model). Potential within a homogeneous half-space (dashed line) is shown for reference. Profile of potential outside, but parallel to, the fault zone is not shown because values beyond 10 km are within 1% of values within the fault zone.

Greater enhancement of both the potential and the electric field are seen landward of the ocean (Figure 2, Table 5), and the equipotential surface of the conductive ocean is clearly seen (Figure 2). Channeling of the current by the ocean onto land enhances the electric field four-fold over the homogeneous response, much larger than does the fault model. As an extreme case, the electric field across the ocean-continent boundary (with a 3000:1 resistivity contrast) is 1290% larger than the corresponding homogeneous field (Table 5). Clearly, channeling by the ocean results more favorable conditions for SES generation at the hypocentral region than does a fault.

The electric fields in Table 5 are the result of a 1 A source. The amount of current needed to generate an SES can be estimated by comparing these numbers to SES magnitudes. Varotsos et al.[4] show that SES can be as large as 250 mV/km 5(2.5×10^{-4} V/m), but we will use a more representative SES amplitude of 10^{-4} V/m for the calculations. Current levels needed to generate this nominal amplitude at a

Table 5 - Electric Field Variations (V/m)

Distance	Homog.	Fault	Ocean	Ocean-Cont.
80 km	2.56×10^{-8}	2.64×10^{-8}	1.31×10^{-7}	3.57×10^{-7}
120 km	1.12×10^{-8}	1.50×10^{-8}	5.50×10^{-8}	

distance of 80 km from the source range from 763 A for the ocean model to 3788 A for the fault model (Table 6). The current levels range from 1818-6663 A for a source at a distance of 120 km (Table 6). As expected, these current levels are lower than are needed in the homogeneous model but never more than an order of magnitude lower. The lowest current needed to generate an SES of 10^{-4} V/m is 281 A for a voltage dipole straddling the large resistivity contrast between the ocean and land (Table 6). This contrast is much larger than any reasonable contrast on land, so the current needed for this contrast should be considered a lower bound on current needed.

Table 6 - Current Levels Needed to Generate SES of 10^{-4} V/m

Distance	Homog.	Fault	Ocean	Ocean-Cont.
80 km	3908 A	3788 A	763 A	281 A
120 km	8929 A	6663 A	1818 A	

These current levels are at least an order of magnitude higher than the 100 A estimated by Utada[26] to create an SES of 10^{-4} V/m in his conceptual model. In the most favorable situation with the SES site straddling a 3000:1 resistivity contrast (the ocean:continent contact), our estimates of the necessary current are still 3 times larger than the 100 A from the earlier estimates. Varotsos et al.[4] speculated that 10-100 A current generation might be reasonable, based on laboratory observations of piezostimulated currents. However, we are not sanguine about the generation of 20-90 times more current actually needed in the two models we present here. This problem of current generation will be exacerbated by the presence of pore fluids in real rocks. Laboratory measurements to date have been made only on dry rocks[25], and we speculate that much smaller current densities will result in saturated rocks because of cancellation of the piezostimulated currents by mobile charge carriers in pore fluids. The generation of current levels in Table 6 by dislocation models and piezostimulated currents becomes even more problematic under these conditions.

Quantitative 3-D modeling of the potential due to a current source embedded in a heterogeneous medium reveals that favorable geologic conditions can indeed enhance the electric field over that in a homogeneous earth. The conductive ocean layer appears to be more efficient at enhancement than does a fault zone. However, the enhancement in either case is less than a factor of 10, which is considerably less than the factor of 100 assumed by previous researchers[4,26]. Generation of sufficient current in the hypocentral region to produce an SES of 10^{-4} V/m at a distance of approximately 100 km appears to be problematic, based on our simulations. The potentials within and outside the fault zone at distances in excess of 30 km are within 1 % of each other, indicating that the selective sensitivity of sites is not likely explained in models with distant sources. We conclude that the models of sources close to the sensitive sites[21,22] are more plausible at this point.

4 Conclusions

Statistical tests of the VAN results show that the formal significance is dependent on parameters selected by the evaluator, including uncertainties in location, magnitude, and time. Other selectable parameters include the choice of catalog for evaluation, conversion formulae between different magnitudes, and regions of sensitivity for each of the SES sites. The presence or absence of aftershocks also affects the formal significance of the results, with lower significance levels associated with the declustered catalog. Overall, we still find the VAN results to be highly significant. This conclusion contradicts some previous research, but this apparent discrepancy may be related to differences in how VAN was scored and to the use of the VAN locations with uncertainties as an approximation to the sensitive regions. Many of these differences may be rendered irrelevant by the use of cost functions which would score the prediction rate of the VAN method with continuous weighting functions rather than binary ones (i.e., success or failure). An obvious extension of the statistical testing would be the use of decision theory to evaluate the success of the VAN method by weighting successes, failed predictions, and surprises.

While much work on the mechanism remains, quantitative modeling of responses in heterogeneous structures supports earlier assertions[21] that serious difficulties arise in the creation of SES by current sources in the hypocentral region. Current levels required at the hypocentral region are 1-2 orders of magnitude larger than those predicted from piezostimulated or dislocation currents. Additionally, the marked contrast between sensitive and insensitive sites is suggestive of a locally generated signal close to the sensitive sites; our modeling shows that adjacent sites distant from the source should exhibit similar electric fields regardless of the site resistivity. Focused studies of the regions within 10 km of the sensitive sites may be fruitful in the search for the causative mechanism.

Acknowledgments

This work was supported partially by grant #14-08-0001-G1808 from the United States Geological Survey.

References

1. P. Varotsos and K. Alexopoulos, *Tectonophysics*, **110**, 73 (1984a).
2. P. Varotsos and K. Alexopoulos, *Tectonophysics*, **110**, 99 (1984b).
3. P. Varotsos and M. Lazaridou, *Tectonophysics*, **188**, 321 (1991).
4. P. Varotsos, K. Alexopoulos, and M. Lazaridou, *Tectonophysics*, **224**, 1 (1993).
5. S.K. Park, M.J.S. Johnston, T.R. Madden, F.D. Morgan, and H.F. Morrison, *Rev. Geophys.*, **31**, 117 (1993).
6. K. Nomicos and P. Chatzidakos, *Tectonophysics*, **224**, 39 (1993).
7. S.K. Park, Proc. 12th IAGA Workshop on Electromagnetic Induction in the Earth, Brest, France, 179 (1994).
8. T. Nagao, M. Uyeshima, and S. Uyeda, *EOS Trans. Am. Geophys. Union*, **75(suppl.)**, 454 (1994).
9. D.D. Jackson, *Geophys. Res. Lett.*, **22**, in press (1995).
10. K. Meyer, P. Varotsos, K. Alexopoulos, and K. Nomicos, *Tectonophysics*, **120**, 153 (1985).
11. K. Hamada, *Tectonophysics*, **224**, 203 (1993).
12. R.L. Aceves, S.K. Park, and D.J. Strauss, *Geophys. Res. Lett.*, **22**, in press (1995).
13. Y.Y. Kagan, *Geophys. Res. Lett.*, **22**, in press (1995).
14. F. Mulargia and P. Gasperini, *Geophys. J. Intl.*, **111**, 32 (1992).
15. M. Wyss and A. Allman, *Geophys. Res. Lett.*, **22**, in press (1995).
16. P. Reasenberg, *J. Geophys. Res.*, **90**, 5479 (1985).
17. P. Varotsos, K. Eftaxias, F. Vallianatos, and M. Lazaridou, *Geophys. Res. Lett.*, **22**, in press (1995).
18. E. Dologlou, *Tectonophysics*, **224**, 175 (1993).
19. F. Mulargia and P. Gasperini, *Geophys. Res. Lett.*, **22**, in press (1995).
20. P.B. Stark, *Geophys. Res. Lett.*, **22**, in press (1995).
21. P. Bernard, *J. Geophys. Res.*, **97**, 17531 (1992).

22. N. Gershenzon and M. Gokhberg, *Tectonophysics*, **224**, 169 (1993).
23. L. Slifkin, *Tectonophysics*, **224**, 149 (1993).
24. P. Varotsos and K. Alexopoulos in *Thermodynamics of Point Defects and their Relation with Bulk Properties*, ed. by S. Amelinckx, R. Gevers, and J. Nihoul (North-Holland, Amsterdam, 1986).
25. V. Hadjicontis and C. Mavromatou, *Geophys. Res. Lett.*, **21**, 1687-1690 (1994).
26. H. Utada, *Tectonophysics*, **224**, 153 (1993).
27. J. Xhang, R.L. Mackie, and T.R. Madden, *Geophysics*, submitted (1995).

RE-EXAMINATION OF STATISTICAL EVALUATION OF THE SES PREDICTION IN GREECE

KAZUO HAMADA

NIED(National Research Institute for Earth Science and Disaster Prevention)
Tsukuba-shi 305 Japan

Earthquake Prediction, based on continuous measurements of the electric fields of the Earth, has been systematically carried out by Varotsos and co-workers in Greece. Predictions were issued by telegrams to Greek authorities. Statistical re-evaluation of these predictions was made by comparing telegram prediction information with earthquake parameters determined by the National Observatory of Athens. The space-time domain selected for this re-evaluation was 19.2-26.3° E, 35.7-41.1° N, from February 26, 1987, to July 5, 1989. Successfully predicted earthquakes are defined here as those that occurred within 22 days of the prediction, within 100 km of the predicted epicenter and with a magnitude difference (between predicted and true ones) not greater than 0.7. Twelve predictions which include expected magnitude $Ms \geq 5.3$ were issued in the space-time domain selected. Following our criteria, nine were successful, three were false. The magnitudes ML (Athens) of the predicted earthquakes varied from 4.0 to 5.5. For earthquakes with ML (Athens) ≥ 5.0, the ratio of the predicted to the total number of earthquakes is 5/8 (63 %) and success rate of the prediction is 5/12 (42 %). With a confidence level greater than 95 %, it is rejected that five successful prediction for earthquakes with $ML \geq 5.0$ out of twelve can be explained by a random model of earthquake occurrence.

1 Introduction

Continuous measurements of the electric field of the Earth have been carried out at various sites in Greece since the beginning of 1983 by Varotsos and co-workers [1]. The data are continuously transmitted through telephone lines on real time bases to the central station, Glyfada in the suburbs of Athens.

Varotsos and Alexopoulos [2,3] reported that transient variations in the electric field of the Earth were detected before the occurrence of an earthquake. These precursory phenomena they named seismic electric signals (SES). Earthquake predictions were actually achieved based on SES [4]. Quite recently the SES prediction was accepted by Greek authorities in March 1994. However, even now, there is a criticism that successful predictions can be explained by a random model of earthquake occurrence and there is no connection between SES and earthquakes.

The aim of this paper is to show the relationship between SES and following, larger earthquakes from the statistical view point within the selected space-time domain: for 861 days from 1987 through 1989 in and around Greece.

The author had presented the paper entitled "Statistical evaluation of the SES predictions issued in Greece: alarm and success rates" [5], which clearly showed statistical significance for earthquakes with Mb(USGS) $\geqq 5.0$. This conclusion was based on comparison of prediction information with earthquake catalogue published by USGS. The present paper shows re-examination of statistical significance in a space-time domain similar to the previous examination, comparing prediction information with an earthquake catalogue published by National Observatory of Athens this time. This catalogue is expected to be better than the USGS catalogue as far as Greek earthquakes are concerned.

As far as the author knows, any precursory phenomena ever observed in Japan have not in general been useful in predicting earthquakes, depending upon time, location, magnitudes and other complicated circumstances. Therefore, the author keenly desires to continue this kind of examination for a much longer period up to the present. Unfortunately, the number of SES observation stations was reduced from 18 to 4 in November 1989 just after a few months later than the present examination period. Such very incomplete operational condition lasted until the end of 1994. The author intends to evaluate the SES prediction for a much longer period after recovering of operational condition for SES observation. Particularly the author pays attention to the predictions for May 13, 1995 event with Mw=6.5 (USGS) and June 15, 1995 event with Ms=6.5(USGS) which are the two largest events since 1984 in Greek regions.

2 Data

Earthquake prediction data based on SES are given in Table 1, which is the same as that in the previous examination by Hamada [5], except for expression of predicted epicenters. In the present table, locations are given by latitude and longitude instead of azimuth and distance from Athens, which were obtained through personal communication with Varotsos. Parameters of earthquakes which actually occurred are taken from the catalogue published by the National Observatory of Athens in which magnitude scale is ML(Athens).

3 Definition of successfully predicted earthquakes

Definition of successfully predicted earthquakes is the same as in the previous paper [5]. That is, those that occurred within 22 days of the prediction being issued, within 100 km of the predicted epicenter and differing from the predicted magnitude by no more than 0.7. In addition to the above definition, the following supplementary rules are adopted. If there is more than one earthquakes which satisfies the prediction condition, the largest one is taken as the predicted one. Furthermore, an earthquake can only be predicted once. Two or more predictions correspond to two or more earthquakes. Without such supplementary rules, predicted earthquakes could not be uniquely determined. The definition and rules adopted here are made only for a simple statistical test and real problems of the prediction which essentially include social aspects are not taken into consideration.

4 Prediction information and successfully predicted earthquakes

Table 1 shows prediction information that is the same as Table 1 of the paper by Hamada [5]. Prediction information was omitted here if there was ambiguity that could not be evaluated quantitatively. There were 24 predictions from February 26, 1987 to July 5, 1989. The Ms, which is used in prediction, is estimated to be ML(Athens)+0.5 in Varotsos and Lazaridou [6]. Earthquake magnitudes can be troublesome because of the variety of magnitude types and the scatter in their measurement. However, the present examination, in which magnitude scales Ms and ML are used, are simpler than the previous examination [5] in which magnitude scales $Ms, ML,$ and Mb(USGS) were used. Table 1 also shows predicted events based on our criteria described before with quantitative difference between predicted and true events, where the relationship $Ms= ML$(Athens)+0.5 is taken into account.

5 Evaluation of the SES predictions

For statistical examination of earthquake occurrences, the space-time domain was selected as: 19.2-26.3 ˚ E, 35.7-41.1˚ N and the 861 days from February 26, 1987, to July 5, 1989. Figure 1 illustrates all epicenters of earthquakes with ML(Athens)$\geqq$4.0 within the selected space-time domain, which are taken from the catalogue published by National Observatory of Athens.

Table 2 summarizes the results of the prediction in each magnitude range in the case of expected $Ms\geqq$5.0. In this case 1, twenty four predictions were issued; 18 were successful while 6 failed.

Since prediction seem to be more successful for larger earthquakes, predictions which include expected $Ms\geqq$5.3 , were examined and the results are given in Table 3. In this case 2, total number of prediction was 12 ; 9 were successful while 3 failed. More importantly, five out of eight events whose actual $ML\geqq$5.0 were predicted in both cases, even though the number of prediction issued was only a half in case 2 which include expected $Ms\geqq$5.3. Without relationship between predicted Ms and ML of true events, these facts are not able to be explained. In other words, twelve predictions in case 2 are selected not randomly but really depending upon magnitude out of 24 predictions in case 1.

5.1 Statistical significance of successful prediction for earthquakes with ML $\geqq$5.0

We consider the probability of successful prediction when $ML\geqq$5.0 in Table 3 based on a random model of earthquake occurrence. Assuming that 8 events with

$ML \geqq 5.0$ occur in the same area in a 861-day period and the prediction is always issued for that area, the probability to detect one $ML \geqq 5.0$ event by one prediction for 22 days is estimated to be 0.185, since the probability is estimated to be $1-exp(-22 \times 8/861)$. This value is obviously larger than a real one because that the prediction was not issued for the whole area but within one or two 100-km-radius areas.

Assuming that an event E occurs with a probability p in a single trial, then the probability $Psum$ that the event E will occur r times or more in n trials is derived as follows:

$$Psum = \sum_{r=r}^{n} P_r \qquad (1)$$

$$P_r = {}_nC_r p^r q^{n-r}, \quad q=1-p, \quad {}_nC_r : combination$$

If a successful prediction is assumed to occur with a probability of 0.185 in one trial, then the probability that a successful prediction will occur 5 times or more in 12 trials is 5.0 %.

6 Conclusion

Regardless of an area in the examined space for which the prediction is issued, with a confidence level larger than 95 %, it is rejected that five successful predictions for earthquakes with $ML \geqq 5.0$ out of 12 predictions which include expected $Ms \geqq 5.3$ can be explained by the random model of earthquake occurrence. The present re-examination confirmed the major conclusion by Hamada [5]. That is, it suggests a physical connection between SES and subsequent earthquakes, at least an event with a magnitude $ML(\text{Athens}) \geqq 5.0$.

7 Acknowledgement

The author expresses to Professor Toshiyasu Nagao of Kanazawa University his warm gratitude for very substantial help given during the analytical procedures reported in this paper.

References

1. P. Varotsos, K. Alexopoulos, K. Nomicos, and M. Lazaridou, In:O.Kulhanek(Editor), Seismic Source Physics and Earthquake Prediction Research, *Tectonophysics*, 152, 193(1988).
2. P. Varotsos and K. Alexopoulos, *Tectonophysics*, 110, 73(1984a).
3. P. Varotsos and K. Alexopoulos, *Tectonophysics*, 110, 99(1984b).
4. P. Varotsos, K. Alexopoulos and M. Lazaridou, *Tectonophysics*, 224, 1(1993).
5. K. Hamada, 1993. *Tectonophysics*, 224, 203(1993).
6. P. Varotsos and M. Lazaridou, *Tectonophysics*, 188, 321(1991).

Table 1:   Prediction information and predicted earthquakes.
DD, MM, YY, Lat., Lon., *Ms* indicate date, month, year, latitude(° N), longitude(° E), and magnitude
Ms of expected earthquakes, respectively.
ML, dM, dT, dst indicate magnitude *ML*(Athens), magnitude difference *dM= ML- Ms*+0.5, *dT*=time
lags between predicted and true events (day), and distance(km) between predicted and
true events, respectively.
O:true prediction; ✕:false prediction

	Prediction issued			Predicted earthquakes						
	DD/MM/YY	Lat.	Lon.	*Ms*	DD/MM/YY	Lat.	Lon.	*ML*	*dM*	*dT*	*dst*
O1	26/02/87	37.9	20.3	6.5	27/02/87	38.37	20.42	5.4	-0.6	1	53
✕2	27/04/87	36.7	21.5	5.5	Failure						
O3	13/06/87	38.0	21.5	5.2	21/06/87	37.13	21.41	4.2	-0.5	8	97
✕4	01/02/88	39.3	25.4	5.0	Failure						
O5	10/03/88	40.2	20.8	5.0	26/03/88	40.09	19.85	4.9	0.4	16	82
	or	38.3	21.0	5.0							
O6	02/04/88	38.0	20.9	5.0	12/04/88	37.79	20.32	4.0	-0.5	10	56
	or	36.0	21.4	5.5							
✕7	03/04/88	38.9	23.8	5.0	Failure						
O8	07/04/88	38.8	21.1	5.0	24/04/88	38.84	20.33	4.5	0.0	17	67
	or	40.2	20.8	5.0							
O9	28/04/88	37.9	20.3	5.0	09/05/88	37.69	19.93	4.4	0.1	11	40
	or	39.0	20.5	5.0							
O10	15/05/88	37.9	20.3	5.3	18/05/88	38.35	20.47	5.3	0.5	3	52
	or	40.0	21.0	5.0							
O11	21/05/88	37.9	20.3	5.3	22/05/88	38.35	20.54	5.0	0.2	1	54
	or	40.2	20.8	5.0							
O12	30/05/88	37.9	20.3	5.4	02/06/88	38.27	20.36	4.5	-0.4	3	41
	or	40.2	20.8	5.0							
O13	04/06/88	37.9	20.3	5.0	06/06/88	38.30	20.48	4.5	0.0	2	47
✕14	10/06/88	36.7	22.2	5.1	Failure						
O15	21/06/88	37.9	20.3	5.0	26/06/88	38.32	20.40	4.0	-0.5	5	47
O16	10/07/88	37.1	21.2	5.2	23/07/88	36.85	21.87	4.0	-0.7	13	66
O17	13/07/88	38.0	23.0	5.0	16/07/88	37.42	22.86	4.5	0.0	3	66
O18	10/09/88	38.0	21.0	5.8	22/09/88	37.99	21.11	5.0	-0.3	21	10
	or	39.9	21.3	5.3							
O19	30/09/88	38.0	21.0	5.3	16/10/88	37.90	20.96	5.5	0.7	16	12
	or	40.0	21.0	5.0							
O20	03/10/88	38.0	21.1	5.3	15/10/88	37.47	21.58	4.3	-0.5	12	72
	or	38.0	23.0	5.0							
✕21	21/10/88	38.0	21.0	6.4	Failure						
	or	40.5	20.4	5.5							
✕22	02/03/89	40.1	21.0	5.0	Failure						
	or	37.9	20.3	5.4							
O23	03/06/89	37.9	20.3	5.5							
	or	40.2	20.8	5.0	04/06/89	39.67	20.16	4.0	-0.5	1	80
O24	13/06/89	38.0	21.5	5.2	18/06/89	38.41	20.59	4.0	-0.7	5	92
	or	40.2	20.8	4.8							

Table 2: Case 1. expected $Ms \geqq 5.0$
Total number of predictions issued=24.
Successful predictions=18 (75 %), Failure=6.
Total time length in which the prediction was issued: 373 days=43% of all.
NEQ: the number of earthquakes in the examined space-time-domain.
NPE: the number of predicted earthquakes.
RPE: rate of predicted earthquakes to all earthquakes.

Magnitude range	NEQ	NPE	RPE
ML(Athens) $\geqq 4.0$	122	18	15%
ML(Athens) $\geqq 4.5$	33	10	30%
ML(Athens) $\geqq 5.0$	8	5	63%

Table 3: Case 2, which include expected $Ms \geqq 5.3$
Total number of predictions issued=12 (sequential numbers 1,2,6,10,11,
12,18,19,20,21,22, and 23 in Table 1).
Successful predictions=9 (75 %), Failure= 3.
Total time length in which the prediction was issued: 213 days=25% of all.
NEQ, NPE and RPE as in Table 2.

Magnitude range	NEQ	NPE	RPE
ML(Athens) $\geqq 4.0$	122	9	7%
ML(Athens) $\geqq 4.5$	33	6	18%
ML(Athens) $\geqq 5.0$	8	5	63%

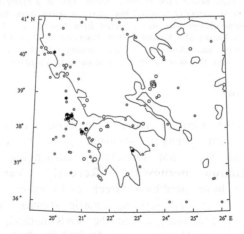

Fig. 1: Epicentral distribution in the selected space-time domain.
Period: February 26,1987-July 5, 1989.
Small circles: $4.0 \leqq ML < 4.5$; medium size circles: $4.5 \leqq ML < 5.0$; and large circles: $5.0 \leqq ML$.
Solid circles indicate predicted 10 events with $ML \geqq 4.5$ which are shown in Table 1.
The predicted 5 events with sequential numbers 1, 10, 11, 12, and 13 in Table 1 are concentrated around a location 38.3 ° N, 20.4 ° E.

ANOMALOUS CHANGES IN GEOELECTRIC POTENTIAL PRECEDING FOUR EARTHQUAKES IN JAPAN

T. NAGAO

Department of Earth Science, Kanazawa University, Kanazawa, 920-11, Japan

S. UYEDA

Earthquake Prediction Research Center, Tokai University, Shimizu, 424, Japan and Geodynamics Res. Inst. / Dept. Geol. & Geophys., Texas A&M University, College Station, Texas 77843, USA

Y. ASAI, Y. KONO

Department of Earth Science, Kanazawa University, Kanazawa, 920-11, Japan

Whether the geoelectric potential field exhibits anomalous changes before earthquakes is a matter of long-standing debate. Recently, many positive results have been reported for Greek earthquakes by the VAN group. To check the exportability of their idea, similar observations have been carried out in Ishikawa Prefecture, central Japan, facing the Japan Sea. Two long span geoelectric potential monitoring systems have been in operation since March, 1991. During the observation period, three M>5 earthquakes (focal depth <50 km) occurred in central Japan. Anomalous geoelectric potential changes were recorded before each of these events. This is a really good sign for VAN method, because it is generally believed that Japan is noisy from the electrical point of view. An anomalous change was also recorded from June 13 to 25, 1993 prior to the devastating M=7.8 Earthquake Southwest Off Hokkaido of July 12, 1993. While it is not certain that these anomalous changes are causally related to the earthquakes, it is worthwhile to continue the observations in a more comprehensive program.

1 Introduction

Since the mid-19th Century there have been numerous reports that geoelectric potential changes occurred prior to earthquakes [e.g. Rikitake[1]]. Most were not sufficiently convincing to be used for practical prediction purposes. Recently, Varotsos and his colleagues[2,3,4,5] have predicted Greek earthquakes using geoelectric potential changes they call seismic electric signals (SES). Success of their method, now commonly called the VAN-method, has been shown to be statistically significant (e. g., Hamada[6]). Park et al.[7] reviewed electromagnetic precursors, stating "The relationship between the SES at sensitive sites and earthquakes in specific regions thus appears to be well defined and repeatable".

VAN-type geoelectric potential monitoring began in Japan in 1987 because it seemed worthwhile to test the method in other parts of the world. If we can confirm their method, it is the best example of exportability of the VAN method. Because, it is well known that Japan is one of the noisiest countries of all over the world in electrical point of view. Some encouraging results have been obtained, but none are entirely convincing[8,9,10,]. In this report, recent results from our Hokuriku network are presented.

2 Method

In 1991, we started geoelectric potential monitoring in the Hokuriku region, central Japan (Figs. 1a and 1b) at two networks. Each network consists of six long-span dipoles using the earthing electrodes at Nippon Telephone and Telegraph Corporation (NTT) exchanges (Figs.1c and 1d). Geoelectric potential data are sampled digitally every 20 seconds and recorded at Kanazawa University. Fig. 1b shows the epicenters of all Japan Meterological Agency Magnitude >4.0 that occurred in central Japan from March 1, 1991 to July 31, 1993. There were three JMA M>5 earthquakes (focal depth <50 km) in the region during our observation period:

EQ 1:West Gifu Prefecture, 20h16m JST, July 30, 1992,
    JMA M=5.3, focal depth = 9 km ,
EQ 2:Off Noto Peninsula , 22h27m JST, February 7, 1993,
    JMA M=6.6, focal depth = 25 km,
EQ 3:Ohtaki-Mura ,Nagano Prefecture, 05h18m JST, April 23,
    1993, JMA M=5.1, focal depth = 8 km.

Figs. 2a and 2b show the hourly mean values, for the total period of monitoring, of the geoelectric potential differences over the six dipoles of the Komatsu network before and after eliminating the magnetotelluric and short period disturbances using BAYTAP-G program[11,12]: the magneto-telluric components were first eliminated by using the transfer function obtained from the three-component geomagnetic data at the Kakioka Geomagnetic Observatory, JMA and the remainder were high-cut filtered to minimize AIC parameter[11]. Fig. 2c shows the records from the Suzu network after the same procedure (the raw Suzu records are not shown for the sake of brevity).

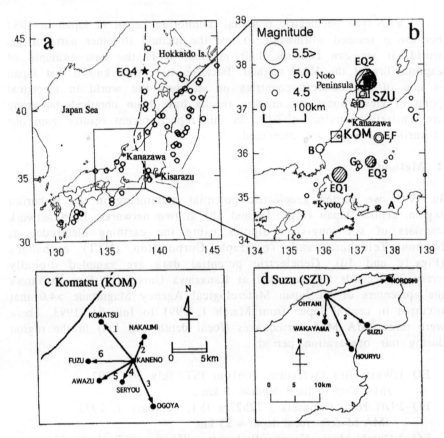

Fig. 1a   Japanese region.   Seismicity is shown by open circles representing M≥7.0 earthquakes (1885-1988).   Lines in the offshore area :plate boundaries.   Dashed line through eastern Japan Sea :   recently proposed North America-Eurasia convergent plate boundary [17].   Black star: epicenter of the M=7.8 Off Southwest Hokkaido Earthquake of July 12, 1993.   Asterisks: cities.

Fig. 1b   Central Japan.   Open circles: all earthquakes with JMA M ≥ 4 (March 1, 1991 - July 31, 1994).   EQ 1, EQ 2, and EQ 3:   three JMA M> 5.0 events.   Rectangles KOM and SZU : areas in Figs. 1c and 1d.   A-G: events with 5.0≥M≥4.5.

Figs. 1c and 1d.   Dipole configuration at Komatsu (KOM) and Suzu (SZU) networks.   Black dots: NTT electrodes.   Increase (decrease) in potential at electrodes pointed by arrows means negative (positive) change on the records in Fig. 2.

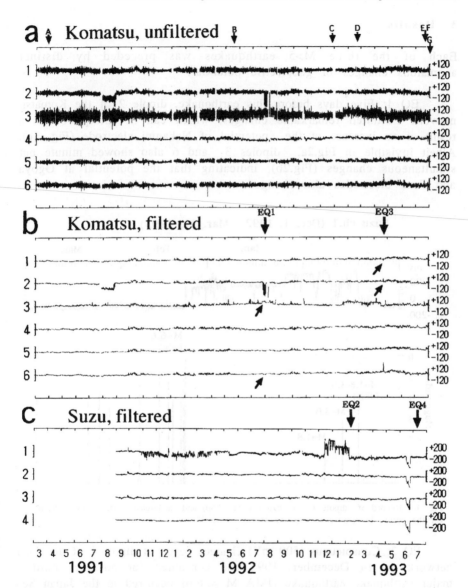

Fig. 2 Continuous records of hourly mean values of geoelectric potential differences of dipoles at Komatsu (a: raw data and b: after cleaning) and Suzu (c: after cleaning) networks. EQ 1 - EQ 4 are indicated by arrows. Numbers: dipole numbers in Figs. 1c and 1d. Top arrows A-G: occurrence times of 5.0≥M≥4.5 events.

3   Results

Each of the three M>5 earthquakes was preceded by distinct anomalous changes of geoelectric potential:

EQ 1: Five days before this earthquake, dipole 2 of the Komatsu network exhibited remarkable changes, which can be interpreted as a result of an increase of the   potential of Nakaumi electrode (Fig. 1c). Albeit invisible in Fig.2a,   dipoles 3   and 6 also showed minute but simultaneous changes (Fig.2b), indicating that the potential at Ogoya and Fuzu electrodes also changed.

Suzu ch.1 (Dec. 1, 1992 - Mar. 31, 1993)

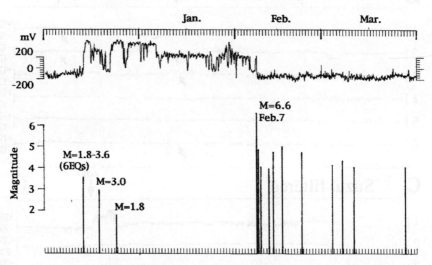

Fig. 3   Record of dipole 1 of Suzu network (top) and earthquake occurrences (bottom) for the period December 1, 1992 to March 31, 1993.

EQ 2: Conspicuous (500 mV) changes began at dipole 1 of the Suzu network on 14 December, 1992 and continued for 56 days until a major earthquake earthquake (JMA M = 6.6) occurred in the Japan Sea off Suzu city at the northern tip of Noto Peninsula on February 7, 1993. (Fig. 2c). The changes can be explained if the geoelectric potential changed at Noroshi electrode, closest to the epicentral area (Figs. 1b and 1d). Fig. 3, with an expanded time scale, shows that the anomalous changes started almost simultaneously with some M3.6 and smaller earthquakes, which are clearly foreshocks. The geoelectric changes started two days after the onset of foreshock activity and lasted until the main shock. Fig. 3 indicates also that there was a short two-day hiatus in the geoelectric activity, followed by a peak immediately preceding the main shock. Fig. 4, with a more expanded time scale, shows that the rise in the geoelectric potential at dipole 1 started at about 9 hours before the shock and decayed about 2 hours after the shock. Most of the other more rapid changes, seen also at other dipoles, are of magnetotelluric origin since they were also seen at distant stations (See for example the bottom figure in Fig. 4, where simultaneous records of six dipoles at the Kisarazu network (Kinoshita et al.,[8] see Fig.1a for location) are shown.).

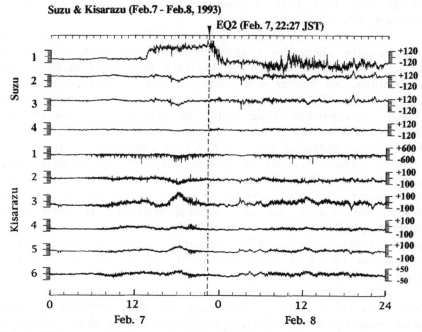

Fig. 4   48 hour records of the very day of the Off Noto Peninsula Earthquake (EQ 2) at Suzu (top)   and Kisarazu (bottom) networks.

EQ 3: An anomalous geoelectric potential change began at dipoles 1, 2 and 6 of the Komatsu network at 10h50m JST, April 21, 1993 and ended at 02h10m of the next day (Fig. 2b). The changes were apparently caused by a potential rise at Komatsu electrode and a simultaneous fall at Nakaumi and Fuzu electrodes. The elimination of short period and magnetotelluric disturbances was needed to extract these signals (Fig. 2a and 2b). At 05h18m, April 23, 27 hours after the end of the potential change, EQ 3 with JMA M=5.1 occurred at Ohtaki-Mura, Nagano Prefecture (see Fig. 1b for epicenter).

EQ 4: On June 13 ,1993, all four dipoles of the Suzu network began to show an anomalous change. The geoelectric potential change at Ohtani electrode, which is common to all dipoles, was probably responsible. The change ended on June 25. On July 12, the devastating JMA M=7.8 Off Southwest Hokkaido Earthquake (EQ4) occurred, causing a huge tsunami as well as serious destruction[13,14]. Although uncertain at this stage, our measured geoelectric potential change may have been a precursor to that giant event, with a precursor time of 27 days.

The epicentral distance of nearly 550 km may be considered unrealistically large for a geoelectric signal to reach. Such a long distance of SES transmission has not been experienced in Greek VAN observations, but neither have magnitude 7.8 earthquakes. Varotsos et al.[5] state that SES of earthquakes with magnitude close to 6.0 can be clearly distinguished from noise at distances of 150-200 km. Recently, SES activity recorded at their Ioannina Station (IOA) on May 3, 1994 was correlated with the M:6.2-6.4 earthquake of May 23 that occurred between Southern Peloponese and Crete (P. Varotsos: private communication). The SES apparently traveled more than 400 km along the western Hellenic subduction zone. In Greece, it had been demonstrated earlier that the apparent resistivity is lower in the direction parallel to the western Hellenic subduction zone than in the perpendicular direction[15]. Bernard[16] proposed some specific mechansims by which precursory electric signals may be observed at great distances.

The epicenter of the Southwest Off Hokkaido Earthquake and the Suzu network are both located along the same proposed convergent plate boundary where incipient subduction of the plate forming the Japan Sea (Eurasia Plate) is suspected under Northeast Japan[17] (North American Plate). Existence of a subterranean high conductivity zone has been demonstrated along subduction plate boundaries such as the Japan Trench and Nankai Trough[18] (Utada, 1987). Putting all these pieces of circumstantial evidence together, we speculate that the seismic source region and our observation sites were favorably linked for electrical signal transmission.

4 Discussion and conclusion

The geoelectric potential changes we have observed were certainly above the noise level, but they were not the same as VAN's ordinary SES which are of much shorter duration.   Our changes measured prior to EQ2 and EQ4 may be of similar nature as that of the so called preseismic Sobolev effect[19] or the Gradual Variation of Electric Field (GVEF)[20].

There was a sizable potential change during August ,1991 (Komatsu network, Figs 2a and 2b) which was not followed by an earthquake. We suggest that this change was probably noise because the records with finer time resolution showed that the change was composed of numerous changes with rectangular forms. As for the long lasting changes in the Suzu records in channel 1, from late September 1991 to early May 1992, it can be noticed, though not clear in Fig.2 scale, that similar disturbances appeared in all channels of both Suzu and Komatsu records. The cause of these long-period changes are being investigated.

In contrast to EQs 1-4, we found no clear signals prior to the earthquakes with $5.0 \geq M \geq 4.5$ (see A-G in Figs. 1b and Fig.2).

It may be too early to conclude that the changes we observed were causally related to the earthquakes. Nevertheless the temporal correlation of the four geoelectric potential changes with the four earthquakes is encouraging enough for us to continue observations on a more expanded scale.

References

1. T. Rikitake in *Earthquake Prediction* (Elsevier, Amsterdam, 1976).
2. P. Varotsos and K. Alexopoulos, *Tectonophys.*,**110**, 73 (1984).
3. P. Varotsos and K. Alexopoulos, *Tectonophys.* ,**110**, 99 (1984).
4. P. Varotsos and M. Lazaridou, *Tectonophys.*, **188**, 321 (1991).
5. P. Varotsos et al., *Tectonophys.*, **224**, 1 (1993).
6. K. Hamada, *Tectonophys.*, **224** , 203 (1993).
7. S. Park et al., *Rev. Geophys.*, **31**, 117 (1993).
8. M. Kinoshita et al., *Bull. Earthq. Res. Inst.*, **64**, 255 (1989).
9. M. Uyeshima et al., *Bull. Earthq. Res. Inst.*, **64**, 487 (1989).
10. T. Kawase et al., *Tectonophys.*, **224**, 83 (1993).
11. M. Ishiguro et al., *Proc. Inst. Stat. Math.*, **32**, 71 (1984).

12. Y. Tamura et al., *Geophys. J. Int.*, **104**, 507 (1991).
13. I. Nakanishi et al., *EOS*, **74**, 377 (1993).
14. Y. Tanioka et al., *EOS*, **74**, 377 (1993).
15. Lazaridou-Varotsou and D. Papanikolaou, *Tectonophys.*, **143**, 337 (1987).
16. P. Bernard, *J. Geophys. Res.*, **97**, 17,531 (1992).
17. K. Nakamura, Bull. Earthq. Res. Inst., 58, 711(1983).
18. H. Utada, *PhD Thesis*, (University of Tokyo, 1987).
19. G. A. Sobolev, *Pageoph*, **113**, 229 (1975).
20. P. Varotsos and K. Alexopoulos, in *Thermodynamics of Point defects and Their Relation with Bulk Properties*, (North-Holland, Amsterdam, 1986).

Part V
Some Related Experimental Programmes

BEHAVIOUR OF THE ELECTRIC POTENTIAL DURING THE ACTIVITY OF AFTERSHOCKS OF THE M7.2 EARTHQUAKE, JAPAN WITH SPECIAL REFERENCE TO SES OF VAN

Y. HONKURA, H. TSUNAKAWA, and M. MATSUSHIMA

Department of Earth and Planetary Sciences, Tokyo Institute of Technology, Tokyo 152, Japan

Observations of the electric potential were made at three sites near the Nojima fault which showed a displacement of about 1 m at the time of an earthquake of M7.2 in Japan. Analyses of the data obtained during the aftershock activity revealed some characteristics of the electric field associated with an inhomogeneous resistivity structure in the vicinity of the fault, including changes similar to seismic electric signals (SES) and gradual variations (GVEF) in the VAN method. Detailed examination of these characteristics suggests an interpretation of *selectivity* which is different from the interpretation of the VAN group.

1 Introduction

An earthquake of M7.2 occurred on 17 January 1995 in the Kobe area, Japan, and caused severe damage including the loss of more than 5,000 lives. It was most unfortunate that very few kinds of systematic observations for earthquake prediction, including observations of the electric potential, have been undertaken in the vicinity of the focal area. In an attempt to learn something from this earthquake, we established, three days after the mainshock, three temporary stations for electric potential measurements near the Nojima fault, which showed a displacement of about 1 m at the time of the mainshock. The main purpose of our observations was to try to examine whether any anomalies can be detected in the electric potential in association with aftershocks.

Detailed examination of the data which we have obtained seems to suggest useful information on whether any seismic electric signals (SES) as reported by the VAN group[1,2] appeared in association with some major aftershocks. In addition to such a basic problem, we can also use the data to investigate what is the physical mechanism of 'selectivity' in the VAN method. In this respect, we should remember that an inhomogeneous crustal structure must be taken into account and the present observation area is in fact inhomogeneous as manifested by the presence of an active fault; this fault was activated and gave rise partly to the M7.2 earthquake.

The aim of this paper is to discuss some aspects of the VAN method, based on our own data and suggest an alternative interpretation of 'selectivity', which has remained a mysterious phenomenon.

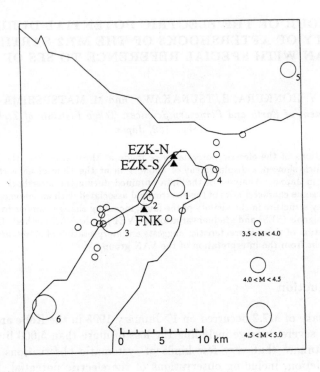

Figure 1: Locations of electric potential observation sites (triangles) near the Nojima fault in the northwestern part of Awaji Island. Circles represent major aftershocks of the M7.2 earthquake during the electric potential observation period.

2   Observations of the Electric Potential

Three temporary observation sites are shown by triangles in Fig. 1. These sites are all located near the Nojima fault which runs in the northwestern part of Awaji-shima Island. In particular, the site EZK-N is located beside the fault; the other two sites are about 1 km to the southeast of the fault. The distance between EZK-N and EZK-S is only about 1 km, but it is important that each is located on the opposite side of the fault from the other.

At all the sites we set up four channels; two channels for the NS component and also two channels for the EW component. The electrode distance is everywhere 40 m, and hence the respective two channels for the NS component and also for the EW component can be considered identical. Although we used stable electrodes of Pb–PbCl$_2$ type,[3] they are still subject to disturbances, for instance, by rainfall.

We can discriminate signals from disturbances of electrodes by examing the simultaneous data at the two identical channels; the observed records represent signals only if both the records for the identical channels are identical. This criterion is essentially the same as the $\Delta V/L$ check for short dipoles in the VAN method. However, we were unable to establish long dipoles as in the VAN method.

Major aftershocks, which occurred in the vicinity of the sites during the observation period, are also shown in Fig. 1. The earthquake No.6 is the largest, but its magnitude is 4.7 which is rather small compared with the mainshock (M7.2).

3 Data and Analyses

3.1 Characteristics of Noises

All the observation sites are located on the island where no large-scale noise sources can be found, and hence we expected that the condition for electric potential measurements should be rather good. Contrary to our expectation, however, severe noises were found at all the sites; such regional noises turned out to stem from the DC-operated train systems on the mainland. The sea surrounding the island acts in fact as a conducting path between the mainland and the island.

It is interesting to note, however, that noise currents tend to flow in a preferential direction depending on sites. For instance, Fig. 2 shows noise values of the NS component plotted against the EW component at respective sites. The least squares fit indicates that noise currents tend to flow in the N51.7°E direction at EZK-N. This direction turns out to coincide with the local strike of the Nojima fault (see Fig. 1), indicating that noise currents tend to flow along the fault at EZK-N.

Such a preferential direction depends strongly on observation sites, as demonstrated in Fig. 2; at EZK-S, N38.0°W is the preferential direction and it is nearly perpendicular to the preferential direction at EZK-N. We should remember here that these two sites are only 1 km distant from each other. This behaviour, however, can be well accounted for theoretically, if there is a resistivity boundary in a direction of the fault strike and also if EZK-N is located on the conductive side and EZK-S on the resistive side.[4] A similar phenomenon has been well known also in magnetotellurics.[5]

The preferential direction at FNK, shown in Fig. 2, is nearly the same as that at EZK-S. This implies that FNK is also located on the resistive side, as understood in Fig. 1.

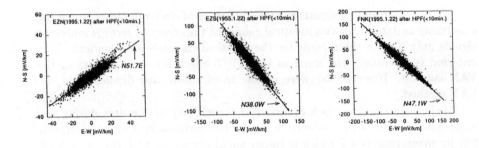

Figure 2: Preferential directions of noise currents at EZK-N (left), EZK-S (middle), and FNK (right).

3.2   Gradual Variations (GVEF in VAN)

There are many reports of gradual variations of the electric potential prior to earthquake occurrences; time duration ranges several hours to a few months.[6] For example, the electric potential difference between two electrodes with a span of 50 m underwent a gradual decrease a few months prior to the M7.0 earthquake.[7] In recent years the VAN method also detected such a gradual variation of the electric potential in some cases, usually before a significant seismic activity.[1,2]

In view of these, we now examine whether any gradual variations can be found in the data. For this, we applied a program package called BAYTAP-G,[8,9] which separates the trend from tidal and irregular constituents. Fig. 3 shows the trend and also the tidal and irregular constituents, together with the original data, for the N38.3°W and the N51.7°E components at EZK-N. Steps appearing in the original data are adjusted in the trend. Arrow symbols represent aftershocks shown in Fig. 1 and bars indicate precipitation at Sumoto located on the Awaji-shima Island.

The trend is found to be sometimes disturbed, mostly by precipitation. Here we note a gradual decrease in the N51.7°E component, which started on 13 February, 5 days prior to the largest aftershock (M4.7) during the observation period. A similar gradual variation can also be seen in both the N38.0°W and the N52.0°E components at EZK-S, as shown in Fig. 4. At this site, the variation started on 12 February. On this day, however, there was slight precipitation amounting to 10 mm or so, but if we compare this case with the other more typical cases of precipitation, such a marked variation in the electric potential is unlikely to be simply due to precipitation. Fig. 5 shows the trend at FNK, but very unfortunately the system became down in the morning of 14 February.

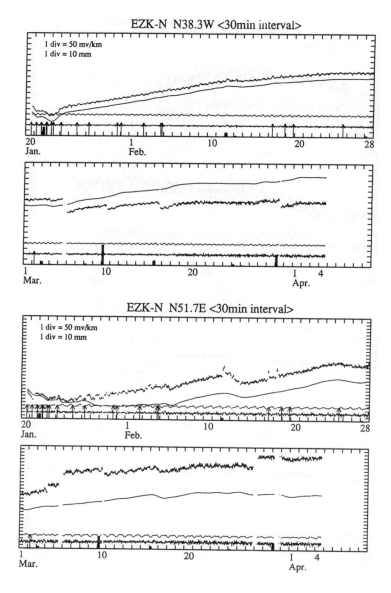

Figure 3: Electric potential data at EZK-N for the N38.3°W and N51.7°E directions. Trend, tidal and irregular components are shown, respectively, together with the original data. Arrows represent major aftershocks, and bars precipitation at Sumoto.

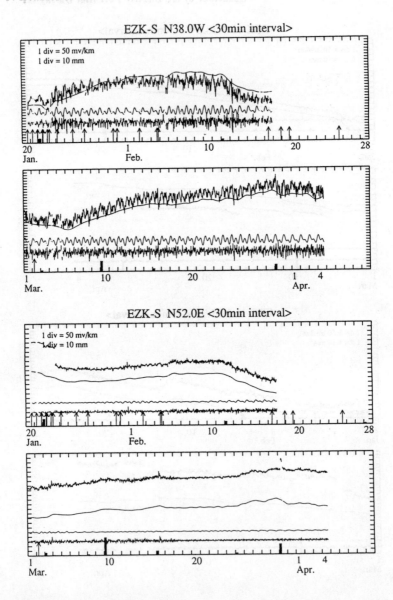

Figure 4: Electric potential data at EZK-S for the N38.0°W and N52.0°E directions. Trend, tidal and irregular components are shown, respectively, together with the original data. Arrows represent major aftershocks, and bars precipitation at Sumoto.

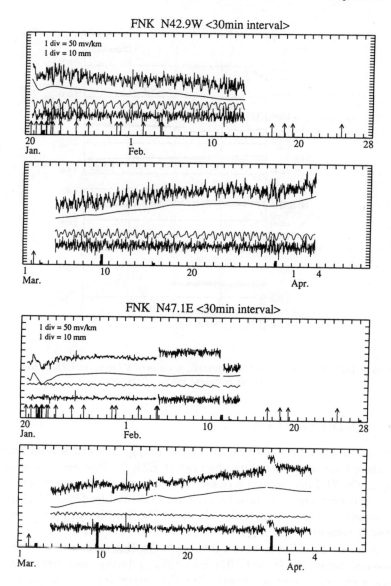

Figure 5: Electric potential data at FNK for the N42.9°W and N47.1°E directions. Trend, tidal and irregular components are shown, respectively, together with the original data. Arrows represent major aftershocks, and bars precipitation at Sumoto.

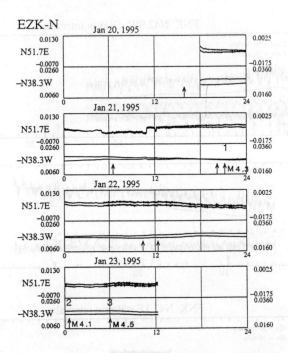

Figure 6: Daily data of the electric potential, in units of V/40 m, at EZK-N for Jan. 20–23. Arrows represent major aftershocks.

3.3   SES-Like Changes

During the aftershock activity, an abrupt change followed several hours later by abrupt recovery was often observed at EZK-N. One example is shown in Fig. 6. On 21 January such a change, together with a change of shorter duration, appeared and it seemed to have been followed by M4.3, M4.1 and M4.5 earthquakes; their epicenters are shown in Fig. 1 with earthquake numbers 1, 2 and 3, respectively.

Another example is shown in Fig. 7. In this case, two similar changes appeared before the largest aftershock (M4.7) during the observation period. Such a change ceased to appear after this aftershock, except the one at the beginning of March. The aftershock activity also decreased.

The changes described in this subsection may be classified as SES's in the VAN method. It should also be noted that the SES-like changes appeared only at EZK-N and did not appear at EZK-S which is only 1 km to the south of EZK-N. In this sense, the concept of 'selectivity' in the VAN method may

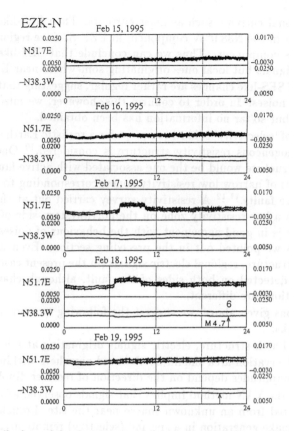

Figure 7: Daily data of the electric potential, in units of V/40 m, at EZK-N for Feb. 15–19. Arrows represent major aftershocks.

be considered to apply to the present case and EZK-N may be considered a sensitive site. However, we do not regard SES-like changes at EZK-N as real SES's for the reasons which will be described in the next section.

4 Discussion and Concluding Remarks

We now discuss the nature of SES-like changes observed at EZK-N. We first point out that they appeared at the channel for the N51.7°E component; this direction coincides with the fault strike and hence regional noise currents turn out to tend to flow along the fault. If SES-like changes originate in the focal areas of the aftershocks shown in Fig. 1, the changes should follow the behaviour of regional currents such as noise currents. Then SES-like changes should also appear in the N38.0°W component at EZK-S, since regional noises did appear in this component. Thus we can conclude that SES-like changes are not regional signals but local ones originating somewhere near EZK-N.

The shapes of SES-like changes are rather regular, suggesting that they are possibly artificial noises. In order to confirm this, however, we must identify the noise source, but so far no information has been obtained.

The concept of 'selectivity' is hard to accept for a realistic Earth structure, even if an inhomogeneous resistivity structure is considered.[10] One notable inhomogeneous structure would be the one associated with active faults; there are some examples of narrow low resistivity zones, corresponding to fractured zones, along active faults.[11,12] A resistivity survey carried out at the Nojima fault showed such a low resistivity zone on the northwestern side of the fault line.[13] This result is in good agreement with the behaviour of noises at EZK-N and EZK-S, as we pointed out in the preceding section. Even in such an inhomogeneous structure, regional electric signals (in the present case, regional noises) could be detected on both sides of the fault, although characteristic preferential directions are different.

For the reasons given above, we suggest the following two possibilities as concluding remarks.

(1) In a realistic Earth structure, electric signals originating at a location far away from the observation area should be observed anywhere within the area, although signal amplitudes depend on the direction of receiver dipole.

(2) Selectivity in the VAN method implies that signals observed at a sensitive site may be emitted from an unknown source near the site through an effect related to earthquake generation in a specific (selective) remote area.

Acknowledgments

We are thankful to Sir James Lighthill for his reading of the manuscript of this paper.

References

1. P. Varotsos et al., *Tectonophysics*, **224**, 1 (1993a).
2. P. Varotsos et al., *Tectonophysics*, **224**, 269 (1993b).
3. G. Pitiau and A. Dupis, *Geophys. Prospect.*, **28**, 792 (1980).
4. G. V. Keller and F. C. Frischknecht, *Electrical Methods in Geophysical Prospecting* (Pergamon Press, Oxford, 1966).

5. T. Rikitake and Y. Honkura, *Solid Earth Geomagnetism* (Terra Sci. Publ., Tokyo, 1985).
6. Y. Honkura, in *Current Research in Earthquake Prediction I*, ed. T. Rikitake (D. Reidel, Tokyo, 1981).
7. S. Koyama and Y. Honkura, *Bull. Earthq. Res. Inst. Univ. Tokyo*, **53**, 939 (1978).
8. M. Ishiguro et al., in *Proc. 9th Inter. Symp. Earth Tide*, 283 (1981).
9. Y. Tamura et al., *Geophys. J. Int.*, **104**, 507 (1991).
10. Y. Honkura and Y. Kuwata, *Tectonophysics*, **224**, 257 (1993).
11. Electromagnetic Research Group for the Active Fault, *J. Geomag. Geoelectr.*, **34**, 103 (1982).
12. Y. Ogawa et al., *J. Geomag. Geoelectr.*, **46**, 403 (1994).
13. T. Mogi, personal communication (1995).

SUBSURFACE LONG WAVE ELECTRIC FIELD CHANGES AS A CANDIDATE FOR AN EARTHQUAKE PRECURSOR

Yukio Fujinawa

National Research Institute for
Earth Science and Disaster Prevention (NIED),
Tennodai 3-1, Tsukuba-shi,
Ibaraki-ken 305, Japan

Kozo Takahashi

Communications Research Laboratory (CRL),
Nukuikita 4-2-1, Koganei-shi, Tokyo, 184, Japan

We have been studying electromagnetic anomalies in the radio wave frequencies preceding earthquakes to assess whether they are useful in earthquake prediction. Changes in vertical electric fields have been measured by using a specially designed antenna by means of deep borehole casing pipe of several hundred meters in length. Researches over several years indicate that the instrument is robust to the presence of civil and natural noise, and that the phenomenon provides useful information in VLF and ULF ranges which can be applied to short-term and imminent earthquake predictions from a few weeks to a few hours before. These phenomema have the property of Sensitivity Selection Rule extensively used in Greece. Recent observational results also suggest that a practical and deterministic method of earthquake prediction may be possible allowing the pulse-like signals of the VLF range to be used for short-term and imminent earthquake prediction.

1 Introduction

At present, a practical method of earthquake prediction which can be used universally has not been devised. In Japan extensive research has been continuing for more than these thirty years without any success in predictions issued to the public before the occurrence of earthquakes. This situation aroused severe criticism including doubts of the possibility of prediction, as strongly advocated by Geller[1].

Despite such a pessimistic atmosphere on earthquake prediction, we have nonetheless experienced many major earthquakes causing serious damage. The worst of these was the 1995 Hyogoken–nanbu earthquake which caused a loss of over 5,500 lives and economic losses totalling more than 100 billion dollars.

Such a serious disaster made Japan feel that earthquake prediction should be more actively pursued. Consequently a great many concerns were directed to the methods monitoring electric field changes, especially the VAN method which have been reported to have very large rates of successes of prediction including release of warnings[2].

There have been accumulating evidences that electromagnetic phenomena of long wave frequency precede earthquakes[3, 4]. However it is also to be noted that there is lack of consensus even about the reality of the phenomena. Nevertheless, we are now in a position to evaluate the electromagnetic phenomena as one of the candidates for precursory phenomena to be used as a means of prediction.

The assessment should be preferably conducted following the Guidelines of the earthquake precursors proposed by IASPEI[5] including 1) definition of the anomalous changes, 2) efficieint method of observation, 3) dominant mechanism of the phenomenon, 4) propagation to the observation point, and 5) prediction related probabilities.

We have been observing subsurface electric field changes using a specially designed antenna to reduce background noises in developed urban environments. Here, we will present our method of observation, and several apparent anomalous phenomena, which seem to offer possibilities of use in a deterministic method of predictions.

2 Observations

2.1 Instruments

The electromagnetic anomalous field changes in the radio wave frequency of less than several kHz have only recently been claimed to be candidates for short–term and/or imminent precursors of earthquakes and volcanic eruptions. This is since the pioneering works of Gokhberg[6]. We have not yet obtained enough knowledge of the phenomena, so that it is natural that there are many measurement methods in this field of work[4].

We have been observing the vertical component of the subsurface electric field strength. The antenna used is a folded dipole micro–antenna with a reflector in the ground as illustrated by Fig. 1. A vertical steel pipe several hundred meters long was used as the linear element of the monopole antenna. Surrounding the pipe, earthed wires were buried to a depth of about 1 m and used as the reflector.

The casing–pipe antenna is taken to be a bare–metal sensor to allow electric field measurement in a conductive medium. Characteristics of the antenna were estimated through the methods of King et al[7] based on the Hallen integral

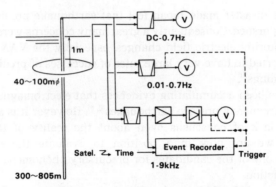

Fig. 1 Schematic diagram of borehole antenna system
and block diagram of measurement.

equation.

Appropriate values for parameters of the model result in the evaluation of the antenna characteristics[8]. The antenna length normalized by wave length is found to be 0.064 ~ 1.9 for 1 kHz in the case of georesistivity $\delta_1 = 1000\ \Omega \cdot m$ and relative dielectic constant $\varepsilon_1 = 8.0$ for homogeneous surrounding sediment. It indicates that the directivity pattern of the antenna radiation is similar to that of a micro dipole in the ULF range, and that of a half-wave length dipole in the VLF range when using pipes several hundred meters long. Consequently we can infer that the antennas detect nearly the vertical component of the electric field in the ULF and VLF ranges.

2.2 Observations

Continuous observations using the borehole antenna have been conducted since March 1989 at Tsukuba (NIE), March 1990 at Izu–Oshima Island (OSM), March 1991 at Hasaki (HAS) and Awano (AWN), March 1992 at Chikura (CKR) and Kofu (KFU), and since March 1993 at Katsuura (KAT) (Fig. 2). The channels observed are "dc" band (0 ~ 0.7 Hz), the ultralow frequency (ULF) band (0.01 ~ 0.7 Hz), and the very low frequency (VLF) band (1 ~ 9 kHz) (Fig. 1). All data are recorded with strip–chart recorders at the observation sites, and are also transmitted to Tsukuba (NIE) by means of a personal computer telecommunication system. From September, 1994 we installed a continuous monitoring recording instrument with a time resolution of about 0.05 sec for rectified VLF signals enabling us to count the number of pulses. Furthermore, a triggering instrument (Fig. 1) is also installed at KAT to record waveforms of signals in the VLF band.

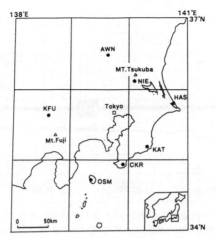

Fig. 2   Sites at which underground electric field changes
were observed using the borehole antennas.

2.3   Comparison with horizontal dipole

On Izu–Oshima island in Sagami Bay, we installed a borehole antenna using a 300 m length casing pipe. This also comprises a roughly 200 m long horizontal dipole with another electrode 100 m deep[9]. Figure 3 shows a typical record, which has three kinds of signals. In the figure, V denotes the vertical component, H the horizontal component, and H' the high pass signal of H. Signal A is larger in the vertical component, B is larger in the horizontal component, and C has amplitudes similar in both components for the amplifier gains chosen. This sample indicates that the borehole antenna does provide signals independent to those of the horizontal dipole. Anomalous signals were observed at the time of a small volcanic eruption occurred on October 4, in 1990 at Mt. Mihara on Izu–Oshima island[9]. The anomalous signals at that occasion are supposed to dominate in the vertical component.

Electrical currents of civil origin are thought to be primarily in the horizontal direction rather than in the vertical direction. And, the telluric electric field has slight vertical component at the horizontally layered medium. So that it would be reasonable to assume that the vertical measurements have a large S/N ratio.

From our limited experience we suspect that the vertical component measurement is more suitable in detecting anomalous signals in environments with a considerable degree of civil noise. However, we have not tried to find extremely sensitive sites as they have done in Greece[2], so we should not be in too much of a hurry to come to any conclusions.

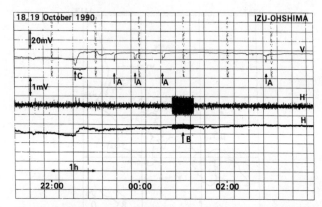

Fig. 3   Typical signals recorded by a borehole antenna for the vertical component measurement and using the horizontal dipoles at Izu–Oshima. Signals A, B, C are relatively dominant in either vertical (V) or horizontal (H and H') and both component, respectively.

3   Intermittent Field Changes

We do not know a priori what the characteristics of anomalous phenomena are. What we can do is to pick up any characteristics of the phenomena from the observational data, and then evaluate the usefulness of the phenomenon through a statistical analysis of the data. Intermittent field changes in the VLF band were observed during the extensive swarm activity in June and July 1989[10].

The upper part of the monitoring records shown in Fig. 4 indicates that clear anomalous VLF signals occurred just before the largest earthquake in the swarm activity in 1989 off the coast of Ito in the Izu–peninsula. Meanwhile the anomalous pattern is discriminated from the normal state in that there is systematic increase of the pulse amplitude with rather periodic undulation. Moderate–scale activity resumed on 26 May 1993, east of the Izu Peninsula. The spatial extent of the latest swarm activity was about half the activity in 1989, but these swarms occurred in nearly the same region and at nearly the same focal depth of less than 10 km. The duration of the former seismic activity was about two weeks and that of the latter was about one week. The largest earthquakes of the former and latter swarms had magnitudes of 5.5 and 4.8, respectively, indicating that the latter activity was less intense than the former.

The lower part of Figure 4 shows the same record about half a day before the third peak of the seismic swarm activity, in May and June of 1993. We can see that the very similar intermittent patterns of field changes were observed at

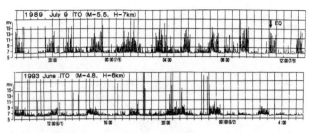

Fig. 4　Intermittent electric field changes at during seismic swarms east of Izu Peninsula in 1989 (upper) part and in 1993 (lower).

both cases of similar crustal activities. Both appearance and duration of the anomalous field changes provide strong evidence that the intermittent VLF electric field change precede certain kinds of earthquakes.

Inspecting records over two years from the beginning of observation at Tsukuba in June, 1989 we picked up eleven instances of similar intermittent VLF changes. This phenomenon is characterized as fluctuations with periods of $1 \sim 2$ hours and intervals of about half an hour.

In order to compare with earthquakes the time window for earthquake occurrence in relation to the anomalous electric field changes was set at three weeks taking account of previous works including those of the VAN group[2]. Windows of earthquake focal parameters are magnitude window $M \geq 4.5$, and the focal depth $H \leq 100$ km. Those thresholds were decided intuitively considering that larger and shallower earthquakes are more likely to produce the preceding electromagnetic changes. It was found that all the depicted eleven cases of anomalous electric field changes preceded earthquakes occurring in central Japan. The precursory time T ranges from $0.4 \sim 15$ days with an average of 2.6 days.

Figure 5 shows the epicenter distribution of earthquakes included in the selected windows of space, time, magnitude and focal depth. The larger symbols correspond to larger magnitudes as denoted in the figure, and the types of syombol indicate the range of the focal depth. The earthquakes with open symbols did not follow the anomalous changes, while the filled ones followed the anomaly. As can be seen from this figure earthquakes in the ocean off Ibaraki $(36° \lesssim \phi \lesssim 37°, 140.5° \leq \lambda)$ and in south central Japan $(\phi \leq 34°)$ were generally not preceded by the anomalous change. There is an exception, however, in the sea around Izu–Oshima Island; where almost all the earthquakes with parameters within the windows were accompanied by the anomalous field changes.

It is not clear why earthquakes in this region were so sensitively observed at

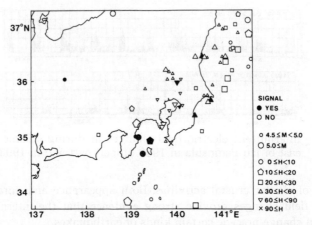

Fig. 5   Epicenter distribution for the earthquakes included in the chosen windows. Open symbols indicate earthquakes not preceded by anomalous changes and filled symbols indicate earthquakes preceded by anomalous changes.

Tsukuba, about 160 km away. The properties might be similar to the selectivity rule for the SES (seismic electric signal) of the VAN group[2].

We do not yet have enough data to estimate prediction related probabilities, so here we can only claim that the intermittent VLF field changes precede shallow earthquakes with moderate magnitudes around Japan in ways that may be useful to earthquake prediction.

4   Imminent Changes before Earthquakes

A great earthquake occurred at 22:23 (JST), October 4, 1994 east of Hokkaido causing huge disasters in the Kuril island and eastern part of Hokkaido. The greatest earthquake for 16 years in the world gave rise to conspicuous ground surface displacement of several tens of centimeters in the eastern most part of Hokkaido observed by an extensive GPS network of GSI (Geographical Survey Institute, Japan)[11], but there was no discernible precursory crustal displacement to be detected even by the kinematic mode GPS measurements at every 2 ~ 3 minutes.

We could not find any apparent anomalous electric field changes before the great earthquake in the slow–speed record concerning characteristics of long period amplitude changes in any of the three frequency bands as had been frequently observed before major earthquakes.

By contrast, the continuous monitoring record of the VLF band rectified

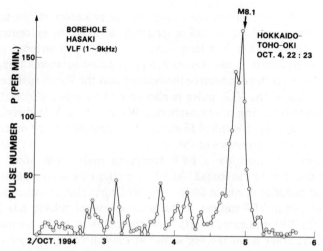

Fig. 6 Number of pulse with amplitude larger than 0.134 mV are counted for 1 minute at every hour from 58^m to 59^m at Hasaki using the fast–speed strip chart recording. It is obvious that the evolution of the pulse number change is closely related to the imminent preparation and occurrences of the great Hokkaido–Toho–Oki earthquake.

data at Hasaki clearly shows that conspicuously larger pulses started to appear in the record from about October 1st, increased in number on October 3rd and 4th just before the earthquake occurrence. In order to quantify that time evolution the number of pulses with amplitude larger than 0.134 mV was tentatively counted every hour to be adopted as a characteristic of the phenomena. Figure 6 shows that the number of pulses increased gradually from October 1st on, but the rate of increase changed greatly from about 17:00 October 4, about 5 hours before the earthquake, and attained its peak about 20 minutes before the great earthquake occurrence. The pulse number quickly attained its background level several hours after the great event.

The characteristics of the pulse number change in time strongly suggests that the VLF pulses were closely related to the occurrence of the great Hokkaido –Toho–Oki earthquake. It is also to be noted that almost the same result was obtained independently by Prof. K. Oike (private communication) near Kyoto about 1,400 km apart from the epicenter and about 500 km apart from our observation site at Hasaki. Oike observed the signals by means of the ball type antenna measuring atmospheric electric field changes in the frequency ranges of $1 \sim 10$ kHz (VLF) and 163 kHz (ELF)[12]. These results suggest that the VLF radio waves of pulse like waveform are caused in the processes of the preparatory stage of the great earthquake, and propagated far from the epicenter reaching more than a thousand km. The long distance propagation without appreciable attenuation indicates that the signals are propagated as waves trapped in the waveguide between the conductive ionosphere and the Earth surface.

The increase of the VLF pulse is also caused by other effects; the largest portion is due to lightning or atmospherics. We may be able to discriminate the seismic radio wave signals (SRS) from those of atmospheric or other origin on the basis of the characteristics of the SRS.

Waveforms obtained by a FFT analysing instrument show that, SRS observed at the time of the Ibaraki–ken–Oki earthquake with magnitude 6.2, at an epicentral distance of about 50 km, the accompanying conspicuous increase of number of SRS in VLF range is of duration of several miliseconds. Frequency analysis of the SRS indicates nearly uniform intensity in the frequency range chosen. On the other hand the big signal at the time of the major Hyogo–ken nanbu earthquake of magnitude 7.2 about 500 km from the epicenter has very low energy in higher frequency component. It is suspected that the higher frequency components are dissipated during propagation resulting in the difference in comparison with those at the time of the earthquake of about 50 km epicentral distance. Accumulation of knowledges about the waveform and frequency energy spectra of SRS will provide us means to discriminate efficiently the SRS from the atmospherics or other noises.

5 Concluding Remarks

We have been studying subsurface electric field changes in order to assess the phenomenon as a candidate precursory phenomenon for short–term and/or imminent earthquake prediction.

Observations are limited within a time of several years, a region of central Japan, the measurement frequencies of long waves, and physical observables of vertical and horizontal electric field. It is suggested from these experiments that:

1) Vertical electric field component measurement provides profitable data even in high civil noise background.

2) Intermittent increase of VLF signal amplitude is a precursor of shallow and larger than moderate earthquakes with a considerable probability. The phenomenon is characterized by a selectivity rule for observation sites[2].

3) Repeated occurrence of seismic swarm gave rise to similar intermittent changes of electric field in the VLF band providing confirmative evidence of the reality of the phenomena connected with crustal activity.

4) Selectivity cannot be explained by the difference of the background noise levels.

5) Number of VLF signal per unit time is suggested to be a profitable characteristic to be used in the imminent prediction.

In order to make an assessment of the phenomena in line with the Guideline[5] a systematic research of several years is necessary. One important field of work is to make an experiment to locate the source region and determine propagation path of the phenomena utilizing various techniques[13), 14), 15)] to provide more definite knowledge as to the relation between the phenomena and earthquake, to infer which mechanisms are dominant in the numerous models proposed[4] and to examine the possibility to estimate other two items of earthquake prediction, that is, place and magnitude.

Acknowledgements

We wish to acknowledge the encouragement of Drs. Shigetsugu Uehara and Kozo Hamada. We are much indebted to Drs. Ryuji Ikeda, and Shoji Sakata of NIED, Mr. Yoshiro Takamatsu of Nittetsu Yosetsu Co., Ltd., and Mr. Hideo Minowa of Tecs Co., Ltd. for their help with the observations. We also thank Ms. Mayumi Yoshida for her help in analyzing data.

References

1. R. Geller, *Nature* **352**, 275 (1991).
2. P. Varotsos and M. Lazaridou, *Tectonophys.* **188**, 321 (1991).
3. S.K. Park et al, *Rev. Geophys.* **31**, 117 (1993).
4. M. Hayakawa and Y. Fujinawa (eds.), *Electromagnetic Phenomena Related to Earthquake Prediction* (Terra Science Pub. Tokyo, 1994).
5. M. Wyss, *Evaluation of Proposed Earthquake Precursors* (Amer. Geophys. Union, Washington, D.C., 1991).
6. M.B. Gokhberg et al, *J. Geophys. Res.* **87**, 7824 (1982).
7. W.P.R. King et al, *Lateral Electromagnetic Waves* (Springer–Verlag, 1992).
8. Y. Fujinawa et al in *Technical Note of the National Research Institute for Earth Science and Disaster Prevention* No.**166**, ed. Y. Fujinawa (1995).
9. Y. Fujinawa et al, *Geophys. Res. Lett.* **19**, 9 (1992).
10. Y. Fujinawa and K. Takahashi, *Nature*, **347**, 376 (1990).
11. GSI in *Rep. Coordinating Committee for Earthquake Prediction* **54**, (1995).
12. K. Oike and T. Ogawa, *J. Geomag. Geoelectr.* **38**, 1031 (1986).
13. M. Hayakawa et al, *Phys. Earth and Planet. Inter.* **77**, 127 (1993).
14. K. Takahashi and Y. Fujinawa, *Phys. Earth and Planet. Inter.* **77**, 33 (1993).
15. T. Yoshino et al, *Ann. Geophysicae* **3**, 727 (1985).

NOTES ON GENERATION AND PROPAGATION OF SEISMIC TRANSIENT ELECTRIC SIGNALS

Yuji ENOMOTO

Mechanical Engineering Laboratory,
Agency of Industrial Science and Technology, MITI
Namiki 1-2,Tsukuba, Ibaraki 305 Japan

Although much attention in this critical review of VAN has focused on statistical assessments of VAN's prediction, it is evident that the seismic electromagnetic activities have been observed not only by the VAN method, but also by the other different methods. Important matters are vital to understanding the nature of seismic electromagnetic signals (SEMS) in greater detail, and to searching for sensitive methods of observation that may be applicable in populated and industrialized areas to detect seismic precursor signals and, if possible, use these to detect imminent seismic precursor signals. From this point of view, this paper discusses the following three points, with emphasis on the comparability of the VAN signals (SES):
1) Possible existence of SEMS, as suggested by laboratory experiments on fracture-induced electromagnetic signals,
2) Working hypotheses regarding various observations of seismic electromagnetic signals, including those from VAN,
3) Possible source mechanism for SEMS.

1 Laboratory experiments on fracture-induced electric signals

Laboratory experiments which simulate the precursor stage leading to final fracture of the Earth's crust (earthquake) are important for understanding the nature of SEMS. Various experiments have been conducted to detect electromagnetic signals associated with the fracturing of rocks. The results, however, have not always been consistent. A typical difference is that some experiments attributed the origin of SEMS to piezoelectric effect of quartz crystals in quartz-bearing rocks such as granite, while others did not. For electric signals propagated through rock media, the relation of signal intensity to the distance from the focal area to an observing site has not been confirmed.

To resolve some of these uncertainties, we conducted experiments to detect transient electric signals induced by mechanical and thermal stimulation of rocks using a new charge detector that can detect rapidly fluctuating electric charge signals. This feature allows it to record the burst-like electric currents emitted from mechanically or thermally stimulated areas of rocks. The following four series of experiments were conducted.

The first series of fracture experiments of rock subjected to uniaxial compression, as shown in the inset of Fig.1, showed that pronounced transient electric signals could be detected in the final stage of accelerated deformation leading to the final failure[1]. Typical results for granite are given in Fig.1. It should also be noted that the signals have been detected not only from quartz-bearing rocks such as granites but also from quartz-free rocks like serpentinite, although the intensity of signals from quartz-free rocks was about one-fourth as large as those of from granites.

In a second series of experiments, a long rectangular bar made of granite was subjected to a guillotine-type fracture at one end while a copper electrode was positioned at another end to detect electric signals[2]. Figure 2 shows the transient electric current as a function of the distance r (mm) of the detector electrode from the fracture zone.

The following conclusions can be drawn from Fig.2,

1) The transient electric current generated due to the fracture is $10^{-10} \sim 10^{-8}$ A/cm^2. This value is comparable to that estimated by Prof. Varotsos, as necessary to cause detectable SES at the ground level[3].

2) The current signal intensity is proportional to $1/r$ in these measurements.

3) The current intensity for coarse-grained granite having electrical resisitivity of 8.4 x 10^6 ohm cm is 17.5 times larger than that of fine-grained granite

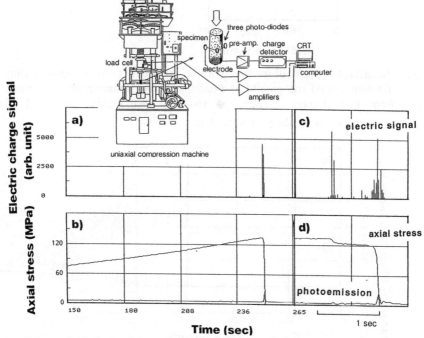

Fig. 1   Schematical view of experimental set-up (inset) and typical records of (a) electric charge signal and (b) axial compression stress and photoemission signals from fracturing granite[1]. Time scale is magnified in (c) and (d).

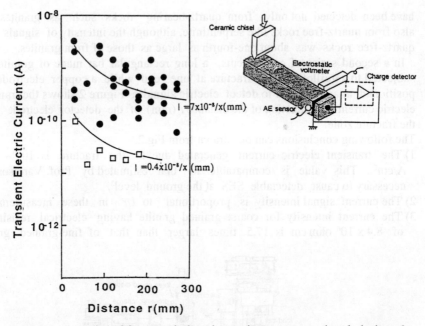

Fig.2 Integrated intensity of fracture induced transient current signal during the fracture period of granite as a function of the distance r of the electrode from the fracture zone[2]. ● coarse-grained granite and ☐ fine-grained granite in their natural state.

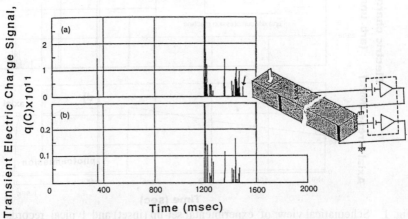

Fig. 3 Typical measurements at room temperature of fracture induced transient current at an electrode bias zero at (a) 5 cm and (b) 105 cm from the fracture zone.

having electrical   resistivity of 0.49 x 10^6 ohm·cm. This suggests that the difference between the current intensity of coarse-grained granite and that of fine-grained   granite is comparable   to   the   difference   in electric conductivity.

A third series of experiment was devised to determine whether the signal is detectable in the field-scale. To this end, a larger bar of coarse grained granite of 3 x 3 x 110 cm in size was used. The signals were detected at two positions 5 cm and 105 cm distant from the fracture point. It is noted, as shown in Fig. 3, that the signal caused by a fracture more than 3 cm square is detectable even 100 cm from the fracture zone.

If a test 100,000 times larger in scale were conducted, it would involve a fracture of the earth's crust of 3 km square, perhaps comparable to the earthquake of a magnitude 4~4.5 on the Richter scale. The implication of 1) in the second series of experiment suggests that the transient   electric current in the focal zone would 10^{-10}~10^{-8} A/cm^2 x (3x105 cm)^2 = 9~900 A.

The results of the third series of experiments indicate that the transient electric current can propagate through rock media over a distance of about 100 km from the focal zone. This estimate is simple but encouraging for the conduct of   field measurements to detect seismic electric signals. By this method as described later, anomalous electric signal could actually be detected prior to the earthquakes which occurred near Tsukuba, north of Tokyo.

Although there is a dispute as to whether SES detected by the VAN method is real or not, the author believes from these   laboratory experiments that detection of SES is promising way of detecting the precursor stage to faulting.

2 Working hypothesis for various observations of SEMS

Seismic electromagnetic phenomena may appear in various different forms, such as electric currents, radio waves, electrostatic induction, electrification and   electric discharge/lightning. This means that various types of observation may be used to detect the phenomena. Several methods of observation have, in fact, been made to detect seismic precursory electromagnetic signals, as schematically shown in Fig. 4[4].

The VAN method has been used to observe the electrical potential difference between two horizontal electrodes at a frequency of 0.1 Hz[5]. The underground electric field variation of vertical components has been observed   by   Fujinawa and Takahashi using a bore-hole antenna at VLF and ULF ranges[6]. Antennas   on   the ground have been utilized to detect the radio waves by several investigators[7-9]. In contrast, Enomoto and Hashimoto have observed the earth current of highly fluctuating components, effectively at frequencies higher than   1   MHz[10]. This observation is related to   laboratory experiments as described above. Figure 5 shows a typical record in   which   anomalous   signals   appeared before a

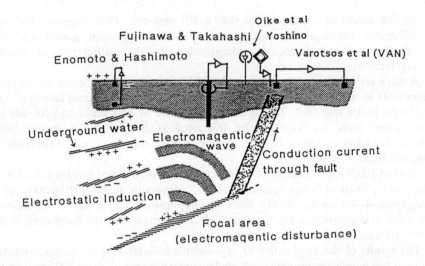

Fig. 4 Schematical view of various detection for SEMS and that of propagation mode of electromagnetic disturbance from a focal zone[4].

magnitude 4.7 earthquake that occurred near Tsukuba, northeast of Tokyo.

All of the methods described above have detected anomalous electromagnetic signals (AES) associated with earthquakes. It should be noted that the duration and the time lag of the anomalous signals that have been detected, are very similar to those of the SES observed by VAN method. At present, however, no seismic anomalous signals have been detected simultaneously on more than one instrument, including VAN. A programme to examine in more detail the comparability of our method with that of VAN and to better understand the nature of SES and SEMS will be carried out in Japan.

3 Generation and propagation mechanisms of SEMS

The observations described above are related to a working hypothesis regarding the propagation/generation mechanisms of seismic electro-magnetic activities, so, the conditions may be better understood by comparing these working hypotheses. This will probably clarify VAN approach.

In general, there are three ways that the seismic electromagnetic disturbances generated in the focal area are propagated through crust media: 1) electric conduction current, 2) electromagnetic wave propagation, and 3) electrostatic induction or electromagneticinduction.

Professor Varotsos argues for a piezostimulated transient electric current generation, and propagation as electric current through a conducting channel of the earth's crust, which falls into category 1)[11]. Fujinawa and Takahashi's observation is related to an argument that radio waves having frequency below 3 kHz suffer less

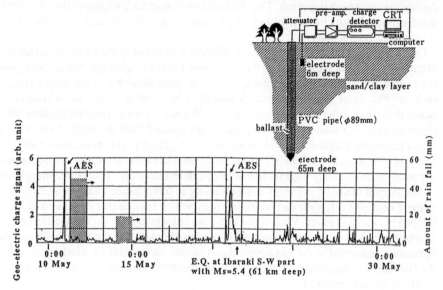

Fig. 5 Electric charge signal as measured 10 - 20 May 1993[10]. An earthauake with Ms =5.4 and with an epicenter at (36 03'N, 139 54"E) occurred at a focal depth of 61 km at 11:36 JST on 21 May 1993. AESs indicate anomalous electric signals appeared prior to the earthquake.

attenuation when they are propagated through the earth's crust[6].

The ground surface may be transiently electrified after the transmission of some seismic electromagnetic disturbance from the focal zone. Then, during the discharge process, fast fluctuations of elcctric charges at the ground surface may occur. We can probably detect such a highly fluctuating electric charge disturbance.

To date, various models have been proposed to explain the generation mechanism of SES and SEMS. They can be classified into two categories:

1) Mechanical interaction

1-a) piezoelectricity

1-b) piezo-stimulated transient current

1-c) stress-induced movements of charge dislocations or charge carriers.

2) Physicochemical interaction

2-a) electrokinetic effects of fluid flow,

2-b) gas electrification or discharge due to thermally stimulated exoelectron emissions from new cracks.

It should be noted that SES detected by the VAN method have always appeared during the preseismic period. The lack of coseismic SES suggests that the main mechanism should be attributed to physicochemical interactions rather than mechanical interactions.

The mechanism of 2-b) has not been discussed in detail to date and therefore it is discussed briefly in the following[9]. Over a long period of geologic time, rocks have been subjected to irradiation by the α, β, γ rays from decaying radio active species such as ^{238}U. Lone pairs of electrons produced by irradiation are often trapped in lattice defects or at the impurities in minerals. When such rock is thermally stimulated, electrons at higher than critical temperature are emitted from the trapped site. This is called thermally stimulated exoelectrons emission (TSEE). The peak temperature for TSEE of granite is about 400 °C. At the depth near the focal zone, the environmental temperature is probably slightly below the peak temperature. Therefore, stimulation of the trapped centres by energy corresponding to temperature changes of, say, fifty or more degrees centigrade can generate TSEE, which may electrify the gas molecules trapped in cracks or pores. Violent gas flows are induced by the pressure difference of coalescing cracks in the precursor stage to an earthquake and the electrical potential difference generated in the gas environment causes abrupt electric discharges. When the final slip occurs, these differences of the electric potential disappear. Therefore, SEMS is mostly preseismic.

4 Summary

It is evident that the VAN method opens the door for the short-term earthquake prediction, although there is still a dispute on the statistical assesment on VAN's prediction. Further studies are needed to develop an acceptable model for SEMS and then look for more sensitive methods to detect imminent seismic precursor signals.

References

1. Y. Enomoto, T. Shimamoto, A. Tsutumi and H. Hashimoto, *Electromagentic Phenomena Related to Earthquake Prediction*, 253 - 259, 1994.

2. Y. Enomoto, M. Akai, H. Hashimoto, S. Mori and Y. Asabe, *Wear*, 168, 135-142, 1993.

3. P.A. Varotsos and K. D. Alexopoulos, *Thermodynamics of Point Defects and their Relation with Bulk Properties,* North-Holland, Amsterdam,

p 419, 1986

4. Y. Enomoto, J. Japn. Soc. Tribologists, 39, 745 - 751, 1994 [in Japanese]; Japn. J. Tribology, Allerton Press [English Translation, in press].

5. P. Varotsos, and K. Alexopoulos, *Techonophys.*,110, 73-98,1984.

6. Y. Fujinawa and K. Takahashi, *Electromagnetic Phenomena Related to Earthquake Prediction*, ed. M. Hayakawa and Y. Fujinawa, 131-147, 1994.

7. K.Oike and T. Yamada, *ibid.*, 115-130, 1994.

8. M. B. Gokhberg, V. A. Morgounov, T. Yshino and I. Tomizawa, *J. Geophys. Res.*, 87, 7824 - 7828, 1982.

9. A. C. Fraser-Smith, A. Bernardi, R. A. Helliwell, P. R. McGill and O. Villard, Jr., U.S. Geological Survey Professional Paper 1550-C, C17-C25, 1993.

10. Y. Enomoto and H. Hashimoto, *Electromagnetic Phenomena Reltated to Earthquake Prediction*, ed. M. Hayakawa and Y. Fujinawa, 261-269, 1994.

11. P. Varotsos, A. Alexopoulos and M. Lazaridou, *Techtonophy.*, 224, 1 - 37 1993.

IMPLEMENTATION OF VAN TECHNIQUE IN GUATEMALA:
Project description

O. KULHÁNEK

Seismological Department, Uppsala University, Box 2101,
S-750 02 Uppsala, Sweden

This is a brief description of a low-budget project aimed at deployment of 2-3 monitoring VAN stations on the territory of Guatemala to investigate the possibilities of the VAN technique for short-term earthquake prediction in the country. With respect to the high seismicity and relatively low level of electrification, Guatemala seems to be the right area to examine the "exportability" of the VAN technique.

1   Introduction

The isthmus of Central America is a region which is systematically exposed to most kinds of rapid-onset natural disasters like earthquakes, volcanic eruptions, floods, landslides and windstorms. The impact of these calamities upon population and further socio-economic development is often catastrophic. Many regions, Central America included, meet the disasters by giving preferential treatment to disaster response. Once the disaster occurrs, massive aid is, of course, indispensable and of great value in reducing human suffering. Nevertheless, I feel that we should focus on and give more support to disaster prevention and mitigation. When limiting ourselves to earthquakes, this can be done in two ways. Firstly, by assessment of the seismic hazard followed by seismic risk mitigation through development planning, modification of infrastructure and implementation of building codes and zoning provisions. Secondly, by introducing prediction methods which include accurate and reliable assessments of the time, place and size of a future severe earthquake. Whereas the former approach has already been successfully applied in many countries, the latter is still at the stage of research.

In Greece, during the last 15 years, a group of physicists (VAN), headed by Prof. Varotsos, developed a technique based on seismo-electric precursors which proved to be successful in predicting larger earthquakes in Greece. Paper by Varotsos et al[1] provides all relevant information. The technique is realatively inexpensive and would, therefore, be extremely useful also for massive use in developing countries with seismic danger. The "exportability" of the VAN technique is, however, one of the questions still much debated. The technique has already been tested e.g. in Japan and France, but in spite of certain success the results are, so far, by no means comparable with those achieved in Greece. Besides, Japan is a highly electrified country and hence to discriminate seismo-electric signals against electric noise is often a problem in itself and France is a country with rather low seismicity.

This short note summarizes motivations and brief decription of projected implementation of the VAN technique in Guatemala. The project is scheduled for the period 1995-2000, commencing July 1, 1995. With respect to the seismicity and low level of electrification, Guatemala should be the right area to test the VAN technique in a new seismo-tectonic environment.

2 Seismotectonics of Guatemala

The narrow strip of land that unites the two Americas is characterized by the interaction between four major tectonic plates, namely the North American plate, the Caribbean plate, the Nazca plate and the Cocos plate. All types of plate boundaries are present in this region. Due to continuous relative motion between the plates and accumulation of stresses along the corresponding plate boundaries, Central America exhibits high seismic activity manifested by numerous large earthquakes (Fig. 1). On average, 4-5 earthquakes with magnitude 6 or greater hit the region every year. Consequences of such earthquakes (e.g. San Salvador, 1986; Managua 1972; Guatemala, 1976) in terms of losses of lives, destruction and environmental degradation are catastrophic.

In Guatemala, the limits of the Caribbean, Cocos and North American plates meet. As a consequence of this "triple junction" the seismic activity of the country is concentrated mainly on two source zones. Firstly, on the subduction zone trending parallel to the Pacific margin and secondly, on the mountain range with dominantly E-W trend. The mountain range includes several distinct and seismically highly active faults (Polochic, Montagua, Jocoban) which separate the North American and Caribbean plates (Fig. 1). The main seismic danges derives from shallow inland events. In the past, the country has experienced several

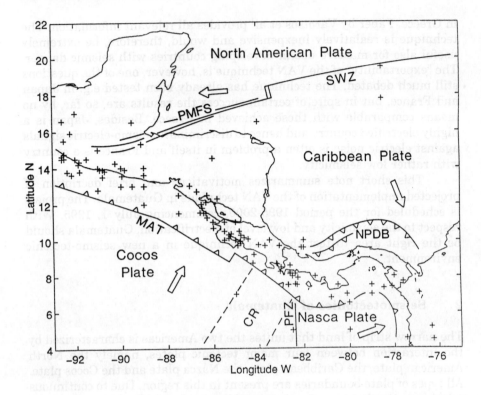

Figure 1. Seismicity and main tectonic features in Central America according to Rojas et al[2] and Villagran[3] (modified). Epicentres (crosses) are shown only for large earthquakes, M≥7, from this century. PMFS = Polochic-Motagua-Jocoban fault system, SWZ = Swan fracture zone, MAT = Middle America trench, CR = Cocos ridge, PFZ = Panama fracture zone, NPDB = North Panama deformed belt. Arrows show the direction of plate motion. Triangles indicate thrust faulting.

disastrous earthquakes of this type. The latest shock, with catastrophic consequences, occurred on February 4, 1976, M=7.3, along the eastern margin of the Montagua fault. Ground brakage was observed along a well defined line extending for 230 km. Displacement is mainly sinistral with the horizontal offset ranging up to about 300 cm and averaging about 100 cm. The 1976 earthquake affected a large area of Guatemala including the capital. There were 26.000 deaths and 1.2 million injuries; the economic losses amounted to US $ 5.000 million.

3 Experimental test of the VAN-prediction method in Guatemala

In March 1995, a low-budget project entitled "The establishment of the VAN earthquake prediction technique in Guatemala" has been launched. It is a joint venture of three Guatemalan organizations (Univeridad Francisco Marroquin; Universidad del Valle; Instituto Nacional de Sismologia, Vulcanologia, Meteorologia y Hidrologia) and the Seismological Department, Uppsala University. The chief investigator is Dr. Juan Carlos Villagran of the Universidad Francisco Marroquin. The co-investigators include specialists in physics, engineering and seismology.

The main objectives of the project are strightforward: deployment of 2-3 monitoring VAN stations on the territory of Guatemala to investigate the potential of the VAN technique for earthquake prediction in the country. Provided that positive results will be achieved, we expect to be able also to contribute to the debate on VAN's "exportability", since the seismo-tectonic and geological character of Guatemala is quite different from that in Greece. With respect to the enormous suffering caused by several major earthquakes in the past, we feel that it is of vital importance for Guatemala (and other Central American countries) to search for and to implement reliable techniques for earthquake predictions. These should complement on-going works on seismic hazard assessments and seismic risk mitigation.

Due to the intrinsic nature of the VAN technique, the project has been designed for a five-year period, 1995-2000. We hope that during this time interval we shall be able to find sensitive sites for monitoring Seismic Electric Signals (SES), establish the proper dipole configurations at each site and collect sufficient data to be able to start with the task of earthquake prediction, including the problem of "site selectivity". All work will be performed in close collaboration with the VAN group in Athens. The proposed time schedule is as follows:

First year
- Initial training phase, including visit to Athens and consultations with Prof Varotsos
- Initial site selection based on analysis of geological and seismic data
- Site evaluation with respect to the ability to monitor SES

Second year
- Visit by Prof. Varotsos to Guatemala

- Selection of the 2nd site
- Evaluation of the ability to monitor SES at the 2nd site
- Data collection at the 1st site and systematic data analysis
- Initial evaluation of results with respect to site selection and data acquisition
- Determination of need for subsequent stations

Third year
- Data collection and analysis at the 1st and 2nd station
- Selection of the third site, depending on the results from previous project phases
- Evaluation of the 3rd site

Fourth year
- Data collection and analysis at all operating stations
- Initial evaluation of results

Fifth year
- Data collection and analysis at all operating stations
- Final evaluation and publication of results

The costs of the project are covered partly by the Universidad Francisco Marroquin, Guatemala and partly by the Swedish Agency for Research Cooperation with Developing Countries, SAREC, Sweden.

References

1. P. Varotsos et al, *Tectonophys* **224**, 1 (1993).
2. W. Rojas et al, *Rev. Geol. Amer. Central* **16**, 5 (1993).
3. M. Villagran, *MSc Thesis*, Univ. Bergen (1995).

Part VI
Reactions to the
Review Meeting

A SEISMOLOGIST'S VIEW OF VAN

HIROO KANAMORI

Seismological Laboratory, California Institute of Technology, Pasadena CA 91125, USA

Although a rigorous statistical assessment of the VAN earthquake prediction method is difficult to make, the correlation between the Seismo-Electric Signal (SES) activities observed with the VAN method and large earthquakes in Greece appears significant. This correlation suggests that SES is a manifestation of regional geophysical activities caused by flow of gas and water in the Earth's crust. Such flow is likely to occur in an area of active extensional tectonics such as Greece. It would be advisable to try to understand the physics of the VAN method by enhancing the existing programs in Greece with not only electric, but also seismological, hydrological and other geophysical investigations. It would be unwise, however, just to export the VAN method to another country as a ready-made prediction tool. If the method is to be investigated in other countries, well designed seismological and geophysical investigations would be required before its usefulness for earthquake prediction in the region can be fully evaluated.

1 Introduction

The VAN (Varotsos-Alexopoulos-Nomicos) method measures telluric electric self-potential with a combination of short and long dipoles and uses it for prediction of earthquakes. In the following "VAN" refers to the method, or the research group consisting of Varotsos and his colleagues. The signal is called Seismo-Electric Signal (SES), and has been reported to appear precursory to earthquakes [e.g. Varotsos and Lazaridou, 1991, Varotsos, et al. 1993]. The following is my personal view of the VAN method that I gained from the literature, the Royal Society conference held in London on May 11 and 12, 1995, and the information FAX provided by Dr. Varotsos.

In general there are many electric noise sources in the crust, and most (about 98%) of the observed electric events are noise, and only about 2% of them are of tectonic origin and are relevant to earthquakes. This extremely high noise level seems to have been one of the reasons for skepticism about the very basis of the VAN method. However, the noise-removal procedure that was developed by VAN and was subsequently verified by Nagao et al. [1995] seems to have made a convincing case that SES is indeed of tectonic origin. The following discussion is based on this premise.

2 Test of VAN Predictions of Small Earthquakes

Whether a prediction is successful or not is judged by the difference in magnitude, ΔM, location (measured by the distance, Δr), and time (Δt) between the predicted event and the actual event that occurred. The following are three sets of commonly used criteria for successful VAN predictions.

a) For a single SES, $\Delta M=\pm 0.7$, $\Delta r=100$ km, and $\Delta t=11$ days.
b) For an SES activity, $\Delta M=\pm 0.7$, $\Delta r=100$ km, and $\Delta t=22$ days.
c) For a gradual variation of electric field (GVEF), $\Delta M=\pm 0.7$, $\Delta r=100$ km, and Δt=several weeks.

Considering the gradual and continuous nature of SES activities, the uncertainty in magnitude and location, and the finite source dimensions of earthquakes, these values cannot be defined very rigidly. However, the ambiguities in these parameters have caused some difficulties and confusion in the statistical tests performed by many investigators. Here I use these criteria only for purposes of discussion and will not attempt to use them rigorously for statistical assessment of predictions. Thus my judgement is inevitably subjective.

The magnitude used by VAN for predictions is denoted by M_S(ATH). It is difficult to define this magnitude rigorously, but it is roughly equal to $M_L+0.5$ where M_L is the standard local magnitude determined by the Athens Observatory. Since M_L is approximately the same as m_b used in PDE of the National Earthquake Information Services for m_b around 5, I will assume that M_S(ATH)$=M_L+0.5=m_b+0.5$ in the following. For earthquakes in the magnitude range of 5 to 6, both m_b and M_L are approximately equal to the internationally used scale like M_S and M_w. Thus, M_S(ATH) is approximately 0.5 unit larger than that of the common magnitude scales. This means that when VAN predicts M_S(ATH)=6, that means that an earthquake with 5.5 on the commonly used scale is actually predicted. For very large earthquakes, significantly larger than 6, M_S(ATH) seems to become comparable to M_S and M_w. Some investigators may disagree with this interpretation in detail. Unfortunately this ambiguity of M_S(ATH) has been a very confusing issue, but is inevitable for historical reasons, and is in no way VAN's fault.

Most of the statistical tests have been done for M_S(ATH)≥ 5.0. Considering $\Delta M=\pm 0.7$ and the relation between M_S(ATH) and m_b, this means that an earthquake with M_S(ATH)≥ 4.3 or $m_b \geq 3.8$ needs to be included in the statistics. This is a very small magnitude, and obviously the PDE catalog is incomplete at this magnitude. A rough estimate indicates that about 120 events occur every year in the whole area of Greece (the total area, $S=1.9 \times 10^5$ km^2). We then expect, on the average,

$$\lambda = 120 \times (10^4 \pi / 1.9 \times 10^5) \times (22/365) = 1.2 \text{ events}$$

in the 22-day interval in a predicted area (a circle with a radius of 100 km), if seismicity is spatially uniform. This is a very high rate, and translates to a probability $p=1-\exp(-\lambda)=0.70$ (70%) of having at least 1 event during the 22 day period by chance, and predictions of such events would not be very meaningful.

However, this argument is subject to large uncertainties. For example, 1) some of the earthquakes are the aftershocks of a larger earthquake and should not be counted as the independent earthquakes in the above statistics; 2) the seismicity is not uniform in Greece. If the seismicity in the target area of prediction is higher than the average in Greece, λ and p would be even larger; if the seismicity of the area is lower than the average, λ and p would be lower; 3) the catalog and the magnitude are very uncertain for these small events, and λ is subject to large uncertainty anyway. For these reasons, depending on the assumptions used for statistics, the conclusion regarding whether the VAN success rate is better than that expected by chance varies so much that the debate on the statistics for small earthquakes (i.e. as small as $M_S(ATH)=5$) seems inconclusive.

3 Large Earthquakes

The statistics for large events is hampered by the small number of such events. To have sufficiently large number of events for making meaningful statistical arguments, it would be important to look at the data over a long period of time. Unfortunately, most of the statistical arguments that have been made so far are for a relatively short period of time [e.g. 1/1/1987 to 6/30/1989, Hamada, 1993] which contains only a few large events. Unfortunately the results for a longer time period were only briefly presented by Dr. Uyeda and Dr. Varotsos at the conference in response to my request, and I have not been able to make a thorough assessment of them myself. According to Dr. Varotsos, for the time period 1987 to 1995, out of 13 $M_S(ATH) \geq 5.8$ earthquakes, 9 were successfully predicted, 3 were missed, and 1 prediction was unsuccessful (the predicted parameters were incorrect). There were only a few false alarms (1 or 2, an announcement issued on January 11, 1991, for an earthquake near Thessaloniki can be considered a false alarm). According to Dr. Uyeda, for the time period of 1/1/1983 to 9/30/1993, out of 10 events with $m_b \geq 5.5$, 6 events were successfully predicted, 1 event was missed and 3 predictions were unsuccessful. The criteria used for successful predictions in these assessments are not exactly the same as those mentioned above, and other investigators might give a somewhat lower success rate using a similar but a more stringent set of criteria. For large earthquakes, the frequencies of the events and predictions are relatively low, so that the chance of having a "successful prediction" by chance is very small. Considering the various uncertainties and the qualitative nature of prediction, I do not feel that giving a definite number for success rate is meaningful at present, but my judgement from examining the individual cases for these large events is that the SES's and the following large earthquakes are indeed causally related in most cases. This statement represents my personal judgement, and is admittedly subjective.

In this context, the recent sequence of VAN predictions and large earthquakes that occurred just before and after the conference is intriguing. The summary is as follows (the names and locations of the stations are in Figure 1).

(1) A prediction was issued on the basis of an SES activity recorded on 4/6/1995 at ASS (FAX from Varotsos dated 4/7/95). This is a double prediction (two

possibilities were presented); one of the events was predicted near Thessaloniki with M_S(ATH)=5.8.

An earthquake with M_S(ATH)=6.0 (NEIS m_b=5.2) occurred on 5/4/1995 50 km ESE of Thessaloniki (Figure 1). Since Δt=28 days, this prediction could be called a failure with the conventional statistical argument. Nevertheless, since this area is not a particularly active area, this prediction appears very unique.

(2) A prediction was issued on the basis of an SES activity at IOA on 4/6/1995 (FAX dated 4/27/1995). The prediction was: M_S(ATH)=5, about 200 km west of Athens. This is a small event, and is not a matter of interest here. Then, two new SES activities were recorded at IOA on 4/18 and 4/19, 1995, and a double prediction was made. This prediction was slightly revised on 4/30/1995. Of the 2 predictions, a prediction of M_S(ATH)=5.5 to 6.0 at a location a few tens of km NW of IOA is considered more consistent with the observed SES.

An earthquake with M_S(ATH)=6.6 (M_w=6.5, NE-SW striking normal fault) occurred on 5/13/1995, about 100 km NE of IOA (Figure 1). The delay time is Δt=24 days, and the errors are Δr=100 km and ΔM=0.6. The seismic activity in this area has been very low, and apparently no earthquake with this magnitude has occurred for at least 1,000 years. Again, the conventional statistical evaluation with the standard set of criteria would call this prediction a failure because of the large Δt.

(3) A prediction was issued on the basis of an SES activity at VOL recorded on 4/30/1995 (FAX dated 5/19/1995). This prediction is somewhat qualitative but is most intriguing. First of all, on the basis of the amplitude and characteristics of the SES activity, VAN stated that the predicted event would be comparable to the 5/13 event in size but at a different location. They excluded the area near station VOL, and the selectivity areas for stations IOA, ASS and KER, which left a relatively small portion of Greece. Even more interesting is that VAN did not exclude the area around station GOR because the station was not functioning.

An earthquake with M_S(ATH)=6.2 (M_w=6.4, E-W striking normal fault) occurred on 6/15/1995 about 40-50 km south of GOR near the edge of the IOA selectivity area (Figure 1). Since Δt=46 days, and the exact location was not given in the prediction (because of the insufficient calibration), the conventional statistical evaluation for this event would be either negative or impossible to make. Nevertheless, the qualitative statements in the FAX sent by Varotsos on May 19 are very suggestive of this earthquake. This example suggests to me that the SES is a real tectonic signal related to the regional seismic activities.

(4) Another interesting aspect of this series of predictions is that, even after the occurrence of one of the largest earthquakes in Greece on May 13 in a previously inactive area, VAN had another outstanding prediction for another earthquake with a comparable magnitude. An earthquake with a "comparable" magnitude did occur on June 15. Now (at the time of this writing, i.e. 6/19/1995) there is no outstanding prediction of large earthquakes in Greece. If no large earthquake occurs in the near future without further SES activities, this sequence of events would enhance the

significance of the relationship between SES and VAN predictions.

The critic of the VAN method would argue that these predictions are vague, and they were double predictions in two of the three cases. Also, with the standard set of criteria, none of them could be claimed as a success (if Δt is extended to 50 days, however, two of them would be successes). Although this criticism may be valid, I feel that the relationship between the SES that led to the predictions and the ensuing earthquakes is striking enough to convince me that SES is a significant signal associated with earthquake sequences.

4 Nature of VAN Predictions

My interpretation of SES activity and the earthquake that follows it is that both SES's and earthquakes are a manifestation of a regional geophysical process. The mechanism of SES is not well understood, and several mechanisms have been proposed: e.g. 1) piezoelectric effects; 2) electro-kinetics (streaming potential); 3) migration of dislocation; 4) dehydration of hydrous minerals. The role of gas and fluid flow in the crust appears important in some of these mechanisms. The absence of SES during coseismic period is often viewed as a peculiar aspect of SES, but if SES is caused by fluid flow rather than stress change, it may not be that strange.

If a large region enters into some critical state with increased flux of gas and water, one might experience a gradual increase of geophysical activities (increased seismicity, ground water anomaly (Radon, or other gasses, water level etc), electric, magnetic and electro-magnetic (radio frequency) noise etc) associated with a significant seismic event. The recent activities in Thessaloniki (m_b=5.2) and in northern Greece (M_w=6.5), and some earlier earthquakes were apparently preceded by significant regional seismic activity. Also Dr. Sudo mentioned at this meeting that some SES's were preceded by small earthquakes. In this scenario, SES or SES activities may be due to sudden influx of gas or water into the crust. Then water may eventually diffuse into a nearby fault zone and weaken the fault to cause an earthquake. This diffusion is slow and is responsible for the delay of an earthquake after detection of SES. This kind of process is most likely to occur in areas of active extensional tectonics such as Greece.

If this is the case, the VAN method may not be very different from what seismologists have been doing for years (monitoring seismicity, Radon, etc). For example, during the 1965 Matsushiro swarm in Japan, warnings for a possible occurrence of a large earthquake were issued to local residents on the basis of increased seismicity [Rikitake, 1979]. The difference is that, instead of monitoring seismicity, VAN uses electric self-potential for warning. Then why does VAN look far more successful than others? This could be really due to the extensive work of Dr. Varotsos and his colleagues on site selection and selectivity studies and Varotsos' ability of accurately remembering the details of the past activities and patterns.

5   Discussion and Conclusions

As far as I know no one has done as extensive electric measurements as VAN in seismically active areas. If someone does as much work as Varotsos and his colleagues do, it may be possible to make VAN type predictions if the geophysical and social environments are similar to those of Greece.

Geophysical environment is an interesting factor. If the above interpretation is correct, the method is more likely to work in areas of active extensional tectonics or of swarm type seismicity (like Matsushiro), but would be less successful in other areas where large earthquakes tend to occur abruptly without obvious foreshock activity. I believe that VAN is monitoring a significant geophysical process, but because of the insufficient instrumentation (seismological, electrical, and geophysical (magnetic, ground water etc)), it is unclear at present what the process is.

Social environment is an important factor to consider when we discuss whether the method is socially beneficial or not. At present, VAN predictions are uncertain as much as ±0.7 unit in M, and about 100 km in location; also the time window is somewhat vague (e.g. a few weeks). For example, if a VAN prediction of an $M=6$ earthquake is made for somewhere between Los Angeles and San Diego, then either an $M=6.7$ earthquake in Los Angeles or an $M=5.3$ earthquake in San Diego would qualify as the predicted event. Also, VAN predictions occasionally involve two predictions for the same SES. This type of uncertain prediction could be useful for the places where economical and social environments can tolerate this much of uncertainty. However, most industrialized communities would require far more accurate and definitive predictions. The local government must decide whether this kind of prediction is useful or not on the basis of their social and economical environments.

If the process is regional, as suggested above, some uncertainty is probably inevitable. Also, the VAN method may be able to predict the onset of an earthquake, but the event can cascade into a much larger event. Thus a definitive prediction of M would be also difficult. These uncertainties could be reduced with more stations and more experience about selectivity, but to what extent they can be reduced is yet to be seen. Considering the extensive heterogeneity of the crust and the indeterministic nature of earthquake process, I personally feel that deterministic predictions with small errors would be difficult. All of these questions remain to be investigated further.

Although the validity of the VAN method is still debated in Greece, in view of the large amount of work that has been done in Greece, it would be advisable to try to understand the physics of the VAN method by enhancing the existing programs in Greece with not only electric, but also seismological, hydrological and other geophysical investigations. It would be important to study the nature and the physical mechanisms of earthquakes in the region with modern broadband instruments. We would need more detailed knowledge of elastic and electric structures of the crust to understand the SES selectivity, which seems to be a critical element of the VAN method.

Regarding the question of exportability of the VAN method, it would be unwise to just "export" the method to another country as a ready-made tool for earthquake prediction; it would be very naive to think that one can solve the problem of earthquake prediction just by introducing the VAN method. If the VAN method is to be investigated in other countries, well designed seismological and geophysical investigations would be required before its usefulness for earthquake prediction in the region can be fully evaluated. It would require years of hard work in eliminating the noises, establishing all the calibrations and selectivity, and accumulating experiences, as Varotsos and his colleagues have done.

Acknowledgements

This is contribution #5569, Division of Geological Sciences, California Institute of Technology.

References

Hamada, K., Statistical evaluation of the SES predictions issued in Greece: alarm and success rates, *Tectonophysics*, **224**, 203-210 (1993).

Nagao, T., M. Uyeshima, and S. Uyeda, An independent check of VAN's criteria for signal recognition, *Geophys. Res. Lett., submitted* (1995).

Rikitake, T., *Theory of Earthquake Forecast and Warning (in Japanese)*, Center for Academic Publications Japan, Tokyo, 1-371 (1979)

Varotsos, P., K. Alexopoulos, and M. Lazaridou, Latest aspects of earthquake prediction in Greece based on seismic electric signals, II, *Tectonophysics*, **224**, 1-37 (1993).

Varotsos, P., and M. Lazaridou, Latest aspects of earthquake prediction in Greece based on seismic electric signals, *Tectonophysics*, **188**, 321-347 (1991).

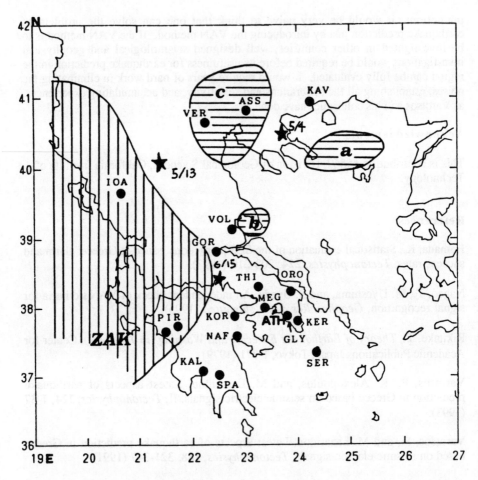

Figure 1: The locations of the stations (solid circle) and the earthquakes of May 4, 1995, May 13, 1995, and June 15, 1995 (star). Shaded areas indicate "selectivity" areas.

UNDERWATER ELECTRIC FIELD SENSORS

G.H. BACKHOUSE

Kernovel Ltd., 7 Park Road, Wadebridge, Cornwall, PL27 7NQ, UK

Underwater electric field sensors can be used to give better rejection of unwanted noise

Underwater electric field sensors can be used to supplement land-based electric field sensor measurements. The lower background noise levels in the sea provide a more stable environment for measurements. Underwater sensor electrode systems originally developed for military use, are now available for geophysical measurements.

Rugged silver/silver chloride electrodes, with very low source impedances and self-noise levels, have been developed to measure nanovolt per metre electric field signals. The proprietary electrode element construction, together with appropriate encapsulation, allows the sensors to be deployed in the sea for long periods without significant degradation in performance. Electrode pair differential voltage drift is around 50 microvolts per year, and 1/f noise is very low. A sensor with one metre electrode spacing can be made with a noise level of 5 nanovolts per metre, at 10 milliHertz. The sensor measurement frequency range is from DC to a few kiloHertz.

The sensors have well-defined source impedances, and, as a result, give very high rejection of common-mode signal noise. Electrode spacings are often one metre or less, and so temperature gradient effects are minimal. The sensors are relatively easy to install on the sea bed, even when they are attached to a heavy base unit, because of their small size.

A typical device has three mutually-orthogonal electric field sensor axes. A three-axis magnetic sensor can be included, together with a pressure sensor. The combined measurements can be used to reduce the influence of magnetic field variations, and tidal flow-induced voltage gradients. The sensors can be made with data recording equipment built-in, or data can be transmitted along a cable, on the sea bed, to a land-based data recorder.

It appears that the gradual variation in electric field, (GVEF), will give the largest signal and the longest warning period before an earthquake. One of the areas of future work should be the development of systems which can detect this signal, albeit in the presence of very large interfering signals from many other sources.

The choice of measurement site plays a most important part in determining the ultimate performance of a sensor system, and no effort should be spared in finding a quiet site, even if it is not in a convenient location.

SOME PERSONAL CONCLUSIONS FROM THE MEETING

C.W.A. BROWITT

British Geological Survey, West Mains Road,
Edinburgh EH9 3LA, Scotland

This short note suggests the potential value of involving scientists from many countries in collaborative investigations around the VAN sensitive areas in Greece.

It is likely that electrical and electromagnetic methods will contribute to earthquake prediction, and scoping studies presented at this meeting have shown that appropriate signal generation mechanisms could exist.

In the general case (that is, for all types of earthquakes), it is unlikely that one method or precursor type will be a unique earthquake predictor.

The VAN method appears to require considerable experience on the data acquisition side; particularly in tuning the instruments and finding geologically appropriate sensitive sites. It may be this problem and a lack of such experience by other groups which is leading to a slow take-up of the ideas outside Greece.

If the method is more generally applicable then this slow take-up is a tragedy and particularly ironic during the IDNDR where the thrust is to exchange and build experience and to transfer technology rapidly on humanitarian grounds.

I advocate, therefore, that in order for other workers to gain experience and to "calibrate" themselves and their instruments, the sensitive sites identified in Greece become open laboratories for other groups to work with, and alongside, the VAN group to this end. There will, of course, be logistical difficulties but in view of what is at stake and with the full co-operation of all parties, these can and must be overcome.

My own position on the general applicability of the method is neutral: the region and tectonic setting of experiments to-date are parochial from a Global view. That neutrality, which is widespread, could be quickly challenged through adoption of the above recommendation.

A BRIEF LOOK BACK AT THE REVIEW MEETING'S PROCEEDINGS

JAMES LIGHTHILL

University College London, Department of Mathematics, Gower St., London WC1E 6BT, UK.

This fourth paper in part VI is a short summary of parts I to V that follows the general lines of a talk, attempting such a summary, which was actually given at the review meeting.

Before inviting Academician Keilis-Borok, Director of the International Institute for Earthquake Prediction in Moscow, to offer his broad reactions to the discussion at this review meeting, I may perhaps briefly recall some essential aspects of the presentations that have been given. So this is your chairman's backward look at the papers as a whole, to be balanced at the end by his forward look towards some needs, generally agreed at the meeting, for future research in this field.

Some essential aims of the meeting, as stated at the outset by Seiya Uyeda, were to fill gaps in knowledge about VAN as well as to eliminate misunderstandings about details of the method. Later in his introduction, he outlined the nature of the case for reviewing VAN in 1995. He reminded us (as Bob Geller would later do in still greater detail) of the accumulated series of failures of attempted "mechanics-based" predictive approaches, such as ground-deformation measurements; in particular, modern data seem not to exhibit premonitory tilt.

In the electromagnetic field, he looked forward to papers by Drs Fujinawa and Enomoto, respectively outlining recent records of VLF and HF signals that are considered ` with hindsight' to have been precursors to some earthquakes. But then he pointed out that d.c. Seismic Electrical Signals (SES) had, in the meantime, been studied for over a decade by the VAN group; which, moreover, had used them not merely in a retrospective mode but as a practical means for providing earthquake warnings in advance.

Against this background he recalled that each SES station included large numbers of dipoles (pairs of buried electrodes between which the electrical potential difference was measured) including many short dipoles of length around 100 m, and many long dipoles of several kilometres' length. Rejection of noise played an absolutely vital role in SES determination; magneto-telluric noise (of largely ionospheric origin) being eliminated as a by-product of its appearance at all SES stations, while local noise from known industrial or community sources was eliminated by appropriately combining signals from short and long dipoles with carefully chosen relative placings. After such noise elimination, the true SES (including both the north-south and east-west components of electric field) could emerge.

Certain misunderstandings about VAN were linked to the supposition that approaches had been essentially unchanged since the method was first proposed in 1981. On the contrary, developing experience in analysing output from SES stations had, as would have been expected, brought about a progressive refinement in the classification of those SES which may be regarded as possibly useful precursors of substantial earthquakes. A key distinction was between any isolated SES peak and the so-called SES Activity (with a closely spaced sequences of very many peaks); the nature of any forecast being different, especially as regards the time window, for these two SES types. (A third interesting type is the Gradual Variation of Electric Field.)

Professor Uyeda indicated too the care which had proved necessary in the selection of specially sensitive areas for the location of SES stations; interpreting them, essentially, as geologically heterogeneous areas with the basement close to the surface. Finally, he posed a question to which the meeting would be continually returning: the question of whether records from VAN show the approach to have been specially successful in forecasting the relatively bigger earthquakes.

In the next paper Professor Varotsos himself reminded us first that the VAN approach is by no means a purely empirical method based on observing correlations between SES and subsequent earthquake activity. On the contrary, the method was initiated in response to specific hypotheses about mechanisms by which SES might be generated in response to stress changes, and various modern developments of mechanism concepts along broadly related lines, supported by laboratory experiments, would be described in later papers.

Some initial 1981 ideas about signals of millisecond duration that might precede an earthquake by a few minutes had later been abandoned in favour of the far more fruitful concept of the SES proper: a signal of some minutes' duration which is considered to precede an earthquake by a number of days (up to 11 days for a single SES, or a few weeks for the "big event" following multiple-peaked SES Activity).

Necessarily, a trial-and-error approach had been used to determine the best locations for siting of SES stations. At these stations experience indicated how to use, not only the general characteristics of SES, but specifically their polarity, in epicentre prediction. In certain instances such experience showed the need to forecast two alternative possibilities for epicentral location.

Professor Varotsos drew attention to the 1993 sensitivity map which has been the one consistently used in all VAN predictions since then. He added that, in parallel with the real-time transmission to the VAN centre in Athens of the analog signals from all SES sites, there has since 1992 been independent digital recording of data at each end of the four sensitive sites now used (Assiros, Ioannina, Pirgos and Keratea), allowing useful checks to be made. These considerations highlight a rather surprising contrast between measures taken to refine the VAN method in recent years and the fact that most statistical attempts to evaluate VAN performance have concentrated on the earlier period 1987-89.

Professor Nomicos, also a member of the original VAN group, gave a valuable paper which stressed several techniques essential to noise elimination (including the need for extreme care in the selection of electrode materials to avoid electrode noise). He outlined also the data capture and data transmission techniques which allowed full details of the SES from all stations to be analyzed in real time at the VAN centre in Athens.

Then, in the first of a number of talks devoted to the possible "exportability" of VAN, Dr. Massinon reviewed his experiences since 1989 in setting up a network of SES stations in the alpine region of southeastern France. He reported (i) success in the identification of 5 sites for such stations that would avoid serious industrial or community noise, along with (ii) satisfactory procedures for eliminating magneto-telluric noise. On the other hand, the region's relatively low seismicity had allowed practically no progress in the selection of specially sensitive sites for stations. At the existing sites, some apparent SES precursors had been observed within a distance of around 100 km from subsequent earthquakes, although there had been some false alarms and some failures to predict. (Sadly, while other talks at the review meeting are all represented by papers in this book, a family health tragedy made it impossible for Dr. Massinon to contribute such a paper.)

Next, in the first of a series of papers about possible mechanisms which might underlie observations made in the course of VAN experiments, David Lazarus addressed questions of both the origin of SES as earthquake precursors and their detection over long distances. Some candidate mechanisms for their origin included the possibility of phase transitions such as were known from laboratory experiments to occur at high hydrostatic pressure under conditions of increasing stress. Detection over long distances might call for the presence of extensive strata with high dielectric constant — although the existence of such strata extending over many tens of kilometres was doubted by contributors to the discussion; while others noted that, in any case, stress redistributions culminating in a substantial earthquake may build up over a considerable area. In the meantime, Dr. Lazarus made a specific plea for more spectral data about SES, which could greatly assist attempts at interpretation.

Laboratory experiments related to possible SES mechanisms were described both by Larry Slifkin and by Clare Mavromatou. The ubiquity of crystal defects, including especially the well known edge dislocations, was emphasized by Dr. Slifkin. His experiments demonstrated the large electric potentials that can be generated in a building-up of stress, which multiplies such dislocations as well as causing them to move; the observed electrical signals arising from the fact that relaxation times are then much greater for charge distribution than for stress. The experiments made jointly by Dr. Mavromatou and Dr. V. Hadjicontis used actual rock samples under stress, and demonstrated "stress-induced polarisation"; essentially a phenomenon expected to occur under those conditions, involving substantial rates of change of stress, under which they measured large transient currents.

Another contribution to the "mechanisms" discussion was made by Ken Sudo, whose analysis of 32 large events studied by the VAN group suggested the possible influence of foreshocks on observed SES. Professor Varotsos agreed to attempt his own investigation of whether foreshocks had been significantly correlated to VAN predictions.

Next the meeting heard a highly professional series of papers offering counterarguments against VAN, beginning with a comprehensive lecture in which Bob Geller argued strongly against the possibility of uncovering any effective method for short-term earthquake prediction. By an effective method he meant one which would define "within narrow limits" three features of a predicted earthquake: magnitude, epicentral location, and time of occurrence; and he went on to ask for VAN predictions to take more unambiguous forms in all three of these respects. Moreover, he criticized strongly the wording of very many press articles (especially in Japanese newspapers) which seemed to make exaggerated claims for success for VAN predictions.

Professor Geller's specific arguments against the possibility of short-term prediction began (as I mentioned earlier) with a rather full analysis of the hitherto unsuccessful history of attempted "mechanics-based" predictive approaches. Then he reviewed the failure of so-called "empirical" approaches — in which no attempt was made to understand the link between the precursor and the earthquake. Lastly, he stressed fundamental difficulties in the way of short-term prediction such as may arise from essentially chaotic elements in both the progression and amplification of rupture processes.

Extensive discussion of Professor Geller's paper focused above all on whether limits as narrow as he proposed were indeed essential to the utility of short-term predictions. Meteorological experience had (it was noted) shown how significant savings of lives had followed from the dissemination of 24-hour landfall warnings for Tropical Cyclones even in the days when such warnings had still possessed a mean error of 300 km — even though later reductions of mean error, first to 200 km and very recently to 120 km, had greatly increased the warnings' usefulness. In the meantime, the closed nature of the present meeting might be viewed as an insurance against further possibilities of misleading press reports.

Another lecture casting doubt on the VAN approach was given by Dr. Mulargia. His was the first of a series of papers statistically analysing the time-series of all VAN forecasts made during the three years 1987-89. Essentially, his approach was to ask whether it would be possible to do as well by means of a purely statistics-based forecast that utilised no SES information at all but rested on established knowledge about the temporal and spatial distribution of Greek earthquakes with magnitudes (as recorded at Athens) exceeding 5. His paper used such established knowledge (including data on distributions of aftershock occurrences) to produce forecasts apparently as good as those issued by the VAN group.

On the following morning Max Wyss, speaking as Chairman of IASPEI's Sub-Commission on Earthquake Prediction, brought forward some further counterarguments against VAN — although he did not share Professor Geller's opinion (see above) that short-term prediction was inherently impossible. But he stressed the rigorous approach adopted by the Sub-Commission in evaluating prediction methods.

Above all, the Sub-Commission insisted that the methods evaluated had to be defined precisely, because of the need to study — as Professor Mulargia had done — whether an actual success rate for predictions might have been arrived at by chance. Against the background of the detail required for a statistical evaluation along these lines, he viewed the VAN procedures as too imprecise. Also, he queried the existence of convincing mechanisms for VAN.

In response to a question, Professor Wyss agreed that the VAN group should not be judged on the basis of speculations in their earliest (1981) paper, and he acknowledged that `a hypothesis can be developed by stages'. On the other hand, he regretted that VAN predictions tend to locate each parameter (magnitude, epicenter, time) in a defined interval, yielding a sort of "box" in parameter space — even though a prediction taking the form of a probability distribution might be scientifically more natural and could still lend itself to statistical evaluation.

Professor Wyss also paid extensive attention to important differences between the earthquake magnitudes provided locally by the Seismological Observatory in Athens and those which were arrived at later by standard international processes of investigation. These distinctions were admitted as important by those at the meeting, some of whom pointed out on the other hand that local magnitude had in practice needed to be used in developing VAN methods, which ought therefore to be judged by comparing predicted with observed local magnitudes.

On this latter basis some difficulties arose in predictions where the expected magnitude was 5.0 but a possible error of 0.7 was allowed for. A `box' hypothesis of this type implied that magnitudes of low as 4.3 might be claimed as successes; yet earthquakes of such low magnitudes were rather common in certain areas of Greece — often allowing such `successes' to be ascribed to chance.

Professor Varotsos emphasized in reply to this point that the VAN group have never sought to predict all earthquakes of magnitude as low as 4.3, although he admitted that there were difficulties with predicted magnitudes of 5 because of the error bound implicit in statistical relationships between earthquake magnitude and SES amplitude. Professor Varotsos expressed a preference for the VAN method to be judged on predictions of relatively large earthquakes; say, those with (local) magnitude ≥ 5.8.

Next, Steve Park offered a revised statistical analysis of the VAN predictions made during 1987-89. Before doing so, however, he drew attention to widespread international agreement — expressed for example in the conference at Lake Arrowhead, California — to the effect that crustal processes are generating measurable electric fields; which accordingly (even though controversy on detailed mechanisms remains) must be among the phenomena meriting study in the context of earthquake prediction.

His statistical method, developed jointly with David Strauss, gave results of comparisons with pure statistics-based forecasts which were dependent on the assumptions used about the statistics of aftershocks. For example, VAN successes in forecasts of up to 22 days were calculated as significant at the $P < 0.0007$ level if compared with a pure-chance forecast using no statistical information about aftershocks; however, on two alternative assumptions about aftershocks, this was reduced by either one or two orders of magnitude — in the latter case, merely to significance at the $P < 0.07$ level.

Professor Park reiterated that the analysis included predictions for magnitudes of 5 and above with the same error bound of 0.7 as had been mentioned earlier. Once again, however, pleas were made to pursue future evaluations on a basis revised in two ways (concentration on earthquakes of larger magnitude, and analysis of more data than were contained in the limited 1987-89 time-series).

A reminder needed to be given, moreover, that `earthquake prediction is not a game!' The general public in areas severely threatened by earthquakes looks for guidance above all in the form of warnings of events of relatively larger magnitude. A judgement of whether, historically, VAN has succeeded in providing a useful basis for such warnings may be more important than conclusions emerging from some tidy set of evaluation rules.

The paper by Dr. Hamada addressed this issue — although again only for the 1987-89 period, when 5 out of 8 "large" earthquakes had been correctly forecast. (The VAN group claimed in reply that at least one missed forecast had resulted from the unavoidable absence overseas of their principal forecaster.)

There followed some further papers related to the possible exportability of VAN. Professor Nagao, although emphasizing that Japan is electrically noisy, described two SES stations which had been set up in relatively less noisy locations: Komatsu and Suzu with 6 and 4 dipoles respectively. Only long dipoles had been used, with rather mixed results. Each of two earthquakes had been preceded by a single SES at Komatsu, and a third by SES Activity at Suzu; yet a major disappointment had been the absence of any precursor for the great Kobe earthquake.

The interesting paper by Dr. Honkura described attempts, begun 3 days after the Kobe earthquake, to predict the associated aftershocks using a group of short dipoles. In a necessarily hurried investigation, problems of both electrode noise and noise from electrical railways could not be fully overcome; thus, it remained an open question whether any observed SES were related to aftershocks.

Dr. Fujinawa argued that the electrical-noise characteristics of Japan suggested an abandonment of horizontal dipoles in favour of the use of vertical antennae buried deep into the ground. Measurements with such an antenna of 800 m depth in d.c., ULF and VLF had indicated the existence of VLF precursors for some large earthquakes — including Kobe.

It was generally agreed that VAN might be particularly exportable to regions that combined high seismicity (to allow identification of sensitive areas) with low electrical noise. In this context, a recent visit by Professor Kulhanek to the country of Guatemala (which satisfied both conditions) had created good hopes for the installation of a VAN network there.

Yet another valuable paper on possible SES mechanisms was then contributed by Professor Le Mouël. He agreed with Professor Park that crustal processes do produce measurable electric-field changes; and, in particular, described such changes as observed in a volcanic crater. He outlined work by Pascal Bernard on possible electrokinetic mechanisms for the changes, together with some results of electrical signal measurements made in a large quarry near Paris. He emphasized that electrokinetic effects were associated with fluid motions within porous media under strain. His own interpretation of SES recorded at distances of order 100 km from a subsequent epicenter was based on the possibility that some distant triggering of fluid flow could arise from crustal deformations associated with a coming earthquake.

In discussion it was agreed that more research was needed on details of the geology — including groundwater features — of those sensitive areas in Greece which the VAN groups had identified; where, moreover, it would be important to give further study to electrical properties of the crust as well as to aspects of local seismicity. The announcement that Dr. Massinon would collaborate with the VAN group on installation of an augmented network of seismometers was particularly welcomed.   New studies on social aspects (including training in responses to warnings) of earthquake prediction in Greece also needed to be made.   Some more comprehensive ideas on future research programmes would emerge from the ensuing review paper by Academician Keilis-Borok.

NON-SEISMOLOGICAL FIELDS IN EARTHQUAKE PREDICTION RESEARCH

V.I. KEILIS - BOROK

International Institute of Earthquake Prediction Theory and Mathematical Geophysics, Russian Acad. Sci., Warshavskoye sh. 79, kor. 2, Moscow 113556 Russia

Research in the short-term dynamics of the lithosphere, earthquake prediction included, indicates the following new possibilities for further test and development of the VAN method: Evaluation of statistical significance of predictions and optimization of prediction algorithm, both based on an errors diagram; applications to damage prevention, allowing to realistic estimation of the accuracy of VAN predictions; incorporation of geometry of the fault system into prediction algorithm; integration of VAN signal with other potentially predictive fields; integration with multistage prediction; mathematical modeling of fault system dynamics, and its possible relations to different precursors, VAN - type ones included. These possibilities are relevant not only to the VAN data but to other prediction - related fields; they have been explored so far mainly for premonitory seismicity patterns.

1   Immediate possibilities

1.1   Evaluation of Performance and Statistical Significance of a Prediction Method

Two approaches to evaluation of VAN method are discussed by Nagao et al.[1], Schnirman and Mulargia[2,3]. One is oriented at prediction without "initially fixed unambiguous rules"[1]. This may have its merits in a preliminary exploration of potential precursors, when formalization may lead to premature rejection of a deserving hypothesis. However at its present stage VAN method seems to merit a rigid test such as suggested by Schnirman[2] and Mulargia[3] in order to eliminate the danger of data-fitting and to optimize the rule for declaration of alarms.

Earthquake prediction algorithms, VAN among them, inevitably include some adjustable parameters, which in lieu of an adequate theory can't be determined a priori and have to be chosen retrospectively to improve the success-to-failure scores. This creates the danger illustrated by Figure 1 (after Gabrielov et al.[4]). It shows "prediction" of random numbers by other (independent) random numbers. An apparently good success-to-failure score is obtained by retrospective adjustment of only two parameters: threshold for declaration of alarm (which is 70) and duration of a single alarm (which is 5 time-units). In real prediction methods adjustable parameters include also magnitude thresh-

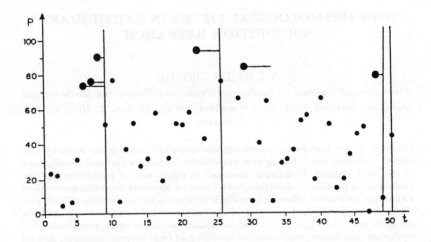

Figure 1: Illustration of the danger of retrospective data-fitting.

olds, size of the territory etc. Methodology allowing to deal with this uncertainty was developed by G. Molchan [5,6,7,8]. It can be briefly summarized as follows.

Errors diagram. Consider the basic test of a prediction algorithm: it is applied to a certain territory; N strong earthquakes occurred there during the time period T, covered by prediction; the alarms declared cover altogether the time τ and have missed $n \leq N$ strong earthquakes. An obvious requirement is that the same set of adjustable parameters is used for the whole period T. The quality of prediction is characterized by dimensionless parameters $n^\circ = n : N$, $\tau^\circ = \tau : T$. The tradeoff between n° and τ° depends on the choice of parameters. In Figure 1, for example, the total duration of alarms can be reduced at the cost of missing one more strong earthquake if the threshold for declaration of alarms is raised to 90 or if duration of a single alarm is reduced to 3.

Evaluation of performance ("quality") of a prediction algorithm requires that this test is repeated with various combinations of adjustable parameters and the results summarized by the errors diagram such as shown schematically in Figure 2 (after Molchan [7]). Different points on the diagram correspond to different combinations of adjustable parameters. Diagonal $n^\circ + \tau^\circ = 1$ corresponds to random predictions: at each step in time an alarm is declared with fixed probability p independent on observations; different values of p correspond to different points on the diagonal. Points above it can be disregarded, since they do not correspond to a meaningful algorithm: reversing the rule for

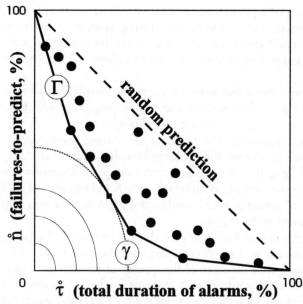

Figure 2: Error diagram for optimization of a prediction algorithm.

declaration of an alarm one would replace them by the points $(1 - n^\circ, 1 - \tau^\circ)$ below the diagonal. The error diagram may allow also for the variation of more essential features of a prediction algorithm, inevitably chosen with a certain degree of freedom: formalization of premonitory phenomena, their scaling, etc.

Performance of an algorithm is characterized by the *error curve* – a lower envelope of the points on error diagram (curve Γ in Figure 2). This curve summarizes the best predictions, which can be achieved by variation of parameters and other adjustable decisions. The whole curve Γ is necessary to compare different algorithms and to use prediction for damage reduction, as discussed in Section 1.2. Being well known in prediction of other phenomena, the error diagrams were just recently introduced into earthquake prediction research[6,7].

Evaluation of statistical significance of predictions. Methodology of evaluation is suggested in the paper by Molchan et al.[5]. It allows for specific formulation of earthquake predictions: they indicate the time interval, during which the magnitude will exceed a certain limit, while in the better developed Kolmogoroff-Wiener prediction concept the value of magnitude would be extrapolated for the next step in time; this is described by as the difference between "horizontal" and "vertical" predictions[5]. Evaluation obviously requires the test of a prediction method with fixed adjustable parameters. Then one can estimate the probability of getting as good or better results under a competing "null hypothesis" that alarms are not connected with strong earthquakes and happen to be successful only by chance. This leads to the following problem [5]:

Let us consider a test with duration T in normalized time-scale $\mathbf{t} = t : T$, so that $0 \leq \mathbf{t} \leq 1$. Two independent sets of random points, $\mathbf{t} = x_1, x_2, \ldots, x_K$ and $\mathbf{t} = y_1, y_2, \ldots, y_K$ are tossed into this interval.

First set represents the times of the strong earthquakes, second – the starting times of alarms; the duration of a single alarm is τ'. By definition x_j is predicted by y_i if $y_i \leq x_j \leq y_i + \tau'$ and there are no points x between y_i and x_j. Let κ be the number of points x_j which were predicted in this sense; ν – the number of points y_i which predicted a point x_i. Since x_i may be predicted by several precursors, $\nu \geq \kappa$.

The problem is to find joint probability distribution for (κ, ν), knowing the distributions of x_i and y_i. We can find then the distribution $F(f)$ of a statistic $f(\kappa, \nu)$, chosen as a measure of the quality of predictions. This measure may be defined in such a way that it is larger for better quality. Then statistical significance of predictions considered is $\{1 - F\{f(\kappa_{\text{obs}}, \nu_{\text{obs}})\}$ where the index obs indicates observed values.

Joint distribution of (κ, ν) is given by Molchan et al. [5] for two cases: distributions are uniform for the times of strong earthquakes and of alarms; or distributions are uniform for the times of strong earthquakes and of main shocks, while each main shock generates an alarm with a probability p. A stronger null hypothesis is that the moments of strong earthquakes are not homogeneoulsy distributed in time, but depend on the time u elapsed since the previous one. Then for many distributions of u the following prediction rule is optimal[7,8]: declare an alarm for $0 < u \leq \varepsilon \bar{u}$ and from $u = .75\bar{u}$ onwards; here $\bar{u}$ is average value of return time $u, \varepsilon \ll 1$. The first interval accounts for the clustering of strong earthquakes, second – for their recurrence time. This rule is optimal for minimax prediction strategy, described in Section 1.2. It will give, on average, $n^\circ = \tau^\circ = .35$, providing that estimation of $\bar{u}$ is sufficiently accurate. Preliminary analysis of VAN predictions by Schirman et al. [2] indicated that they are possibly valid for earthquakes with magnitudes $M \geq 5$ but not for the whole range $M \geq 3$.

1.2   How to Use the Currently Realistic Predictions for Damage Reduction?

Predictions of low accuracy may be used for damage reduction, providing that:
- Prediction indicates the probability of false alarm and specific (but not

necessarily narrow) area, magnitude range, and time interval where a strong earthquake has to be expected.

- Flexible scenarios of response to prediction are prepared.

Then this response may be escalated, de-escalated or retargeted, depending on what specifically is predicted at the moment of decision and on combined risk of false alarm and failure to activate some safety measures.

This approach takes advantage on the diversity of necessary safety measures, dictated by the diversity of possible damage from the earthquakes.

Diversity of damage to population and economy includes: destruction of buildings, lifelines and other constructions; triggering of technological disasters, particularly the fires and release of toxic, radioactive and genetically active materials; triggering of other natural disasters, such as floods, avalanches, landslides, tsunamis etc. As dangerous are the *socio-economic impacts of the earthquakes* such as: disruption of vital services (supply, medical, financial, law enforcement etc.); epidemics; disruptive anxiety of population, profiteering and crime; disruption of economy that is a drop of production and employment, destabilization of prices, credit, stock market, national currency etc. Such impacts may be inflicted also by *undue release of predictions*.

These phenomena are developing in different time-scales from immediate damage to chain reaction, lasting tens of years and spreading regionally if not worldwide.

A hierarchy of diverse safety measures is required by this diversity of damage.

Permanent measures, to be maintained over the decades, include: restriction of land use; building codes; insurance and special taxation; earthquake preparedness of civil defense type such as emergency legislation, installation of stand-by resources, development of scenaria of response to prediction and of post-disaster actions, simulation alarms etc.; earthquakes-related R&D.

Temporary measures, activated in response to an earthquake prediction may include: enhancement of permanent measures, listed above; transfer to emergency legislation, up to martial law; mobilization of post – disaster services; neutralization of the high-risk objects; evacuation of population; monitoring of socio-economic changes etc. These measures are required in different forms on local, provincial, national and international levels. Different measures require different lead times, from seconds to years, to be activated; having different cost they can be realistically maintained for different time-periods, from hours to decades; and they have to be spread over different territories – from selected points to large regions.

Different stages of earthquake prediction are therefore required for damage reduction. Let us remind that these stages are the following (typical duration of alarms is indicated in the brackets):

- Background: maps of the maximal possible magnitude and of the average return time of destructive earthquakes with different magnitudes.
- Long - term .... (10^1 years)
- Intermediate - term .... (years)
- Short - term .... ($10^{-1} - 10^{-2}$ years)
- Immediate .... (hours or less)

No single stage can replace another one for damage reduction and no single measure is sufficient alone. On the other hand many important measures require neither high accuracy of prediction nor large expenses. Accordingly, the practical value of predictions would increase, if different stages of prediction are combined. Quality of predictions may also increase in such case. For example, in VAN predictions, which are the short-term ones, the rate of false alarms might be reduced, if intermediate-term prediction, based on other data, is allowed for.

The framework for decision-making is developed in the papers by Molchan [6,7,8] (see also the early paper by Kantorovich et al. [9]). A point on the error curve (Figure 2) has to be chosen for actual predictions. This choice depends on how predictions are used. The choice is obvious in the two extreme situations – when failures to predict or, to the contrary, false alarms have to be avoided with 100% certainty. In the first case the alarm should be maintained permanently ($n^\circ = 0$, $\tau^\circ = 1$); in the second case no alarms should be ever declared ($n^\circ = 1$, $\tau^\circ = 0$). To choose an optimal strategy in any other situation one has to juxtapose the error curve with the loss function γ depicting the cost of safety measures minus the damage which they prevent.

Let us describe some basic types of loss functions [7,8].

- $\gamma = \gamma(n^\circ, \tau^\circ)$. By definition γ increases with either of the arguments. Consider on the errors diagram the contours $\gamma(n^\circ, \tau^\circ) = d$, assuming that they are convex. When d is too small, the contours will not intersect with the error curve: such small losses are unachievable with the algorithm, for which the error curve Γ is determined. The point, where γ and Γ touch each other determines both minimal achievable loss and the optimal set of free decisions in the prediction algorithm. This optimization requires minimal information on the losses. In the next example more information is required, in the last one it is not used at all, so that Γ represent hazard rather then risk.

- Discounted losses. This measure allows for the following factors:
- Reduction of losses from different alarms and different earthquakes to the common moment of time.
- Initial cost of activation and deactivation of safety measures; accordingly the number of alarms has to be considered, apart from their total duration.

– Possible existence of several levels of safety measures, activated or deactivated according to the type of alarms.

Methodology for minimization of such discounted losses is based on the Bellman - type equation.

• Minimax strategy $\gamma = \max\{n^o, \tau^o\}$. This means that the user of predictions prefers to be on safe side, avoiding underestimation of either error. Minimal value of γ is achieved then at the point on Γ where $n^o = \tau^o$. This measure does not require any information on losses.

What if any safety measures would be justified in response to the VAN predictions in their current state? VAN telegrams, announcing the alarms, provide civil protection authorities with a rare possibility: to estimate retrospectively the effect of different measures, which could be activated on the basis of these telegrams. Possibly, a larger effect could be achieved by relatively more reliable predictions of the strongest earthquakes considered, $M \geq 5$ in Greece [2] and $M \geq 6$ in Japan [1].

2 Further possibilities

We shall discuss here the possible links of VAN precursors and other premonitory variations of electric fields with earthquake prediction problem as a whole. We face in this problem three major phenomena:

• Hierarchical system of interacting blocks and faults, which comprises the earthquake-prone parts of the lithosphere [10,11,12].

• Non-linear dynamics of this system featuring chaos and self organization, with a multitude of destabilizing mechanisms, fluids migration among them [12,13]; this system remains permanently in critical state, even after a large discharge of energy [13,14].

• Scenarios of critical transitions, earthquakes included, within this system; their expression in the observable fields as a set of premonitory phenomena [15]. With all diversity of these fields they are closely interconnected by their common origin – the structure and dynamics of the blocks&faults system. Moreover, modeling of the lithosphere dynamics suggests, that some precursors to critical transitions are not field-specific and even not Earth-specific [14,15]. Accordingly, the experience, gained in the study of premonitory seismicity patterns and in modeling of the lithosphere dynamics is relevant to prediction, based on other fields. Obviously, electric precursors would be closely connected with migration of fluids in the lithosphere, since fluids change electric conductivity of the media.

2.1   Structure of Fault System in Earthquake Prediction

Nucleation nodes. Strong earthquakes are nucleated only within specific areas around the junctions and intersections of faults [10,11]. These areas, called "nodes", are characterized by particularly intensive fracturing and contrasting neotectonic movements, resulting in mosaic patterns of the structure and topography. Geomorphologic concept of the nodes was described by Gerasimov and Rantsman [10], their formal definition – by Alexeevskaya et al. [16].

The nodes are well known in geological prospecting, since ore and oil deposits are concentrated in them. It was established in a series of studies, that the epicenters of strong earthquakes are also confined to the nodes [10,11]; possible explanation is discussed in the next section.

Moreover, nucleation of strong earthquakes is confined not to all the nodes, but only to those with special distinctive features of tectonics, topography and gravity anomalies [11,17].

Particularly relevant to this review are two following features, found in the San Andreas and Big Anatolian faults systems, for the nodes, where the earthquakes with $M \geq 6.5$ can be nucleated [11]: proximity of a hydrothermal reservoir, and a Neogen-Quarternary depression of the node on the background of a general unlift. The access of fluids is indirectly indicated by depression too, since it is associated with relative tectonic tension, favorable for fluids accumulation. Nucleation nodes should be allowed for in earthquake prediction research since precursors may be different within and outside the nodes [18]; in particular precursors connected with fluids and therefore with elecrtic fields may be more explicit within the nodes.

Measures of instability. Two integral parameters named geometric (G) and kinematic (**K**) incompatibility in the blocks&faults system control accumulation of deformations and stress, as well as the tendency of the system to a change of faults geometry, to fracturing and, accordingly, to strong earthquakes [19,20,21]. Vector **K** is deviation from the well known Saint Venant condition of kinematic compatibility; it ensures that relative displacements on the faults can be realized through some absolute movements of the blocks, separated by the faults, without deformation of the blocks.

This condition is not sufficient, however, to ensure, that the above tendency is not accumulating near the fault junctions [19,20,21]. An example is shown in Figure 3 (after Gabrielov et al. [21]). Short arrows show the horizontal components of the absolute movement of a block, long arrows – the horizontal component of a slip rate on a fault. A,B,C,D are the corners of the blocks. Initial position of the blocks is shown on the top, extrapolation of the movements (physically unachievable) - on the bottom. Geometric incompatibility

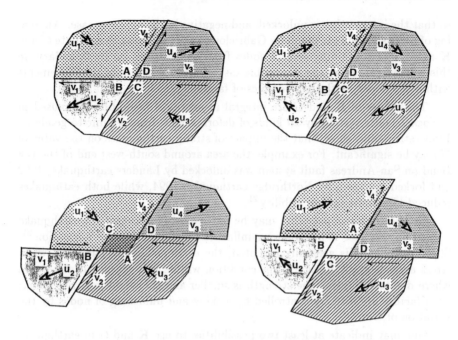

Figure 3: Examples of geometric incompatibility near the faults intersections.

locks up the node on the left and unlocks the node on the right. If all the rates of the blocks motions u_i are the same and, accordingly, the slip rates on each fault v_i are the same on both sides of intersection point, $\mathbf{K} = 0$. However some incompatibility remains near intersection points. In the case shown on the left side of Figure 3 the opposing corners of the blocks tend to move against each other; this is impossible without a change of fault geometry and will result in deformations and fracturing. Note, that fracturing will not restore incompatibiliy, but only transfer it to the newly formed faults of smaller size[20]; eventually this will lead to formation of the nodes, described in previous section. This phenomenon was first described in the paper by McKenzie and Morgan[19] for triple junctions along with a condition, under which "the faults [near a junction] may retain their geometry, as the blocks move".

Geometric incompatibiliy is introduced by Gabrielov et al.[21] as the quantitative measure of deviation from this condition. In the examples in Figure 3 the junction point tends to split into a parallelogram; its area can be represented as Gt^2, if the slip rates are constant. Scalar value G is geometric incompatibility; by definition, it is positive when the corners are pulled apart, so that the intersection is unlocked, and negative in the opposite case. An analog of Stokes formula is found by Gabrielov et al.[21] allowing to estimate G and K in an area containing several nodes from observations on its boundary; in this way we can avoid the impossible task of reconstruction of the movements within the nodes, with their mosaics of faults.

The values of K and G are integral measures of inequilibrium caused in a blocks&faults system by all kinds of deformations, from seismic to geodetic. Pilot estimations show, that the impact of strong earthquakes on the value of G may be significant. For example, the area around south-west end of the Big Bend on San-Andreas fault system was unlocked by Landers earthquake, 1992 and locked up again by Northridge earthquake, 1994, while both eathquakes reduced kinematic incompatibility[21].

Geometric incompatibility may be particularly important for earthquake prediction, because it has a strong influence on the earthquake nucleation[21]. Specifically, as it is generally accepted, the earthquakes are nucleated when stress exceeds static friction. This condition will be first reached in the nodes where due to fracturing the strength is smaller and the stress may rise faster; both factors are directly controlled by locking and unlocking of nodes by the variation of G.

One may indicate at least two possibilities to use K and G in earthquake prediction research: to explore their changes prior to strong earthquakes, thus integrating the above mentioned data on deformations in different time-scales; and to check whether some precursors have different significance for different values of K and G. For example, there are some indications, that alarms should be declared only if the node is locked up[18].

The values of K and G may be estimated simultaneously for different areas, overlapping or nor not. One must note however that contrary to other methods, discussed in this review, the development of methodology for estimation of G has been only started.

2.2   Migration of Fluids, as a Possible Source of Precursors

Fluids in earthquake-prone faults may influence dynamics of seismicity because they control both instability of the fault system as a whole and nucleation of specific earthquakes[12]. This control may be realized by various mechanisms: non-linear filtration of fluids, acting as a lubricator[22]; Rhebinder effect or stress corrosion[23,24]; dissolution of rocks and mass transfer[24]; disturbance of equilibrium of the blocks&faults system[25]. We shall consider now as an example one mechanism of such control.

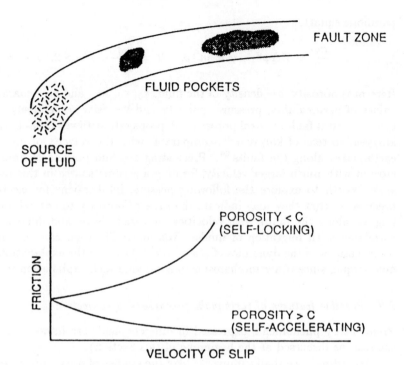

Figure 4: Instability, induced by non-linear filtration of a fluid through the fault zone.

Instability, caused by nonlinear filtration [22]. Let us model a fault zone as a thin porous layer between crustal blocks. Migrating fluids could trigger a seismic slip by reducing the static friction in the fault zone. Strong instability arises, because dynamic friction depends on porosity (Figure 4, after Barenblatt et al. [22]).

When porosity is sufficiently small, friction will increase with the slip rate so that the slip, once started, will self-decelerate; instead of an earthquake a vacillating creep or a slow earthquake will occur, unless the fault is relocked. However, when porosity exceeds certain threshold, the friction will decrease with the slip rate, so that the tiny slips which always occur will self-accelerate, grow, and merge. The porosity may be raised to critical level by infiltration of the fluid itself. Such critical porosity will propagate according to non-linear parabolic equation

$$\frac{\partial \varphi}{\partial t} = V\Delta\varphi^{\alpha}, \quad \varphi = m \cdot \rho, \quad V = \frac{k_0 p_0}{m_0 \eta_0}, \quad \alpha > 2.$$

Here m is porosity, ρ – density of the fluid, k_0, p_0, m_0, and η_0 – characteristic values of permeability, pressure, porosity and viscosity respectively. For the typical crustal faults critical porosity will propagate with velocities of the order km/year to tens of km/year [22], comparable with the velocity of migration of earthquakes along the faults [26]. Preexisting residual pockets of fluid will be crossed with much larger velocity, forming a seismic source in this model. It seems worth to explore the following possible implications for electric field precursors: that they may indicate the raise of porosity to critical level, may migrate along the faults with velocities indicated above, and, finally, may be correlated with migration of fluids. Non-linear filtration can explain many basic features of the dynamics of seismicity [22]; however the explanation will be non unique, since other mechanisms may provide such explanation too [12,23].

2.3   Possible features of earthquake precursors; a conjecture

Premonitory phenomena, expressed in different fields are interconnected (for the reasons discussed at the beginning of this Section).

Accordingly we shall summarize here the studies of premonitory seismicity patterns – both modeling and phenomenology – from a single point of view: what is suggested by these studies for the further quest for non-seisological earthquake precursors?

Where to look for precursors? The generation of the earthquakes is to large extend non-local: *a flow* of earthquakes is generated by *a system* of blocks and faults, rather than each single earthquake – by a single fault [27]. Strictly speaking, no earthquake can be completely separated from the dynamics of the global fault system. Nevertheless the symptoms of the approach of an earthquake may be recognized within a more limited fault system and time-period.

At the intermediate-term stage – within years before an earthquake – the linear dimension of this system is at least 5 to 10 times larger, and sometimes – about 100 times larger [28] than the dimension of an incipient source. Accordingly, the signal of an approaching strong earthquake inconveniently comes not only from a vicinity of its incipient source, but from much wider area, which may include the faults of different nature: strike-slip, thrusts etc. An example is shown in Figure 5 (after Knopoff [29]). Squares show the epicenters of two strongest earthquakes in Southera California: Landers, 1992, $M = 7.6$ (top) and Kern County, 1952, $M = 7.7$ (bottom). Dots outline the area in which the precursors to these earthquakes were formed (in this case the precursors where the peaks of seismic activity in medium magnitude range and dots show the epicenters of corresponding earthquakes). Lines show active faults.

For VAN precursors, as they are reported, these distances are large indeed.

Should the precursors be field-specific? Not necessarily all of them. Some precursors may depend only on geometry of the fault system [18,21] or reproduced by non-linear systems of exceedingly simple design, not even Earth-specific, such as lattices of interacting elements [13,14]. This suggests that some earthquake precursors may reflect the symptoms of instability common for a wide class of non-linear systems [15].

What kind of precursors to look for? If previous conjecture is correct it may be worth to explore the following symptoms of an approaching strong earthquake inferred fron the studies of premonitory seismicity patterns [12,13,15,27]: the field considered becomes more intense, irregular and clustered; range of the spacial correlation in this field and its responce to exitation are increasing. Each of these phenomena may be represented by differently defined precursors. Due to chaotic nature of the lithosphere dynamics such definitions should be sufficiently robust.

Should all precursors be region-specific? Not necessarily. For example, premonitory seismicity patterns in their robust definition are similar (i.e. identical after normalization) in the magnitude range at least from 4.5 to 8.5 and for a wide variety of neotectonic environments, from subduction zones to transform faults to intraplate faults in the platforms, to induced seismicity in mines and artificial lakes [27]. There are no obvious reasons, why a limited similarity may not be encouterd in non-seismological precursors.

How to integrate available data? Methodology of pattern recognition of infrequent events, developed by I. Gelfand school [11,27], proved to be useful at the present "pre-equations" stage, when the search for precursors is essentially a trial-and-error procedure. This methodology allows to cope with complexity and incompleteness of the data when more traditional methods of mathematical statistics are inapplicable due to the limited number of strong earthquakes.

More fundamental alternative is to derive premonitory phenomena from the models of the lithosphere dynamics. Block models, described by Gabrielov et al. [25], deserve attention for this purpose: they may reproduce realistic geometry of faults and, after a feasible modification, – the fluids migration.

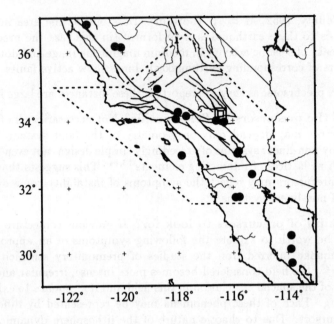

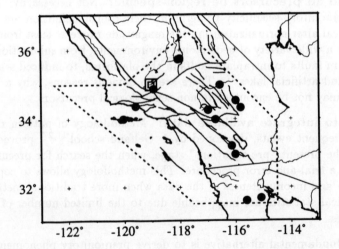

Figure 5: Illustration of large size of the areas, which contribute to formation of premonitory phenomena.

3 Conclusions related to VAN precursors

Two immediate tasks seem unavoidable.

(i) To determine performance of VAN precursors in terms of the errors diagram. This is necessary for their further development, for evaluation of their statistical significance, and for practical application of VAN predictions. Results of the studies[1,2] suggest to consider separately the strongest earthquakes, $M \geq 5$ in Greece and $M \geq 6$ in Japan.

(ii) To analyse, how the previous VAN predictions (telegrams with advance warning) could be used for damage reduction, independently on geophysical considerations. This analysis has to be done by authorities responsible for earthquake preparedness.

Two more tasks are feasible and seem promising.

(iii). To incorporate into VAN prediction the maps of a fault system and of earthquake nucleation zones (see for example, the papers by Gerasimov et al.[10] and Gelfand et al.[11]). This may explain connection between predicted earthquakes and "VAN - sensituive areas".

(iv) To determine basic statistical properties of time series, where VAN signals are detected; this will allow to understand reliability of detection algorithms.

Less immediate but major possibility for further development of VAN - type precursors is (to the author's opinion) to combine them with precursors expressed in different fields and to integrate them with other stages of prediction.

Acknowledgments

The idea of this paper was inspired by Prof. J.Lighthill. It was written, when I was a visiting professor in the Department of the Earth, Atmospheric and Planetary Sciences, Massachusets Institute of Technology. I am grateful to Drs. L. Silver, N.Valette-Silver, S. Sacks, S. Ueda, and P. Varotsos for discussion and advice. Ms. J. Bogunova, Ms. L. Kadlubovich and Ms. O. Matsievskaya transferred the manuscript in copy-ready form with care and speed. The work reviewed in this paper was partly supported by the following grants: International Sciences Foundation (MB3000; MB3300), U.S. National Science Foundation (EAR 94 23818), International Association for the Promotion of Cooperation with Scientists from the Independent States of the Former Soviet Union (INTAS-93-809), International Science and Technology Center (008-94), and Russian Foundation for Basic Research (944-05-16444a).

References

1. T. Nagao, M. Uyeshima and S. Ueda, *Geophys. Res. Lett.* (1995) in print.
2. M. Schnirman *et al., ibid.*
3. F. Mulargia, *This volume.*
4. A. Gabrielov *et al., CERESIS* (Lima, Peru, 1986).
5. G. Molchan *et al., PEPI* **61, N 1-2**, 128-139 (1990).
6. G. Molchan, *Tectonophysics* **193**, 267-276 (1991).
7. G. Molchan, *Computational Seismology and Geodynamics* **2**, 1-10 (AGU 1995).
8. G. Molchan, *Pageoph* (1995) in print.
9. L. Kantorovich *et al., MIT* (1974).
10. I. Gerasimov and E. Rantsman, *Geomorph. Res.* **1**, 1-13 (1973).
11. I. Gelfand *et al., PEPI* **11**, 227-283 (1976).
12. V. Keilis-Borok, *Review of Geophysics* **28**, 19-34 (1990).
13. D. Turcotte, *Fractals and chaos in geology and geophysics* (Cambridge University Press, 1993).
14. W. Newman, A. Gabrielov and D. Turcotte, eds. *Geophys. Monograph* **83** (1994).
15. V. Keilis-Borok, *Physica D* **77**, 193-199 (1994).
16. M. Alexeevskaya *et al., J. Geophys.* **43**, 227-233 (1977).
17. E. Eaton and A. Sadovsky, *Computational Seismology* **18**, 124-127 (Allerton Press, 1987).
18. D. Rundkvist and I. Rotwain, *Computational Seismology* **27**, 201-244 (1994) in Russian.
19. D. McKenzie and W. Morgan, *Nature* **224**, 125-133 (1969).
20. G. King, *PAGEOPH* **121**, 761-815 (1983).
21. A. Gabrielov *et al., submitted to Proc. Natl. Acad. Sci. USA* (1995).
22. G. Barenblatt *et al., Doklady AN SSSR* **269** 831-834 (1983).
23. A. Gabrielov and V. Keilis-Borok, *Pure Appl. Geophys* **121 (3)** 477-494 (1983).
24. N. Pertsov and V. Traskin, *Coll. chemistry and phys.-chem. mechanics* 155-165 (1992) in Russian.
25. A. Gabrielov *et al., PEPI* **61**, 18-28 (1990).
26. E. Vilkovich and M. Schnirman, *Computational Seismology* **14** 27-36 (Allerton Press, 1983).
27. V. Keilis - Borok, ed., *PEPI* **61** (1990).
28. F. Press and C. Allen, *JGR* **B4** 6421-6430 (1995).
29. L. Knopoff *et al., JGR* (1995) in print.

A BRIEF LOOK FORWARD TO FUTURE RESEARCH NEEDS

JAMES LIGHTHILL

*University College London, Department of Mathematics, Gower St.,
London WC1E 6BT, UK.*

A case that, taking into account all the papers in this book, continued research stemming from observations by the VAN group is required, is developed in some detail. International geophysical research in Greece needs to be devoted to features of Greek seismicity relevant to VAN, as well as to electrical properties of the crust and considerations regarding possible pre-seismic migration of fluids. Other work is needed (i) outside Greece on the (still uncertain) "export" potential of VAN, and (ii) in the VAN group itself on further developments in its approach to earthquake prediction.

1 Introduction

This, the book's last paper, is partly based on remarks from the chair that concluded the London review meeting of 11-12 May 1995, but it takes into account too all the texts printed above as they were finalised by authors during the ensuing four months. In these texts, of course, the earthquakes occurring after the meeting (on 13 May in northern Greece and on 15 June in Egion; which were the two largest in Greece for over a decade) are carefully related to the corresponding VAN predictions (those received by myself, for example — along with other interested scientists — on 2 May and on 20 May 1995). It is noteworthy that the distinguished seismologist, Professor H. Kanamori, was influenced partly by these events, as well as by the proceedings of the review meeting (which he had attended in an initially neutral spirit), to give the views he has expressed above in "A seismologist looks at VAN"; suggesting that for the larger earthquakes in Greece the VAN group appears to have usefully identified SES precursors.

But it is for each reader of this book to draw his or her own conclusions from all the different arguments presented in the course of some 300 carefully prepared pages. There will certainly be readers who conclude with Professor Wyss that "the VAN hypothesis has to be rejected", and some of them will be further persuaded by Professor Geller that the goal of short-term earthquake prediction is inherently unattainable. For all of these readers, the set of "future research needs" (related to VAN) on which this final paper aims to concentrate is necessary a null set. Before the meeting, Dr. Geller gave me a solemn (albeit unnecessary) reminder that it would be irresponsible to end the proceedings with recommendations for future research if those were intended merely as a compromise between two opposing points of view which the chairman had not succeeded in reconciling.

The following listing of future research needs, on the other hand, is not at all put forward in such a compromise spirit. On the contrary, it is seen as a natural set of conclusions for an initially neutral geophysicist who attended the meeting (like, for example, Professor Kanamori or Dr. Browitt, both of whom have expressed similar conclusions in their papers in part VI) or who has studied carefully the whole of this book. Furthermore it receives weighty support from the papers of Dr. LeMouël, Director of the Institut de Physique du Globe de Paris, and of Academician Keilis-Borok, immediate past president of the International Union of Geodesy and Geophysics. The list comprises needs for future geophysical studies in Greece, for investigations of VAN's "export" potential, and for developments in earthquake prediction by the VAN group itself.

2 Needs for future geophysical research in Greece

Those initially neutral readers who have been impressed by the VAN group's painstaking accumulation of evidence for SES precursors measured in sensitive areas may agree with the authors just mentioned that there are needs for research by the international geophysics community, in close collaboration with the VAN group, on the nature and extent of these sensitive areas in Greece. Such research needs to have three principal components, related to seismicity, to electrical properties and to displacement of fluids.

2.1 Seismicity

One important question posed by Professor Kanamori is whether the nature of seismicity in Greece involves unusual features that may favour the appearance of SES precursors. Research in this field needs to be well integrated with seismological studies of the active fault system, and will be facilitated by the augmented network of seismometers and other equipment being installed in Greece through collaborative efforts between French and Greek scientists.

In this context, the paper by Academician Keilis-Borok emphasizes the need to consider seismic behaviour not in local terms but in relation to systems of blocks and faults which extend over a large area. Appropriate investigation of essentially non-local processes of slow deformation may throw light on non-local precursors such as the SES observed in sensitive areas.

2.2 Electrical properties

The special electrical properties of such sensitive areas of the earth's crust are moreover of high importance to the understanding of SES precursors. In this context, studies in the Ioannina region as described by Drs. Bernard and LeMouël are an excellent example of what may need to be attempted on a still greater scale.

At the same time, continued research is needed on background electrical noise, and on improved low-noise instrumentation. Furthermore, as Dr. Lazarus has emphasized, some information on SES spectra will be of real value in narrowing the search for mechanisms.

2.3 Displacement of fluids

Amongst all of the potential mechanisms discussed during the review meeting, those associated with a supposed displacement of fluids which may accompany non-local pre-seismic processes of slow deformation appeared of greatest interest to participants. Academician Keilis-Borok emphasized in this context the importance of collective phenomena such as may arise in the course of nonlinear dynamic developments of fluid-containing layers, and some other highly interesting suggestions have been made in papers by Dr. Lazarus and by Drs Bernard and LeMouël.

There is a special challenge in this call to investigate pre-seismic displacement of fluids, because of formidable difficulties in attempting such studies. It is above all this challenge to which the attention of geophysicists all over the world needs to be drawn in the light of the VAN group's findings.

3 The "export" potential of VAN

Even though a fair proportion of readers of this book may agree with the foregoing recommendations for future geophysical research in Greece, many of them may also concur with Professor Kanamori in voicing present doubts about VAN's "export" potential. As he points out, there is as yet insufficient information on whether or not exceptional features of Greek seismicity favour the appearance of SES precursors; and, if so, on whether or not such features are present in other earthquake zones. Some key merits of the programmes of geophysical research suggested above arise, of course, from their potential for a more positive identification of mechanisms that may underlie SES precursors in Greece. Evidently, such identification could help in forming judgements regarding which other earthquake zones may be suited to experiments with VAN.

Various programmes relevant to this issue have been described in part V above. In the context of the International Decade for Natural Disaster Reduction (IDNDR), with its emphasis on the needs of developing countries which are specially prone to natural disasters, the paper by Professor Kulhanek seems particularly helpful. He points out that Guatemala suffers from severe natural disasters — especially, those resulting from its high level of seismicity — while its relatively low level of electrical noise favours the possible introduction of an earthquake prediction system based on SES precursors. Therefore, the willingness of the Swedish authorities interested in IDNDR, with Professor Kulhanek as their principal agent, to support careful attempts at exporting VAN to Guatemala in a way which makes maximum use of experiments by the VAN group and experience on the identification of sensitive areas, seems thoroughly commendable.

In the meantime, there is of course great interest in information about firm intentions to intensify SES research in Japan, notwithstanding (i) its high levels of electrical noise, and (ii) certain suggestions (see above) that further researches in Greece might usefully have preceded an "export" drive on so massive a scale. This is perhaps a programme of research about which it is inherently impossible to hazard forecasts of its results; and yet which, by bringing the great scientific, technological and financial resources of Japan to bear on problems of enormous difficulty, might well lead to outcomes of completely unexpected nature and value.

4 Earthquake prediction by the VAN group itself

Finally, in accordance with the general philosophy regarding future research needs expounded in the Introduction to this paper, it is a pleasure to end this book with an encouragement to the distinguished members of the VAN group to continue to expand their own earthquake prediction programme. This will involve expansions on the instrumental side, with a renewed increase in numbers of VAN stations and the further introduction of low-noise data-acquisition and data-handling equipment. It will involve additional work on the statistical significance of VAN findings, suggested by some of the papers in this volume (for example, that by Dr. Sudo on the possible influence of foreshocks). And above all, in the context of this attempt at not only a critical, but also an interdisciplinary, review of VAN, it may be important to expand the interests of the group into social aspects of their earthquake prediction procedures.

Now I may conclude by offering my very sincere thanks to all participants in the review meeting and to all the contributors to this volume. Collectively, these individuals have been able to lay before the scientific public for the first time (I believe) a volume from which informed judgements can at last be reached about earthquake prediction from seismic electrical signals.